QUANTUM CHROMODYNAMICS AT HIGH ENERGY

Filling a gap in the current literature, this book is the first entirely dedicated to high energy quantum chromodynamics (QCD) including parton saturation and the color glass condensate (CGC). It presents groundbreaking progress on the subject and describes many problems at the forefront of research, bringing postgraduate students, theorists, and interested experimentalists up to date with the current state of research in this field.

The material is presented in a pedagogical way, with numerous examples and exercises. Discussion ranges from the quasi-classical McLerran–Venugopalan model to the linear BFKL and nonlinear BK/JIMWLK small-x evolution equations. The authors adopt both a theoretical and an experimental outlook, and present the physics of strong interactions in a universal way, making it useful for physicists from various subcommunities of high energy and nuclear physics, and applicable to processes studied at all high energy accelerators around the world. A selection of color figures is available online at www.cambridge.org/9780521112574.

This title, first published in 2013, has been reissued as an Open Access publication on Cambridge Core.

YURI V. KOVCHEGOV is Professor in the Department of Physics at The Ohio State University. He is a world leader in the field of high energy QCD. In 2006 he was awarded The Raymond and Beverly Sackler Prize in the Physical Sciences by Tel Aviv University for a number of groundbreaking contributions in the field. The Balitsky–Kovchegov equation bears his name.

EUGENE LEVIN is Professor Emeritus in the School of Physics and Astronomy at Tel Aviv University. He is the founding father of the field of parton saturation and of the constituent quark model. Equations and approaches that bear his name include the Levin–Frankfurt quark-counting rules, the Gribov–Levin–Ryskin nonlinear equation, the Levin–Tuchin solution, and the Kharzeev–Levin–Nardi approach, reflecting only a selection of his many contributions to high energy physics.

CAMBRIDGE MONOGRAPHS ON PARTICLE PHYSICS, NUCLEAR PHYSICS AND COSMOLOGY

General Editors: T. Ericson, P. V. Landshoff

QUANTUM CHROMODYNAMICS
AT HIGH ENERGY

YURI V. KOVCHEGOV

The Ohio State University, USA

EUGENE LEVIN

Tel-Aviv University, Israel

Shaftesbury Road, Cambridge CB2 8EA, United Kingdom

One Liberty Plaza, 20th Floor, New York, NY 10006, USA

477 Williamstown Road, Port Melbourne, VIC 3207, Australia

314–321, 3rd Floor, Plot 3, Splendor Forum, Jasola District Centre, New Delhi – 110025, India

103 Penang Road, #05–06/07, Visioncrest Commercial, Singapore 238467

Cambridge University Press is part of Cambridge University Press & Assessment, a department of the University of Cambridge.

We share the University's mission to contribute to society through the pursuit of education, learning and research at the highest international levels of excellence.

www.cambridge.org
Information on this title: www.cambridge.org/9781009291415

DOI: 10.1017/9781009291446

First published 2013
Reissued as OA 2022

A catalogue record for this publication is available from the British Library.

ISBN 978-1-009-29141-5 Hardback
ISBN 978-1-009-29142-2 Paperback

Cambridge University Press & Assessment has no responsibility for the persistence or accuracy of URLs for external or third-party internet websites referred to in this publication and does not guarantee that any content on such websites is, or will remain, accurate or appropriate.

Contents

Preface

This book summarizes the developments over the past several decades in the field of strong interactions at high energy. This is the first ever book almost entirely devoted to the physics of parton saturation and the color glass condensate (CGC).

Our main goal in this book is to introduce the reader systematically to the ideas, problems, and methods of high energy quantum chromodynamics (QCD). Over the years, these methods and ideas have led to a new physical picture of high energy hadronic and nuclear interactions, representing them as the interactions of a very dense system of tiny constituents (quarks and gluons) having only a small value of the QCD coupling constant. Owing to the high density of gluons and quarks the interactions in such systems are inherently nonperturbative; nevertheless, a theoretical description of these interactions is possible due to the smallness of the QCD coupling. Our main goals in the book are to show how these new ideas arise from perturbative QCD and to enable the reader to enjoy the beauty and simplicity of these emerging methods and equations.

The book's intended audience is advanced graduate students, postdoctoral fellows, and mature researchers from the neighboring subfields of nuclear and particle physics. We assume that graduate student readers are familiar with quantum field theory at the level of a standard graduate-level course based on the textbooks by Peskin and Schroeder (1995) or Sterman (1993). We also recommend that students should have taken a theoretical particle physics course before attempting to read this book. Nevertheless, we have tried to make this book as self-sufficient as possible, and so we refer to the results of quantum field theory only minimally.

The book is structured as follows. In Chapters 1 through 5 we present general concepts and the results of high energy QCD at a level accessible to a graduate student beginning his or her research in the field. Chapters 6 though 9 deal with more specialized topics and are written at a somewhat higher level; now the reader is expected to do more independent calculations and thinking to follow the presentation. Sections marked with an asterisk * can be skipped in the first reading of the book.

The field of high energy QCD has been developing rapidly over the past few decades, generating vast amounts of new and interesting results. It is impossible to fit all the recent advances into a single book: inevitably some important results have had to be left out. We have tried to overcome this shortcoming by incorporating sections on further reading at the ends of most chapters. In these sections we provide the reader with the references needed to further develop his or her understanding of the subject.

At the ends of many chapters we provide exercises for readers. Fairly difficult problems are marked with an asterisk * and very hard problems are marked with a double asterisk **.

In this book we have aimed to bring the reader to the forefront of research on high energy QCD. We would be thrilled if our readers were able to pursue work in the field after reading this book, generating new theoretical ideas and results which ultimately could be compared with experiment.

We would like to thank our colleagues and collaborators Javier Albacete, Ian Balitsky, Jochen Bartels, Jean-Paul Blaizot, Kostya Boreskov, Eric Braaten, Yuri Dokshitzer, Adrian Dumitru, Victor Fadin, Lonya Frankfurt, Dick Furnstahl, Francois Gelis, Asher Gotsman, Ulrich Heinz, Will Horowitz, Edmond Iancu, Jamal Jalilian-Marian, Oleg Kancheli, Dima Kharzeev, Valera Khoze, Boris Kopeliovich, Alex Kovner, Andrei Leonidov, Lev Lipatov, Mike Lisa, Misha Lublinsky, Uri Maor, Cyrille Marquet, Larry McLerran, Al Mueller, Marzia Nardi, Robert Perry, Robi Peschanski, Dirk Rischke, Misha Ryskin, Anna Stasto, Mark Strikman, Lech Szymanowski, Derek Teaney, Kirill Tuchin, Raju Venugopalan, Heribert Weigert for many productive discussions on the subjects covered in the book. This book would not be possible without the intellectual pleasure and constant support of these discussions. Special thanks go to Javier Albacete for preparing Figs. 4.29, 4.30, 4.31, and 6.3, to Anna Stasto for preparing Fig. 4.33, and to Kunihiro Nagano for preparing Fig. 2.7.

Most of all we are grateful to our wives, children, grandchildren, parents, and grandparents for their unwavering love and support and for their great patience during the writing of this book.

YURI V. KOVCHEGOV

EUGENE LEVIN

November 2011

1

Introduction: basics of QCD perturbation theory

Quantum chromodynamics (QCD) is the theory of strong interactions. This is an exciting physical theory, whose Lagrangian deals with quark and gluon fields and their interactions. At the same time, quarks and gluons do not exist as free particles in nature but combine into bound states (hadrons) instead. This phenomenon, known as *quark confinement*, is one of the most profound puzzles of QCD. Another amazing feature of QCD is the property of *asymptotic freedom*: quarks and gluons tend to interact more weakly over short distances and more strongly over longer distances.

This book is dedicated to another QCD mystery: the behavior of quarks and gluons in high energy collisions. Quantum chromodynamics is omnipresent in high energy collisions of all kinds of known particles. There are vast amounts of high energy scattering data on strong interactions, which have been collected at accelerators around the world. While these data are incredibly diverse they often exhibit intriguingly universal scaling properties, which unify much of the data while puzzling both experimentalists and theorists alike. Such universality appears to imply that the underlying QCD dynamics is the same for a broad range of high energy scattering phenomena.

The main goal of this book is to provide a consistent theoretical description of high energy QCD interactions. We will show that the QCD dynamics in high energy collisions is very sophisticated and often nonlinear. At the same time much solid theoretical progress has been made on the subject over the years. We will present the results of this progress by introducing a universal approach to a broad range of high energy scattering phenomena.

We begin by presenting a brief summary of the tools needed to perform perturbative QCD calculations. Since much of the material in this chapter is covered in standard field theory and particle physics textbooks, we will not derive many results, simply summarizing them and referring the reader to the appropriate literature for detailed derivations.

1.1 The QCD Lagrangian

Quantum chromodynamics is an SU(3) Yang–Mills gauge theory (Yang and Mills 1954) describing the interactions of quarks and gluons. The QCD Lagrangian density is

$$\mathcal{L}_{QCD} = \sum_{\text{flavors } f} \bar{q}_i^f(x) \left[i\gamma^\mu D_\mu - m_f \right]_{ij} q_j^f(x) - \tfrac{1}{4} F_{\mu\nu}^a F^{a\mu\nu} \tag{1.1}$$

1

where $q_i^f(x)$ and $\bar{q}_i^f(x)$ are the quark and antiquark spin-1/2 Dirac fields of color i, flavor f, and mass m_f, with $\bar{q} = q^\dagger \gamma^0$. A field $A_\mu^a(x)$ describes the gluon, which has spin equal to 1, zero mass, and color index a in the adjoint representation of the SU(3) gauge group. Summation over repeated color and Lorentz indices is assumed, with $i, j = 1, 2, 3$ and $a = 1, \ldots, 8$. The covariant derivative D_μ is defined by

$$D_\mu = \partial_\mu - ig A_\mu = \partial_\mu - ig t^a A_\mu^a. \tag{1.2}$$

The t^a are the generators of SU(3) in the fundamental representation ($t^a = \lambda^a/2$, where the λ^a are the Gell-Mann matrices). The non-Abelian gluon field strength tensor $F_{\mu\nu}^a$ is defined by

$$F_{\mu\nu} = t^a F_{\mu\nu}^a = \frac{i}{g} \left[D_\mu, D_\nu \right] \tag{1.3}$$

or, equivalently, by

$$F_{\mu\nu}^a = \partial_\mu A_\nu^a - \partial_\nu A_\mu^a + g f^{abc} A_\mu^b A_\nu^c, \tag{1.4}$$

where f^{abc} are the structure constants of the color group SU(3).

We work in natural units, with $\hbar = c = 1$. Our four-vectors are $x^\mu = (t, \vec{x})$, the partial derivatives are denoted $\partial_\mu = \partial/\partial x^\mu$, and the metric in t, x, y, z coordinates is $g_{\mu\nu} = \text{diag}(+1, -1, -1, -1)$.

The Lagrangian of Eq. (1.1) was proposed by Fritzsch, Gell-Mann, and Leutwyler (1973), Gross and Wilczek (1973, 1974), and Weinberg (1973). The form of the QCD Lagrangian is based on two assumptions confirmed by experimental observations: all hadrons consist of quarks and quarks cannot be observed as free particles. The first observation leads to a new quantum number for quarks: color. Indeed, without this quantum number we cannot build the wave functions for baryons. For example the Ω^- hyperon has spin 3/2 and consists of three s-quarks. This means that the spin and flavor parts of its wave function are symmetric with respect to interchange of the identical valence s-quarks. Owing to the Pauli exclusion principle the full wave function of the three identical quarks has to be antisymmetric. If spin and flavor were the only quantum numbers, it would appear that the spatial wave function of the three s-quarks would have to be antisymmetric. However, this would contradict the fact that Ω^- is a stable particle and is, therefore, a ground state of the three s-quark system. The spatial wave function of a ground state has to be symmetric. To resolve this conundrum we need to introduce a new quantum number that should have at least three different values to make the three strange quarks different in the Ω^- hyperon. This quantum number is the quark *color*.

We then need to determine which particle is responsible for interactions between the quarks forming quark bound states, the hadrons. The interactions between the quarks in mesons and baryons have to be attractive, which indicates that they should depend on quark color: if one introduced interactions between quarks using some global (not gauged) non-Abelian color symmetry then one would not be able to obtain attractive interactions between the quark and the antiquark in a meson and between a pair of quarks in a baryon simultaneously, at least not in the lowest nontrivial order in the interaction. One therefore concludes that the non-Abelian color symmetry has to be gauged by introducing a non-Abelian vector boson responsible for quark interactions. Moreover, as we will see below, the high energy scattering data confirms this conclusion as it demonstrates that the particle responsible for quark interactions has spin equal to 1.

The second experimental observation needed for the construction of the QCD Lagrangian, that quarks are never seen as free particles, means that the forces between quarks should be stronger at longer distances to prevent quarks from leaving a hadron. For point-like particles our best chance of getting such forces is by assuming that quark interactions are mediated by a massless particle. For such a particle the lowest-order quark–antiquark interaction potential decreases at long distances roughly as to $1/r$, where r is the distance between the quarks. (Indeed in a full QCD calculation this behavior changes to $\sim r$, that of a confining potential.) Massive particles would give an exponentially decreasing potential, which would have a shorter range than the potential in the massless case. We therefore conclude that the particle responsible for quark interactions is a non-Abelian massless vector boson, a gluon.

However, particle interactions may generate a mass even for a particle that is massless at the Lagrangian level. To protect the zero mass of the gluon from higher-order corrections we have to assume the existence of gauge symmetry in our Lagrangian. Namely, the Lagrangian should be invariant with respect to

$$q(x) \rightarrow S(x)\,q(x), \tag{1.5a}$$

$$\bar{q}(x) \rightarrow \bar{q}(x)\,S^{-1}(x), \tag{1.5b}$$

$$A_\mu(x) \rightarrow S(x)A_\mu(x)S^{-1}(x) - \frac{i}{g}\left[\partial_\mu S(x)\right]S^{-1}(x), \tag{1.5c}$$

where we have defined a unitary 3×3 matrix

$$S(x) - e^{i\alpha^a(x)t^a}, \tag{1.6}$$

where the $\alpha^a(x)$ are arbitrary real-valued functions; summation over repeated color indices a is again implied. The form of the Yang–Mills Lagrangian (1.1) can be derived directly from the gauge symmetry in Eqs. (1.5) (see e.g. Peskin and Schroeder (1995)).

1.2 A review of Feynman rules for QCD

To derive the Feynman rules from the Lagrangian (1.1) we need to define the functional integral (the QCD partition function)

$$Z_{QCD} - \int \mathcal{D}A\,\mathcal{D}q\,\mathcal{D}\bar{q}\,\exp\left\{i\int d^4x\,\mathcal{L}_{QCD}\,(A,q,\bar{q})\right\}. \tag{1.7}$$

One can see that this integral is divergent since its integrand has the same value for an infinite set of fields related to each other by all possible gauge transformations (1.5). However, the values of physical observables are given by the expectation values of operators. For an arbitrary gauge-invariant operator $\mathcal{O}$ we have the vacuum expectation value

$$\langle \mathcal{O} \rangle \equiv \frac{\int \mathcal{D}A \, \mathcal{D}q \, \mathcal{D}\bar{q} \, \mathcal{O} \exp\{i \int d^4 x \mathcal{L}_{QCD}\}}{\int \mathcal{D}A \, \mathcal{D}q \, \mathcal{D}\bar{q} \exp\{i \int d^4 x \, \mathcal{L}_{QCD}\}} \tag{1.8}$$

The divergences caused by integrations over gauge directions in the numerator and in the denominator of Eq. (1.8) cancel each other. Faddeev and Popov (1967) suggested a procedure allowing one to see such cancellations in the most economic way by multiplying the definition (1.7) with the functional integral identity[1]

$$1 = \int \mathcal{D}\alpha \, \delta(\alpha) = \int \mathcal{D}\alpha \, \delta(G(A^\alpha)) \, \det\left(\frac{\delta G(A^\alpha)}{\delta \alpha}\right), \tag{1.9}$$

where the integral runs over all gauge transformations labeled by α^a (see Eq. (1.6)), A^α is a gauge field related to the original one by the gauge transformation defined by α^a, and $G(A) = 0$ is the gauge-fixing condition. (For instance, $G(A) = \partial_\mu A^\mu$ in a covariant gauge.) Let us restrict ourselves to gauges in which the functional determinant $\det[\delta G(A^\alpha)/\delta \alpha]$ is independent of α^a for a given A^α. Using Eq. (1.9) the expectation values of the operators can be written as

$$\langle \mathcal{O} \rangle = \frac{\left(\int \mathcal{D}\alpha\right) \int \mathcal{D}A \, \mathcal{D}q \, \mathcal{D}\bar{q} \, \mathcal{O} \, \delta(G(A)) \, \det\left(\frac{\delta G(A^\alpha)}{\delta \alpha}\right) \exp\left\{i \int d^4 x \, \mathcal{L}_{QCD}\right\}}{\left(\int \mathcal{D}\alpha\right) \int \mathcal{D}A \, \mathcal{D}q \, \mathcal{D}\bar{q} \, \delta(G(A)) \, \det\left(\frac{\delta G(A^\alpha)}{\delta \alpha}\right) \exp\left\{i \int d^4 x \, \mathcal{L}_{QCD}\right\}}, \tag{1.10}$$

where we have relabeled the integration variable A^α as A everywhere except in the determinants, in which one should put $\alpha^a = 0$ after differentiation thus turning A^α into A. The infinities in the numerator and the denominator of Eq. (1.10) are clearly identifiable as being due to the integration over α^a. As nothing else in the integrands of Eq. (1.10) depends on α we can simply cancel the $\mathcal{D}\alpha$ integrations, writing

$$\langle \mathcal{O} \rangle = \frac{\int \mathcal{D}A \, \mathcal{D}q \, \mathcal{D}\bar{q} \, \mathcal{O} \, \delta(G(A)) \, \det\left(\frac{\delta G(A^\alpha)}{\delta \alpha}\right) \exp\left\{i \int d^4 x \, \mathcal{L}_{QCD}\right\}}{\int \mathcal{D}A \, \mathcal{D}q \, \mathcal{D}\bar{q} \, \delta(G(A)) \, \det\left(\frac{\delta G(A^\alpha)}{\delta \alpha}\right) \exp\left\{i \int d^4 x \, \mathcal{L}_{QCD}\right\}}. \tag{1.11}$$

To obtain the Feynman rules we have to put all the A-dependence in the integrands in Eq. (1.11) into the exponents. We start with the delta functions and first note that making the replacement in Eq. (1.11)

$$\delta(G(A)) \to \delta(G(A) - r(x)), \tag{1.12}$$

where $r(x)$ is some arbitrary function of x^μ, would not change the values of the functional integrals in the numerator and the denominator and would therefore leave $\langle \mathcal{O} \rangle$ unchanged. Indeed different choices of $r(x)$ correspond to different choices of the gauge defined by the $G(A) = r(x)$ gauge condition. Thus the replacement (1.12) simply modifies the function defining the gauge condition: $G(A) \to G(A) - r(x)$. Since our initial gauge-defining function $G(A)$ is arbitrary, and as neither of the integrals in the numerator and the denominator of Eq. (1.11) depends on $G(A)$, we conclude that nothing in the numerator

---

[1] In discussing the Faddeev–Popov method we will follow closely the presentations in Peskin and Schroeder (1995) and in Sterman (1993).

or the denominator of Eq. (1.11) changes if we perform the replacement (1.12). Moreover, the resulting expression,

$$\langle \mathcal{O} \rangle = \frac{\int \mathcal{D}A \, \mathcal{D}q \, \mathcal{D}\bar{q} \, \mathcal{O} \, \delta(G(A) - r(x)) \, \det\left(\frac{\delta G(A^\alpha)}{\delta \alpha}\right) \exp\left\{i \int d^4x \, \mathcal{L}_{QCD}\right\}}{\int \mathcal{D}A \, \mathcal{D}q \, \mathcal{D}\bar{q} \, \delta(G(A) - r(x)) \, \det\left(\frac{\delta G(A^\alpha)}{\delta \alpha}\right) \exp\left\{i \int d^4x \, \mathcal{L}_{QCD}\right\}}, \tag{1.13}$$

is independent of $r(x)$ for the same reasons. We can integrate the numerator and the denominator separately over $r(x)$ by multiplying them with

$$1 = N(\xi) \int \mathcal{D}r \, \exp\left\{-i \int d^4x \, \frac{r^2(x)}{2\xi}\right\}, \tag{1.14}$$

where $N(\xi)$ is a normalization function defined by Eq. (1.14) and ξ is an arbitrary number. Multiplying both the numerator and the denominator of Eq. (1.13) by Eq. (1.14), canceling $N(\xi)$, and performing the r-integrals with the help of the delta functions, we obtain

$$\langle \mathcal{O} \rangle = \frac{\int \mathcal{D}A \, \mathcal{D}q \, \mathcal{D}\bar{q} \, \mathcal{O} \det\left(\frac{\delta G(A^\alpha)}{\delta \alpha}\right) \exp\left\{i \int d^4x \left(\mathcal{L}_{QCD} - \frac{1}{2\xi}[G(a)]^2\right)\right\}}{\int \mathcal{D}A \, \mathcal{D}q \, \mathcal{D}\bar{q} \det\left(\frac{\delta G(A^\alpha)}{\delta \alpha}\right) \exp\left\{i \int d^4x \left(\mathcal{L}_{QCD} - \frac{1}{2\xi}[G(a)]^2\right)\right\}}. \tag{1.15}$$

Finally, in order to remove the determinants of Eq. (1.15) into the exponents one introduces the (unphysical) *Faddeev–Popov ghost field* $c^a(x)$, whose values are complex Grassmann numbers (Faddeev and Popov 1967, Feynman 1963, DeWitt 1967). The ghost field is a Lorentz scalar in the adjoint representation of SU(3). With the help of the Faddeev–Popov ghost field we write

$$\det\left(\frac{\delta G(A^\alpha)}{\delta \alpha}\right) = \int \mathcal{D}c \, \mathcal{D}c^* \, \exp\left\{-i \int d^4x \, c^* \frac{\delta G(A^\alpha)}{\delta \alpha} c\right\} \tag{1.16}$$

with c^* the complex conjugate of the c field. Using Eq. (1.16) in Eq. (1.15) we obtain

$$\langle \mathcal{O} \rangle = \frac{\int \mathcal{D}A \, \mathcal{D}q \, \mathcal{D}\bar{q} \, \mathcal{D}c \, \mathcal{D}c^* \, \mathcal{O} \, \exp\left\{i \int d^4x \, \mathcal{L}(A, q, \bar{q}, c, c^*)\right\}}{\int \mathcal{D}A \, \mathcal{D}q \, \mathcal{D}\bar{q} \, \mathcal{D}c \, \mathcal{D}c^* \, \exp\left\{i \int d^4x \, \mathcal{L}(A, q, \bar{q}, c, c^*)\right\}}, \tag{1.17}$$

where we have defined an effective Lagrangian

$$\mathcal{L}(A, q, \bar{q}, c, c^*) \equiv \mathcal{L}_{QCD} - \frac{1}{2\xi}[G(A)]^2 - c^* \frac{\delta G(A^\alpha)}{\delta \alpha} c. \tag{1.18}$$

Now we are ready to derive the Feynman rules for QCD.

In this book we will employ two main gauge choices. One is the Lorenz gauge, defined by the gauge condition

$$\partial_\mu A^{a\,\mu} = 0. \tag{1.19}$$

Inserting $G(A) = \partial_\mu A^{a\,\mu}$ into Eq. (1.18), after some straightforward algebra (see e.g. Peskin and Schroeder (1995)) we end up with

$$\mathcal{L} = \mathcal{L}_{QCD} - \frac{1}{2\xi}\left(\partial^\mu A^a_\mu\right)^2 + \left(\partial^\mu c^{a*}\right)\left(\delta^{ac}\,\partial^\mu + g f^{abc} A^b_\mu\right)c^c. \tag{1.20}$$

Using Eq. (1.20) we can derive the Feynman rules for QCD by substituting the Lagrangian (1.20) into Eq. (1.7) in place of $\mathcal{L}_{QCD}$.

The other gauge choice that we will be using frequently throughout the book is the light cone gauge, defined by

$$\eta \cdot A^a = \eta^\mu A^a_\mu = 0, \tag{1.21}$$

with η^μ a constant four-vector that is light-like, so that $\eta^2 = \eta_\mu \eta^\mu = 0$. One can show that, in the light cone gauge, $\det[\delta G(A^\alpha)/\delta \alpha]$ does not depend on A^μ when we take the limit $\xi \to 0$. From Eq. (1.18) one can see that in this case the ghost field would not couple to the gluon field and so can be integrated out in the functional integrals of Eq. (1.17). Hence there is no ghost field in the light cone gauge. The effective Lagrangian (1.18) in the light cone gauge becomes

$$\mathcal{L} = \mathcal{L}_{QCD} - \frac{1}{2\xi} \left(\eta^\mu A^a_\mu\right)^2 \tag{1.22}$$

(with an implied $\xi \to 0$ limit).

Below we list the Feynman rules for QCD, in the Lorenz and light cone gauges, which follow from the Lagrangians in Eqs. (1.20) and (1.22). We use the standard notation for a product of two four-vectors $u \cdot v = u_\mu v^\mu$ and for the square of a single four-vector $v_\mu v^\mu = v^2$. The Dirac gamma matrices in the standard Dirac representation, which we will use here, are defined by

$$\gamma^0 = \begin{pmatrix} \mathbf{1} & 0 \\ 0 & -\mathbf{1} \end{pmatrix}, \quad \gamma^i = \begin{pmatrix} 0 & \sigma^i \\ -\sigma^i & 0 \end{pmatrix}, \tag{1.23}$$

where $\mathbf{1}$ is a unit 2×2 matrix, $i = 1, 2, 3$, and σ^i are the Pauli matrices

$$\sigma^1 = \begin{pmatrix} 0 & 1 \\ 1 & 0 \end{pmatrix}, \quad \sigma^2 = \begin{pmatrix} 0 & -i \\ i & 0 \end{pmatrix}, \quad \sigma^3 = \begin{pmatrix} 1 & 0 \\ 0 & -1 \end{pmatrix}. \tag{1.24}$$

As usual, we will write $\not{p} = \gamma^\mu v_\mu$. Arrows on the quark and ghost propagators (see below) indicate the flow of the particle number and, in the cases of the quark propagator and the ghost–gluon vertex, they also indicate the momentum flow. As ghost fields do not exist in the light cone gauge, the Feynman rules for ghosts listed below apply only in the Lorenz gauge.

1.2.1 QCD Feynman rules

Quark propagator: $\quad \dfrac{j \qquad p \qquad i}{\xrightarrow{\hspace{2cm}}} = \dfrac{i(\not{p} + m_f)}{p^2 - m_f^2 + i\epsilon} \delta^{ij},$ $\tag{1.25}$

Ghost propagator: $\quad b \;\text{-----}\!\!\xrightarrow{k}\!\!\text{-----}\; a = \dfrac{i}{k^2 + i\epsilon} \delta^{ab},$ $\tag{1.26}$

Gluon propagator: $\quad \overset{b}{\underset{\nu}{}}\!\overset{k}{\text{0000000000000000}}\!\overset{a}{\underset{\mu}{}} = \dfrac{-iD_{\mu\nu}(k)}{k^2 + i\epsilon} \delta^{ab},$ $\tag{1.27}$

where in the Lorenz gauge ($\partial \cdot A^a = 0$)

$$D_{\mu\nu}(k) = g_{\mu\nu} - (1 - \xi)\frac{k_\mu k_\nu}{k^2}; \qquad (1.28)$$

the choice $\xi = 0$ is referred to as the Landau gauge and the choice $\xi = 1$ is called the Feynman gauge. In the light cone gauge $\eta \cdot A^a = 0$ with $\xi \to 0$ one has

$$D_{\mu\nu}(k) = g_{\mu\nu} - \frac{\eta_\mu k_\nu + \eta_\nu k_\mu}{\eta \cdot k}. \qquad (1.29)$$

Quark–gluon vertex:

$$= i g \gamma^\mu (t^a)_{ji}, \qquad (1.30)$$

Ghost–gluon vertex
(Lorenz gauge only):

$$= g(p + k)^\mu f^{abc} \qquad (1.31)$$

Three-gluon vertex
(all momenta flow
into the vertex):

$$= \begin{aligned} -g f^{abc} [&(k_1 - k_3)^\nu g^{\mu\rho} \\ &+ (k_2 - k_1)^\rho g^{\mu\nu} + (k_3 - k_2)^\mu g^{\nu\rho}] \end{aligned} \qquad (1.32)$$

Four-gluon vertex:

$$\begin{aligned} -ig^2 \big[&f^{abe} f^{cde} (g^{\mu\rho} g^{\nu\sigma} - g^{\mu\sigma} g^{\nu\rho}) \\ &+ f^{ace} f^{hde} (g^{\mu\nu} g^{\rho\sigma} - g^{\mu\sigma} g^{\nu\rho}) \\ &+ f^{ade} f^{bce} (g^{\mu\nu} g^{\rho\sigma} - g^{\mu\rho} g^{\nu\sigma}) \big] \end{aligned} \qquad (1.33)$$

The Feynman rules that are standard for all field theories, such as the conservation of four-momentum in the vertices and the inclusion of a factor -1 for each fermion loop or of proper symmetry factors, apply to QCD as well and will not be explicitly spelled out here.

1.3 Rules of light cone perturbation theory

Many calculations in this book will not be performed using the Feynman rules. Instead we will use light cone perturbation theory (LCPT), following the rules introduced by Lepage and Brodsky (1980) (see Brodsky and Lepage (1989) and Brodsky, Pauli, and Pinsky (1998) for a detailed derivation of the LCPT rules). We begin by introducing the light cone notation.

For any four-vector v^μ we define

$$v^+ = v^0 + v^3, \quad v^- = v^0 - v^3. \tag{1.34}$$

With this notation we see immediately that

$$v^2 = v^+ v^- - \vec{v}_\perp^2, \tag{1.35}$$

where we have defined a vector of transverse components $\vec{v}_\perp = (v^1, v^2)$. A product of two four-vectors v^μ and u^μ in light cone notation is

$$u \cdot v = \frac{1}{2} u^+ v^- + \frac{1}{2} u^- v^+ - \vec{u}_\perp \cdot \vec{v}_\perp. \tag{1.36}$$

The metric has nonzero components $g_{+-} = g_{-+} = 1/2$, $g_{11} = g_{22} = -1$. This gives

$$v_- = \frac{v^0 + v^3}{2} = \frac{v^+}{2}, \quad v_+ = \frac{v^0 - v^3}{2} = \frac{v^-}{2}. \tag{1.37}$$

Note also that $\partial_+ = (1/2)\, \partial^-$ and $\partial_- = (1/2)\, \partial^+$.

Light cone perturbation theory is similar to time-ordered perturbation theory, except that the light cone x^+-direction plays the role of time. (For a good presentation of time-ordered perturbation theory see Sterman (1993).) Our discussion of LCPT here will closely follow Lepage and Brodsky (1980) and Brodsky and Lepage (1989). We will work in the particular light cone gauge

$$A^+ = 0, \tag{1.38}$$

which can be obtained from Eq. (1.21) by choosing $\eta^\mu = (0, 2, \vec{0}_\perp)$, in the $(+, -, \perp)$ notation. Of the remaining A^- and $A_\perp^i$ components of the gluon field ($i = 1, 2$), only the transverse components $A_\perp^i$ are independent. The component A^- can be expressed in terms of the $A_\perp^i$ using the equations of motion for the QCD Lagrangian (1.1). The quark field, which we will denote by $q(x)$, dropping the flavor label, is separated into two spinor components q_+ and q_- defined by

$$q_\pm(x) = \Lambda_\pm q(x), \tag{1.39}$$

where the projection operators $\Lambda_\pm$ are given by

$$\Lambda_\pm = \frac{1}{2} \gamma^0 \gamma^\pm \tag{1.40}$$

and the Dirac matrix $\gamma^\pm = \gamma^0 \pm \gamma^3$. Note that, just like any other projection operators, $\Lambda_\pm$ obey the following relations: $\Lambda_+ \Lambda_- = 0$, $\Lambda_\pm^2 = \Lambda_\pm$, and $\Lambda_+ + \Lambda_- = 1$. The two projections q_+ and q_- are not independent and can also be related using the constraint part of the equations of motion. The dependent field operators A^- and q_- are expressed in terms of $A^i_\perp$ and q_+ as (see Lepage and Brodsky (1980))[2]

$$A^- = -\frac{2}{\partial^+}\, \partial_{\perp j} \cdot A^j_\perp + \frac{2g}{(\partial^+)^2}\left\{\left[i\partial^+ A^j_\perp, A^j_\perp\right] + 2q^\dagger_+ t^a q_+ t^a\right\}, \qquad (1.41)$$

$$q_- = \frac{1}{i\partial^+}\, \gamma^0 \left(-i\,\gamma^j_\perp D_{\perp j} + m\right) q_+ \qquad (1.42)$$

where $j = 1, 2$. Next one defines free gluon and quark fields $\tilde{A}^\mu$ and $\tilde{q}$ by

$$\tilde{A}^\mu = (0, \tilde{A}^-, \vec{A}_\perp), \qquad (1.43)$$

in the $(+, -, \perp)$ notation, with

$$\tilde{A}^- \equiv -\frac{2}{\partial^+}\, \partial_{\perp j} \cdot A^j_\perp \qquad (1.44)$$

and

$$\tilde{q} \equiv q_+ + \frac{1}{i\partial^+}\, \gamma^0 \left(-i\gamma^j_\perp \partial_{\perp j} + m\right) q_+. \qquad (1.45)$$

The light cone Hamiltonian H is defined as the minus component of the four-momentum vector, P^-. It can be written as the sum of free and interaction terms:

$$H = P^- = H_0 + H_{int}, \qquad (1.46)$$

where (Lepage and Brodsky 1980, Brodsky and Lepage 1989, Brodsky, Pauli, and Pinsky 1998)

$$H_0 = \frac{1}{2}\int dx^- \, d^2 x_\perp \left(\bar{\tilde{q}}\,\gamma^+ \frac{m^2 - \nabla^2_\perp}{i\partial^+}\,\tilde{q} - \tilde{A}^a_\mu \nabla^2_\perp \tilde{A}^{a\,\mu}\right) \qquad (1.47)$$

is the free part of the Hamiltonian, while the interaction part is given by

$$H_{int} = \int dx^- d^2 x_\perp \left[-2g\,\mathrm{tr}\left(i\partial^\mu \tilde{A}^\nu [\tilde{A}_\mu, \tilde{A}_\nu]\right) - \frac{g^2}{2}\,\mathrm{tr}\left([\tilde{A}^\mu, \tilde{A}^\nu][\tilde{A}_\mu, \tilde{A}_\nu]\right)\right.$$

$$- g\bar{\tilde{q}}\gamma^\mu A_\mu \tilde{q} + g^2\,\mathrm{tr}\left([i\partial^+ \tilde{A}^\mu, \tilde{A}_\mu]\frac{1}{(i\partial^+)^2}[i\partial^+ \tilde{A}^\nu, \tilde{A}_\nu]\right)$$

$$+ g^2\bar{\tilde{q}}\gamma^\mu A_\mu \gamma^+ \frac{1}{2i\partial^+}\gamma^\nu A_\nu \tilde{q} - g^2\bar{\tilde{q}}\gamma^+\left(\frac{1}{(i\partial^+)^2}[i\partial^+\tilde{A}^\mu, \tilde{A}_\mu]\right)\tilde{q}$$

$$\left. + \frac{g^2}{2}\bar{\tilde{q}}\gamma^+ t^a q \frac{1}{(i\partial^+)^2}\bar{q}\gamma^+ t^a q \right]. \qquad (1.48)$$

Quantizing the theory by expanding $A^i_\perp$ and q_+ in terms of creation and annihilation operators while treating the x^+ light cone direction as time, one can construct light cone time-ordered perturbation theory with the help of the light cone Hamiltonian H. The rules of LCPT for the calculation of scattering amplitudes are given in the following subsection (Lepage and Brodsky 1980, Brodsky and Lepage 1989, Zhang and Harindranath 1993, Brodsky, Pauli, and Pinsky 1998).

---

[2] Our notation in Eqs. (1.1), (1.2), and (1.4), and therefore throughout the book, can be obtained from that of Lepage and Brodsky (1980) and Brodsky and Lepage (1989) by making the replacement $g \to -g$.

1.3.1 QCD LCPT rules

1. Draw all diagrams for a given process at the desired order in the coupling constant, including all possible orderings of the interaction vertices in the light cone time x^+. Assign a four-momentum k^μ to each line such that it is on mass shell, so that $k^2 = m^2$ with m the mass of the particle. Each vertex conserves only the k^+ and $\vec{k}_\perp$ components of the four-momentum. Hence for each line the four-momentum has components as follows:

$$k^\mu = \left(k^+, \frac{\vec{k}_\perp^2 + m^2}{k^+}, \vec{k}_\perp \right). \tag{1.49}$$

2. With quarks associate on-mass-shell spinors in the Lepage and Brodsky (1980) convention:

$$u_\sigma(p) = \frac{1}{\sqrt{p^+}} \left(p^+ + m\gamma^0 + \gamma^0 \vec{\gamma}_\perp \cdot \vec{p}_\perp \right) \chi(\sigma), \tag{1.50}$$

$$v_\sigma(p) = \frac{1}{\sqrt{p^+}} \left(p^+ - m\gamma^0 + \gamma^0 \vec{\gamma}_\perp \cdot \vec{p}_\perp \right) \chi(-\sigma), \tag{1.51}$$

with

$$\chi(+1) = \frac{1}{\sqrt{2}} \begin{pmatrix} 1 \\ 0 \\ 1 \\ 0 \end{pmatrix}, \quad \chi(-1) = \frac{1}{\sqrt{2}} \begin{pmatrix} 0 \\ 1 \\ 0 \\ -1 \end{pmatrix}. \tag{1.52}$$

Gluon lines come with a polarization vector $\epsilon_\lambda^\mu(k)$. In the $A^+ = 0$ gauge this vector is given by

$$\epsilon_\lambda^\mu(k) = \left(0, \frac{2\vec{\epsilon}_\perp^\lambda \cdot \vec{k}_\perp}{k^+}, \vec{\epsilon}_\perp^\lambda \right) \tag{1.53}$$

with transverse polarization vector

$$\vec{\epsilon}_\perp^\lambda = -\frac{1}{\sqrt{2}} (\lambda, i), \tag{1.54}$$

where $\lambda = \pm 1$. Equation (1.53) follows from requiring that $\epsilon_\lambda^+ = 0$ and $\epsilon_\lambda(k) \cdot k = 0$.

3. For each intermediate state there is a factor equal to the light cone energy denominator

$$\frac{1}{\sum\limits_{inc} k^- - \sum\limits_{interm} k^- + i\epsilon} \tag{1.55}$$

where the sums run respectively over all incoming particles present in the initial state in the diagram ("inc") and over all the particles in the intermediate state at hand ("interm"). According to rule 1 above, for each particle we have $k^- = (\vec{k}_\perp^2 + m^2)/k^+$. Since the k^- momentum component is not conserved at the vertices the intermediate states are not on the "energy shell" and the light cone denominator in (1.55) is nonzero. Note that the light cone energy is conserved for the whole scattering process: $\sum_{inc} k^-$ is equal to $\sum_{out} k^-$, where "out" stands for all outgoing particles.[3]

4. Include a factor

$$\frac{\theta(k^+)}{k^+} \tag{1.56}$$

for each internal line, where k^+ flows in the future light cone time direction.

5. For vertices include factors as follows (we assume that the light cone time flows from left to right).

Quark–gluon vertex (i and j are quark color indices):

$$= -g \bar{u}_{\sigma' j}(p+q) \, \epsilon\!\!\!/_\lambda(q) \, (t^a)_{ji} \, u_{\sigma i}(p). \tag{1.57}$$

Three-gluon vertex (all momenta flow into the vertex; asterisks denote complex conjugation):

$$\begin{aligned} = -i g f^{abc} \big[& (k_1 - k_3) \cdot \epsilon^*_{\lambda_2}(k_2) \, \epsilon_{\lambda_1}(k_1) \cdot \epsilon_{\lambda_3}(k_3) \\ & + (k_2 - k_1) \cdot \epsilon_{\lambda_3}(k_3) \, \epsilon_{\lambda_1}(k_1) \cdot \epsilon^*_{\lambda_2}(k_2) \\ & + (k_3 - k_2) \cdot \epsilon_{\lambda_1}(k_1) \, \epsilon_{\lambda_3}(k_3) \cdot \epsilon^*_{\lambda_2}(k_2) \big]. \end{aligned} \tag{1.58}$$

Four-gluon vertex:

$$\begin{aligned} = g^2 \big[& f^{abe} f^{cde} (\epsilon_{\lambda_1} \cdot \epsilon_{\lambda_3} \, \epsilon^*_{\lambda_2} \cdot \epsilon^*_{\lambda_4} - \epsilon_{\lambda_1} \cdot \epsilon^*_{\lambda_4} \, \epsilon_{\lambda_3} \cdot \epsilon^*_{\lambda_2}) \\ & + f^{ace} f^{bde} (\epsilon_{\lambda_1} \cdot \epsilon^*_{\lambda_2} \, \epsilon_{\lambda_3} \cdot \epsilon^*_{\lambda_4} - \epsilon_{\lambda_1} \cdot \epsilon^*_{\lambda_4} \, \epsilon_{\lambda_3} \cdot \epsilon^*_{\lambda_2}) \\ & + f^{ade} f^{bce} (\epsilon_{\lambda_1} \cdot \epsilon^*_{\lambda_2} \, \epsilon_{\lambda_3} \cdot \epsilon^*_{\lambda_4} - \epsilon_{\lambda_1} \cdot \epsilon_{\lambda_3} \, \epsilon^*_{\lambda_2} \cdot \epsilon^*_{\lambda_4}) \big]. \end{aligned} \tag{1.59}$$

In addition to the above vertices, which are (up to some trivial factors due to a different convention) identical to the same vertices in the Feynman rules, there are instantaneous terms in the light cone Hamiltonian giving the four vertices below. Again, light cone time flows to the right while the momentum flow direction is indicated by arrows. Instantaneous quark and gluon lines are denoted by regular quark and gluon lines with a short

---

[3] This light cone energy conservation condition does not apply to light cone wave functions, to be discussed shortly, as they represent only part of the scattering process.

line crossing them.

$$
= g^2 \, \bar{u}_{\sigma_2 j}(p_2) \, \slashed{\epsilon}_{\lambda_1}(k_1) \, \frac{\gamma^+}{2(p_1^+ - k_2^+)} \, \slashed{\epsilon}^*_{\lambda_2}(k_2)
$$
$$
\times (t^a \, t^b)_{ji} \, u_{\sigma_1 i}(p_1), \tag{1.60}
$$

$$
= g^2 \, \bar{u}_{\sigma_2 j}(p_2) \, \gamma^+ \, (t^a)_{ji} \, u_{\sigma_1 i}(p_1)
$$
$$
\times \bar{u}_{\sigma_4 l}(p_4) \, \gamma^+ \, (t^a)_{lk} \, u_{\sigma_3 k}(p_3) \, \frac{1}{(p_1^+ - p_2^+)^2}, \tag{1.61}
$$

$$
= -\, g^2 \, \bar{u}_{\sigma_2 j}(p_2) \, \gamma^+ \, (t^c)_{ji} \, u_{\sigma_1 i}(p_1)
$$
$$
\times \frac{k_1^+ + k_2^+}{(k_1^+ - k_2^+)^2} \, i f^{abc} \epsilon^*_{\lambda_2} \cdot \epsilon_{\lambda_1}, \tag{1.62}
$$

$$
= g^2 \, f^{abe} \, f^{cde} \, \epsilon^*_{\lambda_2} \cdot \epsilon_{\lambda_1} \, \epsilon^*_{\lambda_4} \cdot \epsilon_{\lambda_3}
$$
$$
\times \frac{(k_1^+ + k_2^+)(k_3^+ + k_4^+)}{(k_1^+ - k_2^+)^2}. \tag{1.63}
$$

6. For each independent momentum k^μ integrate with the measure

$$
\int \frac{dk^+ \, d^2 k_\perp}{2(2\pi)^3}. \tag{1.64}
$$

Sum over all internal quark and gluon polarizations and colors.

Again, standard parts of the rules, common to both LCPT and Feynman diagram calculations, such as symmetry factors and a factor -1 for fermion loops and for fermion lines beginning and ending at the initial state, are assumed implicitly.

The rules of LCPT are supplemented by tables of Dirac matrix elements in appendix section A.1. These tables are very useful in the evaluation of LCPT vertices.

1.3.2 Light cone wave function

An important quantity in LCPT, which is hard to construct in the standard Feynman diagram language, is the light cone wave function. Its definition is similar to that of the wave function in quantum mechanics. In our presentation of the light cone wave function we will follow Brodsky, Pauli, and Pinsky (1998). Imagine that we have a hadron state $|\Psi\rangle$. In general this is a superposition of different Fock states

$$|n_G, n_q\rangle \equiv |n_G, \{k_i^+, \vec{k}_{i\perp}, \lambda_i, a_i\}; n_q, \{p_j^+, \vec{p}_{j\perp}, \sigma_j, \alpha_j, f_j\}\rangle, \qquad (1.65)$$

where a particular Fock state has n_G gluons and n_q quarks (and antiquarks). The gluon momenta are labeled k_i^+, $\vec{k}_{i\perp}$, with polarizations λ_i and gluon color indices a_i where $i = 1, \ldots, n_G$. (As usual in LCPT $k_i^- = \vec{k}_{i\perp}^2/k_i^+$, as all particles are on mass shell.) The quark momenta are labeled p_j^+, $\vec{p}_{j\perp}$, with helicities σ_j, colors α_j, and flavors f_j where $j = 1, \ldots, n_q$.

The Fock states form a complete basis such that

$$\sum_{n_G, n_q} \int d\Omega_{n_G+n_q} \, |n_G, n_q\rangle\langle n_G, n_q| = \mathbf{1}, \qquad (1.66)$$

where the phase-space integral is defined by

$$\int d\Omega_{n_G+n_q} = \frac{2P^+(2\pi)^3}{S_n} \int \prod_{i=1}^{n_G} \sum_{\lambda_i, a_i} \frac{dk_i^+ \, d^2 k_{i\perp}}{2k_i^+(2\pi)^3} \prod_{j=1}^{n_q} \sum_{\sigma_j, \alpha_j, f_j} \frac{dp_j^+ \, d^2 p_{j\perp}}{2p_j^+(2\pi)^3}$$

$$\times \delta\left(P^+ - \sum_{l_1=1}^{n_G} k_{l_1}^+ - \sum_{l_2=1}^{n_q} p_{l_2}^+\right) \delta^2\left(\vec{P}_\perp - \sum_{m_1=1}^{n_G} \vec{k}_{m_1\perp} - \sum_{m_2=1}^{n_q} \vec{p}_{m_2\perp}\right)$$

$$(1.67)$$

with symmetry factor $S_n = n_G! \, n_Q! \, n_{\bar{Q}}!$. Here n_Q and $n_{\bar{Q}}$ are respectively the numbers of quarks and antiquarks in the wave function, so that $n_q = n_Q + n_{\bar{Q}}$. The delta functions in Eq. (1.67) represent the conservation of the "plus" and transverse components of the momenta, according to rule 1 of LCPT. The incoming hadron has longitudinal momentum P^+ and transverse momentum $\vec{P}_\perp$. We assume that each Fock state is normalized to 1, so that $\langle n_G, n_q | n_G, n_q \rangle = 1$.

Using Eq. (1.66) we can write

$$|\Psi\rangle = \sum_{n_G, n_q} \int d\Omega_{n_G+n_q} \, |n_G, n_q\rangle\langle n_G, n_q|\Psi\rangle. \qquad (1.68)$$

The quantity

$$\Psi(n_G, n_q) = \langle n_G, n_q|\Psi\rangle \qquad (1.69)$$

is called *the light cone wave function*. It is a multi-particle wave function, describing a Fock state in the hadron with n_G gluons and n_q quarks.

Note that requiring that the state $|\Psi\rangle$ is normalized to unity, $\langle\Psi|\Psi\rangle = 1$, and using Eq. (1.68) we can write

$$1 = \langle\Psi|\Psi\rangle = \sum_{n_G, n_q} \int d\Omega_{n_G+n_q} \, |\Psi(n_G, n_q)|^2. \qquad (1.70)$$

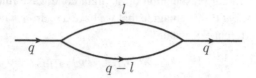

Fig. 1.1. A Feynman diagram in the ϕ^3-theory considered here. The arrows indicate the momentum flow.

We see that each light cone wave function $\Psi(n_G, n_q)$ is normalized to a number less than or equal to 1.

1.4 Sample LCPT calculations

While we expect that the reader has a fluent knowledge of Feynman rules, we realize that it is less likely that he or she is equally fluent with LCPT rules. Therefore, to help the reader become more familiar with LCPT, here we will perform two LCPT calculations. We will first "cross-check" LCPT by calculating a sample scattering amplitude using both the Feynman and LCPT rules and showing that we obtain the same result. We will then set up the rules for calculating light cone wave functions, by considering an example of a basic wave function containing $1 \to 2$ particle splitting.

1.4.1 LCPT "cross-check"

We begin by calculating a simple amplitude in a real scalar ϕ^3 field theory in two ways: using standard Feynman rules and using the rules of LCPT. We will show that the two ways give identical results. This demonstrates that LCPT is indeed equivalent to the standard Feynman diagram approach.

The process we consider is illustrated in Fig. 1.1. We consider a field theory for a real massive scalar field ϕ with Lagrangian

$$\mathcal{L} = \frac{1}{2} \partial_\mu \phi \, \partial^\mu \phi - \frac{m^2}{2} \phi^2 - \frac{\lambda}{3!} \phi^3. \tag{1.71}$$

The contribution of the diagram in Fig. 1.1 (henceforth labeled A) can be written down using the Feynman rules for the real scalar field theory having Lagrangian (1.71) (see e.g. Sterman (1993) on Peskin and Schroeder (1995)):

$$-i\Sigma = \frac{(-i\lambda)^2}{2!} \int \frac{d^4 l}{(2\pi)^4} \frac{i}{l^2 - m^2 + i\epsilon} \frac{i}{(q - l)^2 - m^2 + i\epsilon}. \tag{1.72}$$

Here $1/2!$ is a symmetry factor and m is the mass of the scalar particles.

Working in the light cone variables

$$q^\mu = (q^+, q^-, \vec{q}_\perp), \quad l^\mu = (l^+, l^-, \vec{l}_\perp), \tag{1.73}$$

we write $l^2 = l^+ l^- - \vec{l}_\perp^2$ and $(q - l)^2 = (q^+ - l^+)(q^- - l^-) - (\vec{q}_\perp - \vec{l}_\perp)^2$. Equation (1.72) can now be rewritten as

$$-i\Sigma = \frac{\lambda^2}{4} \int \frac{dl^+ \, dl^- \, d^2 l_\perp}{(2\pi)^4} \frac{1}{l^+ l^- - \vec{l}_\perp^2 - m^2 + i\epsilon}$$

$$\times \frac{1}{(q^+ - l^+)(q^- - l^-) - (\vec{q}_\perp - \vec{l}_\perp)^2 - m^2 + i\epsilon}. \tag{1.74}$$

Now we need to integrate over l^-. In the complex l^--plane the integrand in Eq. (1.74) has two poles,

$$l_1^- = \frac{\vec{l}_\perp^2 + m^2 - i\epsilon}{l^+} \quad \text{and} \quad l_2^- = q^- - \frac{(\vec{q}_\perp - \vec{l}_\perp)^2 + m^2 - i\epsilon}{q^+ - l^+}. \tag{1.75}$$

The l^--integral is nonzero only if these two poles lie in different half-planes. This happens for either (i) $l^+ > 0$, $q^+ - l^+ > 0$ or (ii) $l^+ < 0$, $q^+ - l^+ < 0$. As the incoming particle with momentum q is physical we have $q^+ > 0$, which makes case (ii) impossible to achieve, as there one has $q^+ < l^+ < 0$. We are left with case (i). Closing the l^--integration contour in the lower half-plane we pick up the pole at l_1^-, obtaining

$$\Sigma = \frac{\lambda^2}{2} \int \frac{dl^+ \, d^2 l_\perp}{2(2\pi)^3} \frac{\theta(l^+)\theta(q^+ - l^+)}{l^+(q^+ - l^+)}$$

$$\times \frac{1}{q^- - \dfrac{\vec{l}_\perp^2 + m^2 - i\epsilon}{l^+} - \dfrac{(\vec{q}_\perp - \vec{l}_\perp)^2 + m^2 - i\epsilon}{q^+ - l^+}}$$

$$= \frac{\lambda^2}{2!} \int \frac{dl^+ \, d^2 l_\perp}{2(2\pi)^3} \frac{\theta(l^+)\theta(q^+ - l^+)}{l^+(q^+ - l^+)}$$

$$\times \frac{1}{q^- - \dfrac{\vec{l}_\perp^2 + m^2}{l^+} - \dfrac{(\vec{q}_\perp - \vec{l}_\perp)^2 + m^2}{q^+ - l^+} + i\epsilon}. \tag{1.76}$$

We observe that Eq. (1.76) is identical to what one would obtain for the diagram in Fig. 1.1 if one calculated it using the rules of LCPT from Sec. 1.3 (modified for a scalar particle), as illustrated in Fig. 1.2. Indeed Eq. (1.76) can be obtained by assigning

$$\frac{\theta(l^+)}{l^+} \quad \text{and} \quad \frac{\theta(q^+ - l^+)}{q^+ - l^+} \tag{1.77}$$

for each internal line (LCPT rule 4), including an energy denominator

$$\frac{1}{\displaystyle\sum_{inc} k^- - \sum_{interm} k^- + i\epsilon} = \frac{1}{q^- - \dfrac{\vec{l}_\perp^2 + m^2}{l^+} - \dfrac{(\vec{q}_\perp - \vec{l}_\perp)^2 + m^2}{q^+ - l^+} + i\epsilon} \tag{1.78}$$

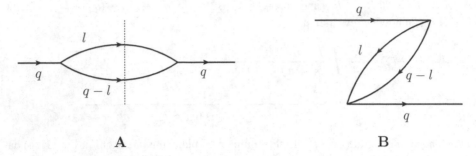

A B

Fig. 1.2. Light cone perturbation theory diagrams in the ϕ^3-theory corresponding to the Feynman diagram in Fig. 1.1. Time flows to the right. The arrows indicate the momentum direction. The vertical dotted line indicates an intermediate state.

for the intermediate state (denoted by the dotted line in Fig. 1.2A), according to LCPT rule 3, and integrating over the internal momentum l with the integration measure

$$\int \frac{dl^+ d^2 l_\perp}{2(2\pi)^3},$$ (1.79)

as prescribed by LCPT rule 6. In LCPT each vertex gives a factor λ (a modification of rule 5 for ϕ^3-theory) and one has to include the symmetry factor $1/2!$ as well. (Scalar particles obviously have no polarization. Neither do they have instantaneous terms.)

We have demonstrated that starting from the Feynman diagram amplitude expression (1.72) we can reduce it to the result that one would obtain by the rules of LCPT. Hence the two approaches in the end give identical expressions for the amplitudes, as expected.

A few words of caution are in order here. In principle the Feynman diagram in Fig. 1.1 corresponds to the two LCPT diagrams A and B shown in Fig. 1.2, which correspond to two different orderings of the vertices (see LCPT rule 1). The two graphs A and B in fact correspond to cases (i) and (ii) considered after Eq. (1.75). Our argument above was simplified by the fact that diagram B in Fig. 1.2 is zero as, according to the LCPT rules, it comes with a factor $\theta(-l^+)\theta(l^+ - q^+)$, which is zero for $q^+ > 0$. The physical meaning of this is quite clear: one cannot generate three particles with positive plus momenta out of nothing (see the lower vertex in Fig. 1.2B). Conversely, three particles with positive plus momenta cannot combine to give nothing (see the upper vertex in Fig. 1.2B). Because of this simplification, we have a one-to-one correspondence between the Feynman diagram in Fig. 1.1 and the LCPT diagram in Fig. 1.2A. In general, each Feynman diagram corresponds to a sum of all the LCPT diagrams with the same topology, including all possible time-orderings and instantaneous terms. A general derivation of an LCPT diagram starting from a Feynman diagram does not simply involve integration over the minus components of the internal momenta; one has to assign each vertex an x^+-coordinate and Fourier transform the diagram (by integrating over the minus momenta) into x^+ coordinate space. One then

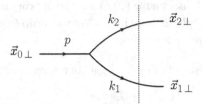

Fig. 1.3. Light cone wave function for a scalar particle splitting into two. The vertical dotted line denotes an intermediate state.

has to integrate over all the x^+-coordinates of the vertices, imposing different orderings: each ordering will lead to a different LCPT diagram.

1.4.2 A sample light cone wave function

Let us calculate, using the rules of LCPT, a sample light cone wave function. The calculation will be instructive, as the wave function we will calculate is similar to certain light cone wave functions that we will use throughout the book. In this calculation we will also illustrate in more detail what is actually meant by the light cone wave function definition (1.69) and will set up the rules for wave function calculations.

The sample wave function is depicted in Fig. 1.3. Again we are working in ϕ^3 real scalar field theory, with the Lagrangian (1.71). The wave function describes a single incoming particle splitting into two. For the scalar field theory only rules 1, 3, 4, and 6 from Sec. 1.3 apply. On top of these rules there is a factor equal to the coupling λ coming from the vertex. In calculating light cone wave functions one has to treat the "outgoing" state on the right of the diagram (the state denoted by the dotted line in Fig. 1.3) as an intermediate state. The reason is that, in describing a scattering process, the light cone wave function is thought of as a part of a larger diagram in which this "outgoing" state in fact undergoes subsequent interactions with other particles and therefore is truly an intermediate state. Our definition of the boost-invariant integration measure (1.67) dictates a slight modification of LCPT rule 4 as well, when calculating light cone wave functions: we treat the incoming lines (the external lines on the left, e.g. line p in Fig. 1.3) as "internal" and include a factor $1/p^+$ for them, while the outgoing lines (the lines on the right, e.g. lines k_1 and k_2 in Fig. 1.3) will be treated as "external" and so will not bring in such factors.

To summarize, when calculating the light cone wave function using LCPT one should follow the rules stated in Sec. 1.3, with the following modifications.

(i) The outgoing state on the right of a diagram is treated as an internal state and brings in an energy denominator according to LCPT rule 3.
(ii) At the same time the outgoing external lines on the right of the diagram bring in only factors $\theta(k^+)$, in modification of LCPT rule 4. (As usual, light cone time flows to the right.)

(iii) The incoming external lines on the left of a diagram bring in factors $1/p^+$, i.e., LCPT rule 4 is extended to apply to those lines. (We will drop $\theta(p^+ > 0)$ as incoming lines always have positive p^+ momentum.)

According to the above-stated rules, the light cone wave function depicted in Fig. 1.3 is

$$\Psi(k_1, k_2) = \frac{1}{p^+} \frac{\lambda}{p^- - k_1^- - k_2^-}$$

$$= \frac{1}{p^+} \frac{\lambda}{\frac{\vec{p}_\perp^2 + m^2}{p^+} - \frac{\vec{k}_{1\perp}^2 + m^2}{k_1^+} - \frac{\vec{k}_{2\perp}^2 + m^2}{k_2^+}}, \tag{1.80}$$

where we have omitted the regulator $i\epsilon$ for simplicity (in fact we will not need it below). Before we simplify this expression, let us note that, as can be seen from Eq. (1.70), the probability of finding such a configuration in a general "dressed" state $|\Psi\rangle$ of the incoming particle is

$$\int d\Omega_2 \left|\Psi(k_1, k_2)\right|^2, \tag{1.81}$$

where, as follows from Eq. (1.67), the phase-space integral for two identical particles is given by

$$\int d\Omega_2 = \frac{2p^+ (2\pi)^3}{2!} \int \frac{dk_1^+ d^2 k_{1\perp}}{2k_1^+ (2\pi)^3} \frac{dk_2^+ d^2 k_{2\perp}}{2k_2^+ (2\pi)^3} \delta\left(p^+ - k_1^+ - k_2^+\right)$$

$$\times \delta^2\left(\vec{p}_\perp - \vec{k}_{1\perp} - \vec{k}_{2\perp}\right)$$

$$= \frac{1}{2!} \int \frac{dk_1^+ d^2 k_{1\perp}}{2k_1^+ (2\pi)^3} \frac{p^+}{p^+ - k_1^+}. \tag{1.82}$$

We see that $k_2^+ = p^+ - k_1^+$ and $\vec{k}_{2\perp} = \vec{p}_\perp - \vec{k}_{1\perp}$. Using these to replace k_2^+ and $\vec{k}_{2\perp}$ in Eq. (1.80) and doing some algebra yields

$$\Psi(k_1, p - k_1) = -\frac{\lambda z_1(1 - z_1)}{(\vec{k}_{1\perp} - z_1 \vec{p}_\perp)^2 + m^2 [1 - z_1(1 - z_1)]}, \tag{1.83}$$

where

$$z_1 = \frac{k_1^+}{p^+} \tag{1.84}$$

is the longitudinal fraction of the original particle's momentum p carried by the particle k_1, which will be identified as a Feynman-x variable in the next chapter. Equation (1.83) gives us the momentum-space two-particle light cone wave function at the lowest order in λ.

Substituting the wave function (1.83) into Eq. (1.81) and using Eq. (1.82) for the phase-space integration measure, one obtains the probability for one particle to fluctuate into two particles:

$$\frac{\lambda^2}{2!} \int \frac{dz_1\, d^2k_{1\perp}}{2(2\pi)^3} \frac{z_1(1-z_1)}{\left\{(\vec{k}_{1\perp} - z_1\vec{p}_\perp)^2 + m^2\left[1 - z_1(1-z_1)\right]\right\}^2} \sim \frac{\lambda^2}{m^2}. \qquad (1.85)$$

Thus the probability of the configuration in Fig. 1.3 is proportional to the coupling constant squared. As the coupling in ϕ^3-theory has the dimension of the mass, the factor m^2 in the denominator of Eq. (1.85) makes the expression dimensionless. We note in passing that the effective dimensionless coupling constant for the perturbative expansion of ϕ^3-theory is λ/m.

It is also instructive to Fourier-transform the wave function (1.83) into transverse coordinate space. The transverse coordinates of the lines are shown in Fig. 1.3. The Fourier transform is accomplished by integrating over the independent transverse momenta, assigning a factor $e^{i\vec{k}_\perp \cdot \vec{x}_\perp}$ for each line, with k the net outgoing momentum carried by the line. For the two-particle wave function (1.83) we have

$$\Psi(\vec{x}_{1\perp}, \vec{x}_{2\perp}, \vec{x}_{0\perp}, z_1)$$

$$= \int \frac{d^2k_{1\perp}\, d^2p_\perp}{(2\pi)^4} e^{i\vec{k}_{1\perp}\cdot\vec{x}_{1\perp} + i\vec{k}_{2\perp}\cdot\vec{x}_{2\perp} - i\vec{p}_\perp\cdot\vec{x}_{0\perp}}\, \Psi(k_1, p - k_1)$$

$$= \int \frac{d^2k_{1\perp}\, d^2p_\perp}{(2\pi)^4} e^{i\vec{k}_{1\perp}\cdot(\vec{x}_{1\perp} - \vec{x}_{2\perp}) - i\vec{p}_\perp\cdot(\vec{x}_{0\perp} - \vec{x}_{2\perp})}\, \Psi(k_1, p - k_1). \qquad (1.86)$$

Substituting Eq. (1.83) into Eq. (1.86) and integrating yields (see Eq. (A.11))

$$\Psi(\vec{x}_{1\perp}, \vec{x}_{2\perp}, \vec{x}_{0\perp}, z_1) = -\frac{\lambda}{2\pi} z_1(1-z_1)\, K_0\left(|\vec{x}_{12}|\, m\sqrt{1 - z_1(1-z_1)}\right)$$

$$\times\, \delta^2(\vec{x}_{0\perp} - z_1\vec{x}_{1\perp} - (1-z_1)\vec{x}_{2\perp}), \qquad (1.87)$$

where $\vec{x}_{ij} \equiv \vec{x}_{i\perp} - \vec{x}_{j\perp}$. Equation (1.87) gives us the $1 \to 2$ splitting wave function shown in Fig. 1.3 in coordinate space. Even though this wave function has been obtained for the scalar ϕ^3-theory case it has a feature valid for theories with higher spin: it contains a delta function insuring that $\vec{x}_{0\perp} = z_1\vec{x}_{1\perp} + (1-z_1)\vec{x}_{2\perp}$. This means that the transverse coordinate positions of the two produced particles are indeed related to each other (Kopeliovich, Tarasov, and Schafer 1999): both the original particle and the two new particles lie on one straight line in transverse coordinate space, and $x_{02} : x_{01} = z_1 : (1 - z_1)$ where $x_{ij} = |\vec{x}_{ij}|$. The transverse coordinate space structure of the wave function (1.87) is illustrated in Fig. 1.4. The same constraint on the transverse plane locations of the produced particles as derived here for the ϕ^3-theory applies to the splittings of particles in quantum electrodynamics (QED) and in QCD.

1.5 Asymptotic freedom

A remarkable property of QCD, known as *asymptotic freedom*, is the fact that the running QCD coupling tends to be small at short distances (corresponding to large values of the

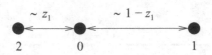

Fig. 1.4. The $1 \to 2$ splitting wave function pictured in transverse coordinate space. The circles represent particles and the numbers label these particles in agreement with the diagram in Fig. 1.3: 0 labels the original particle, while 1 and 2 label the produced particles.

relevant four-momentum squared, $q^2 = -Q^2$ with Q a real number). The running of the QCD coupling constant is given by (Gross and Wilczek 1973, Politzer 1973)[4]

$$\alpha_s(Q^2) = \frac{\alpha_s(\mu^2)}{1 + \alpha_s(\mu^2)\,\beta_2\,\ln(Q^2/\mu^2)}, \tag{1.88}$$

where

$$\beta_2 = \frac{11 N_c - 2N_f}{12\pi} \tag{1.89}$$

with $N_c = 3$ the number of colors and N_f the number of quark flavors. The QCD beta function is given by

$$\beta_{QCD}(\alpha) = -\beta_2 \alpha^2 + O(\alpha^3). \tag{1.90}$$

While $N_f = 6$ in the Standard Model of particle physics, the effective number of flavors relevant for a given physical process depends on the momentum scale Q and may be smaller than six. One can clearly see from Eq. (1.88) that $\alpha_s(Q^2) \to 0$ as $Q^2 \to \infty$: the strong coupling is small at large momenta. Thus quarks and gluons interact *weakly* at asymptotically short distances; this is asymptotic freedom.

Such behavior is in striking contrast with the running of the coupling in quantum electrodynamics (QED), where β_2 is negative, making the QED coupling grow with Q^2 (Landau, Abrikosov, and Halatnikov 1956). The main difference between QED and QCD is in the non-Abelian interactions between the gluons. Owing to these interactions the gluon propagator receives corrections not only from quark loops (which are quite similar to the electron loops in QED) but also from gluon loops. The polarizations of virtual gluons in these loops can be either transverse or longitudinal. The transverse gluon and quark loops generate terms tending to make the QCD beta function positive (and $\beta_2 < 0$). Owing to a large contribution from the longitudinal gluon in the loop, however, the resulting QCD beta function is negative (and $\beta_2 > 0$), leading to asymptotic freedom (see Khriplovich (1969), Gribov (1978), and Dokshitzer and Kharzeev (2004) for more details).

The quantity μ in Eq. (1.88) is an arbitrary scale (known as the renormalization point): physical observables should not depend on its value. In fact Eq. (1.88) can be rewritten as

$$\alpha_s(Q^2) = \frac{1}{\beta_2\,\ln(Q^2/\Lambda_{QCD}^2)} \tag{1.91}$$

---

[4] The QCD beta function was also calculated by 't Hooft but the result was not included in t' Hooft (1972).

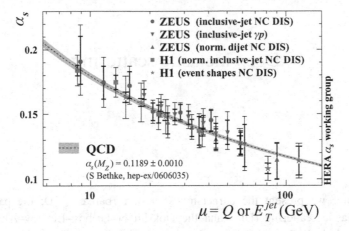

Fig. 1.5. The experimental data on the running QCD coupling from deep inelastic scattering (DIS) experiments at HERA. The dashed line with a band around it is the theoretical prediction for the strong coupling. (Reprinted with permission from H1 and ZEUS collaboration (2008). Copyright 2008 by IOP Publishing.) A color version of this figure is available online at www.cambridge.org/9780521112574.

where $\Lambda_{QCD} \approx 200\text{--}300$ MeV is the fundamental scale of QCD. (The exact value of Λ_{QCD} depends on the renormalization scheme used.) The strong coupling constant $\alpha_s(Q^2)$ becomes large near $Q \approx \Lambda_{QCD}$, leading to strong forces between the quarks and gluons. These strong forces presumably contribute to the confinement of quarks and gluons within hadrons.

For the purposes of this book the most important implication of Eq. (1.88) is that at short distances (large transverse momenta) the strong coupling is small. This small value of the dimensionless running QCD coupling gives the naturally small parameter needed to develop perturbation theory. Therefore the rules are simple: as we probe shorter and shorter distances inside the hadron perturbative QCD calculations become better justified, providing more theoretical control over the problem at hand.

Figure 1.5 shows a compilation of the data on the strong coupling constant determined from deep inelastic electron–proton scattering experiments at a single collider, the Hadron Electron Ring Accelerator (HERA) at the Deutsches Elektronen-Synchrotron (DESY) laboratory in Hamburg, Germany. The dashed line with a narrow band around it in Fig. 1.5 represents our theoretical knowledge of $\alpha_s(Q^2)$, which is based on Eq. (1.88) along with several higher-order corrections (up to three loops). The agreement between theory and data shown in Fig. 1.5 is quite remarkable and is a major triumph in our attempts to understand how QCD works.

2

Deep inelastic scattering

In this chapter we present the cornerstones of perturbative QCD: the parton model of deep inelastic scattering (DIS) and the Dokshitzer–Gribov–Lipatov–Altarelli–Parisi (DGLAP) evolution equations. There exists an extensive literature covering these subjects using Lorentz-covariant Feynman diagram techniques (see the further reading section at the end of the chapter). Here we deviate from the traditional treatment and derive both the parton model and the DGLAP equations using light cone perturbation theory (LCPT). We argue that the light cone approach provides an intuitively clear space–time picture of the scattering process, which is universally applicable for high energy scattering. Owing to this universality, both the LCPT techniques used here and their space–time interpretation will prove very useful in subsequent chapters.

2.1 Kinematics, cross section, and structure functions

One of the simplest scattering processes that occur at short distances is the reaction

$$e + p \longrightarrow e' + X, \tag{2.1}$$

known as deep inelastic electron–proton scattering (DIS). Here e and e' are the incoming and outgoing electron (or positron), p is the proton, and X stands for the other produced particles. The process is illustrated diagrammatically in Fig. 2.1 in the rest frame of the proton. The electron scatters on the proton through the exchange of a virtual photon (denoted γ^*) with a quark in the proton's wave function. The virtual photon usually breaks the proton apart, leading to the production of several new hadrons; these are labeled X in Fig. 2.1. Hence the process is deeply inelastic, which explains its name.

We begin by working in the rest frame of the proton. As shown in Fig. 2.1, the four-momentum of the proton is $P^\mu = (m, \vec{0})$, where m is the proton's mass. The four-momentum of the incoming electron is $p^\mu = (E, \vec{p})$, while the outgoing electron has four-momentum $p'^\mu = (E', \vec{p}')$. Out of the three independent four-momenta P^μ, p^μ, and p'^μ one can construct three Lorentz invariants relevant to the collision dynamics. (Note that $P^2 = m^2$ and $p^2 = p'^2 = m_e^2$, where m_e is the electron's mass; while these masses are indeed Lorentz scalars they do not carry any information about the scattering.) In terms of the virtual photon's four-momentum $q^\mu \equiv p^\mu - p'^\mu$, the three invariants usually employed to describe

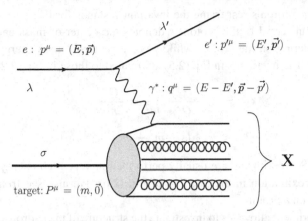

Fig. 2.1. Feynman diagram describing deep inelastic electron–proton scattering. The momentum labels of the lines correspond to the frame in which the target proton is at rest. The wavy line denotes the virtual photon propagator while the corkscrew lines denote the gluons inside the proton.

DIS are

$$Q^2 \equiv -q^2,$$

$$x_{Bj} \equiv \frac{Q^2}{2P \cdot q}, \qquad (2.2)$$

$$y \equiv \frac{P \cdot q}{P \cdot p}.$$

The quantity Q^2 is called the virtuality of the photon, while x_{Bj} is the Bjorken-x variable. In the rest frame of the proton one can easily show that

$$Q^2 = 4EE' \sin^2 \frac{\theta}{2} \qquad (2.3)$$

and

$$y = \frac{E - E'}{E}. \qquad (2.4)$$

Here θ is the electron scattering angle, i.e., the angle between $\vec{p}$ and $\vec{p}\,'$. We therefore see that $q^2 \le 0$ or, equivalently, $Q^2 \ge 0$, which demonstrates that Q is indeed real. In the proton's rest frame the third Lorentz invariant y has a physical interpretation as the fraction of the electron's energy transferred to the proton.

Apart from the three independent invariants in Eq. (2.2) one usually defines other Lorentz-invariant (but not independent) quantities,

$$\nu \equiv \frac{P \cdot q}{m} = E - E',$$

$$\hat{s} \equiv (P + q)^2 = 2P \cdot q + q^2 + m^2, \qquad (2.5)$$

$$s \equiv (P + p)^2.$$

We see that in the proton's rest frame the invariant ν stands for that part of the electron's energy that is transferred to the proton; s denotes the center-of-mass energy squared of the electron scattering on the proton, while $\hat{s}$ is the center-of-mass energy squared of the $\gamma^* + p$ reaction. The invariants in Eq. (2.5) are related to those in Eq. (2.2) via

$$x_{Bj} = \frac{Q^2}{\hat{s} + Q^2 - m^2} = \frac{Q^2}{2m\nu},$$
$$Q^2 = yx_{Bj}(s - m^2 - m_e^2) \approx yx_{Bj}s. \tag{2.6}$$

The fact that DIS experiments are usually performed at very high energy $s \gg m^2 \gg m_e^2$ justifies the approximation in the last line of Eq. (2.6). We also see from Eq. (2.6) that $x_{Bj} \leq 1$ for DIS on a proton.

The DIS experiment allows us to investigate the structure of the hadron at short distances by observing the recoil electron e' in Eq. (2.1). As we will see shortly, a DIS experiment can be thought of as a relativistic electron microscope. We can characterize this "microscope" by its maximal resolution. We will show below that with this DIS microscope we can resolve the sizes of the proton's constituents down to $1/Q$. Thus the physical meaning of the photon virtuality Q^2 is that it is related to the resolution of our "microscope". However, because our microscope is relavistic, we need to introduce one more variable, namely, the time duration of the observation. The number of particles is not conserved in a relativistic system: the number of quarks and gluons inside the proton constantly fluctuates owing to particle splitting and annihilation. Some fluctuations have longer lifetimes while others have shorter lifetimes. Therefore, the number of proton constituents can be different when measured over different observation times. We will show below that the measuring time of the DIS microscope is proportional to $1/x_{Bj}$, so that $t \sim 1/(mx_{Bj})$. This gives one of the two physical interpretations of x_{Bj}.

Using the covariant gauge for the photon propagator we can write the amplitude for the DIS process pictured in Fig. 2.1 as

$$iM_{\sigma,\lambda,\lambda'}(X) = \frac{ie^2}{q^2} \bar{u}_{\lambda'}(p') \gamma_\mu u_\lambda(p) \langle X|J^\mu(0)|P,\sigma\rangle. \tag{2.7}$$

Here λ and λ' are the electron polarizations before and after the interaction and σ is the polarization of the proton (see Fig. 2.1). The initial state of the proton is denoted $|P,\sigma\rangle$, while the final state of the many produced hadrons X in Fig. 2.1 is correspondingly denoted as $|X\rangle$. We define the quark electromagnetic current by

$$J^\mu(x) = \sum_f Z_f \bar{q}^f(x) \gamma^\mu q^f(x), \tag{2.8}$$

where Z_f is the quark's electric charge in units of the electron charge e, $q^f(x)$ is the quark field operator, and the sum in Eq. (2.8) runs over all quark flavors. (All operators in the book are in the Heisenberg representation.)

To calculate the total DIS cross section we need to square the amplitude (2.7), integrate or sum over the final-state quantum numbers, average over the initial-state quantum numbers, divide by the flux factor, and impose energy–momentum conservation (see e.g. Peskin and

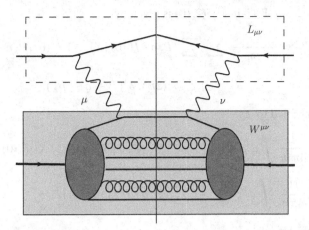

Fig. 2.2. Diagrammatic representation of the DIS cross section calculation as the amplitude squared. The vertical solid line denotes the final-state cut. The rectangular boxes encompass the parts of the diagram contributing to the leptonic tensor $L_{\mu\nu}$ and the hadronic tensor $W^{\mu\nu}$.

Schroeder (1995)). We get

$$\sigma^{ep} = \int \frac{d^3 p'}{(2\pi)^3\,2E\,2E'}\,\frac{1}{4}\sum_{\sigma,\lambda,\lambda'}\sum_{X}|M_{\sigma,\lambda,\lambda'}(X)|^2\,(2\pi)^4\,\delta^4(P+q-p_X). \qquad (2.9)$$

Here p_X denotes the net four-momentum of all the hadrons produced in the scattering process.

Without giving the details of the calculation, which can be found in standard textbooks (Halzen and Martin 1984, Peskin and Schroeder 1995, Sterman 1993), we will write down the following expression for the DIS cross section, which results from substituting Eq. (2.7) into Eq. (2.9):

$$\frac{d\sigma}{d^3 p'} = \frac{\alpha_{EM}^2}{E E' Q^4}\,L_{\mu\nu}W^{\mu\nu}. \qquad (2.10)$$

Equation (2.10) is illustrated in Fig. 2.2, which shows the amplitude from Fig. 2.1 squared. As shown graphically in Fig. 2.2, one can separate the electron and proton contributions to the DIS cross section into leptonic and hadronic parts. Formally, the leptonic part brings in a *leptonic tensor* $L_{\mu\nu}$, while the hadronic part yields a *hadronic tensor* $W^{\mu\nu}$.

From Eqs. (2.7) and (2.9) we can easily see that one defines the *leptonic tensor* by

$$L_{\mu\nu} = \frac{1}{2}\sum_{\lambda=\pm1}\sum_{\lambda'=\pm1}\bar{u}_{\lambda'}(p')\,\gamma_\mu\,u_\lambda(p)\left[\bar{u}_{\lambda'}(p')\,\gamma_\nu\,u_\lambda(p)\right]^*. \qquad (2.11)$$

Summing over the initial and final electron polarizations yields

$$L_{\mu\nu} = \frac{1}{2}\,\mathrm{Tr}\left[(p\!\!\!/' + m_e)\,\gamma_\mu\,(p\!\!\!/ + m_e)\,\gamma_\nu\right]$$
$$= 2\left(p_\mu p'_\nu + p_\nu p'_\mu - p\cdot p'\,g_{\mu\nu} + m_e^2\,g_{\mu\nu}\right), \qquad (2.12)$$

where again m_e is the electron mass.

The *hadronic tensor* $W^{\mu\nu}$ in Eq. (2.10) is given by

$$W^{\mu\nu} = \frac{1}{4\pi m} \frac{1}{2} \sum_{\sigma=\pm 1} \sum_X \langle P, \sigma| J^\mu(0) |X\rangle \langle X| J^\nu(0) |P, \sigma\rangle$$

$$\times (2\pi)^4 \delta^4(P + q - p_X), \qquad (2.13)$$

which can be simplified to

$$W^{\mu\nu} = \frac{1}{4\pi m} \int d^4x \, e^{iq\cdot x} \frac{1}{2} \sum_{\sigma=\pm 1} \sum_X \langle P, \sigma| J^\mu(x) |X\rangle \langle X| J^\nu(0) |P, \sigma\rangle$$

$$= \frac{1}{4\pi m} \int d^4x \, e^{iq\cdot x} \frac{1}{2} \sum_{\sigma=\pm 1} \langle P, \sigma| J^\mu(x) J^\nu(0) |P, \sigma\rangle$$

$$\equiv \frac{1}{4\pi m} \int d^4x \, e^{iq\cdot x} \langle P| J^\mu(x) J^\nu(0) |P\rangle \qquad (2.14)$$

where the last line defines an abbreviated notation for the spin-averaged proton state and m is the mass of the proton.

The strong interaction dynamics in DIS (including nonperturbative contributions) is entirely contained in the hadronic tensor $W^{\mu\nu}$; therefore, it is very hard to calculate $W^{\mu\nu}$ in a "first principles" QCD calculation. However, we can infer more about its structure by noting that conservation of the electromagnetic current (2.8) requires that

$$q_\mu W^{\mu\nu} = 0, \qquad q_\nu W^{\mu\nu} = 0. \qquad (2.15)$$

Imposing the condition (2.15) on $W^{\mu\nu}$ and assuming that the tensor is symmetric one can show that, without loss of generality, it can be written in the following form (see Exercise 2.1 at the end of the chapter):

$$W^{\mu\nu} = - W_1(x_{Bj}, Q^2) \left(g^{\mu\nu} - \frac{q^\mu q^\nu}{q^2} \right)$$

$$+ \frac{W_2(x_{Bj}, Q^2)}{m^2} \left(P^\mu - \frac{P\cdot q}{q^2} q^\mu \right) \left(P^\nu - \frac{P\cdot q}{q^2} q^\nu \right). \qquad (2.16)$$

Here W_1 and W_2 are unknown scalar functions of x_{Bj} and Q^2, called *structure functions*. As $W^{\mu\nu}$ describes the interaction of the virtual photon with the proton, there are only two four-momentum vectors on which it depends: P^μ and q^μ. As $P^2 = m^2$ one can construct only two Lorentz invariants from them that describe the scattering process. We will use x_{Bj} and Q^2 as the two invariants on which W_1 and W_2 depend.

Substituting Eq. (2.16) into Eq. (2.10), after some algebra one can show that the cross section of the reaction $e + p \to e' + X$ in terms of the functions W_1 and W_2 is (for details of the derivation see, for example, the book Halzen and Martin (1984), Chapter 8)

$$\frac{d\sigma^{ep}}{dE'\, d\Omega} = \frac{\alpha_{EM}^2}{4\,E^2 \sin^4\frac{\theta}{2}} \left[W_2(x_{Bj}, Q^2) \cos^2\frac{\theta}{2} + 2W_1(x_{Bj}, Q^2) \sin^2\frac{\theta}{2} \right]. \qquad (2.17)$$

In arriving at Eq. (2.17) we have neglected the mass of the electron m_e, to write

$$d^3 p' = p'^2 \, dp' \, d\Omega \approx E'^2 \, dE' \, d\Omega,$$

where Ω is the solid scattering angle. We have also used Eq. (2.3) to replace Q^2. Equation (2.17) demonstrates that the structure functions W_1 and W_2 can be measured experimentally by studying the angular dependence of the DIS cross section.

Note that the structure functions W_1 and W_2 have the dimension of inverse mass.[1] It is more convenient to define dimensionless structure functions F_1 and F_2, by

$$F_1(x_{Bj}, Q^2) \equiv m W_1(x_{Bj}, Q^2), \tag{2.18a}$$

$$F_2(x_{Bj}, Q^2) \equiv \nu W_2(x_{Bj}, Q^2) = \frac{Q^2}{2m x_{Bj}} W_2(x_{Bj}, Q^2). \tag{2.18b}$$

All the QCD physics in DIS is contained in F_1 and F_2. We will now attempt to calculate these structure functions.

2.2 Parton model and Bjorken scaling

To find the structure functions F_1 and F_2 it is easier to change the frame in which we are working. Instead of the proton's rest frame we will now use a frame in which the proton is ultrarelativistic. Such a frame is usually referred to as the *infinite momentum frame* (IMF) or Bjorken frame. The proton is taken to be moving along the z-axis, and its momentum in this frame is

$$P^\mu \approx \left(P + \frac{m^2}{2P}, 0, 0, P \right) \tag{2.19}$$

in the (P^0, P^1, P^2, P^3) notation. We assume that the proton's momentum is much larger than its mass, $P \gg m$. The virtual photon in the IMF has $q^3 = 0$, so that

$$q^\mu = (q^0, q^1, q^2, 0). \tag{2.20}$$

The part of the DIS process relevant for the calculation of the structure functions, virtual photon–proton scattering, is depicted in Fig. 2.3. Note that, unlike Fig. 2.2, we now draw the proton at the top of the diagram. In fact, in our normal convention a proton at rest (or any other target) is drawn at the bottom of the diagram, while a proton (or any other projectile) moving at high energy is shown at the top of the diagram.

2.2.1 Warm-up: DIS on a single free quark

As a warm-up calculation in preparation for the full *parton model*, let us simply assume that the proton consists of noninteracting quarks and gluons, which we will refer to as *partons*. As we will see below in Sec. 2.3, this is not such a bad approximation as in the IMF the

---

[1] Our single-particle states are normalized such that $\langle p | p' \rangle = (2\pi)^3 \, 2 E_p \, \delta^3(\vec{p} - \vec{p}')$, which allows one to see that the dimension of $W^{\mu\nu}$ in Eq. (2.14) is that of inverse mass.

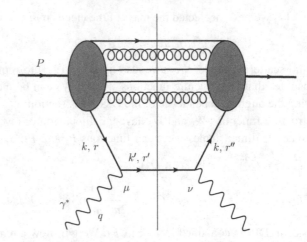

Fig. 2.3. Virtual photon–proton scattering in the IMF.

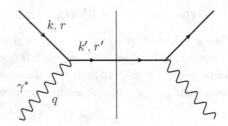

Fig. 2.4. Interaction of a virtual photon with one point-like particle (a parton), as the basic ingredient of the parton model. As usual, the vertical solid line denotes the final-state cut.

typical time scale of the quark and gluon interactions inside the proton is much longer than the time scale of DIS. Hence for the duration of the virtual photon–proton scattering we can assume that the quarks and gluons do not interact with each other. Thus the photon simply interacts with a quark in the proton. To better understand photon–quark scattering let us assume that we simply have one free quark instead of the proton. The diagram giving the cross section of the DIS process is shown in Fig. 2.4.

The hadronic tensor $W_{\mu\nu}$ for the interaction of the virtual photon with the point-like particle (a single quark) has a structure similar to $L_{\mu\nu}$ in Eq. (2.11), namely

$$W_{\mu\nu}^{quark} = \frac{Z_f^2}{2} \sum_{r=\pm1} \sum_{r'=\pm1} \bar{u}_{r'}(k') \gamma_\mu u_r(k) \left[\bar{u}_{r'}(k') \gamma_\nu u_r(k) \right]^* \frac{1}{2m_q} \delta(k'^2 - m_q^2)$$

$$= \frac{Z_f^2}{2} \text{Tr} \left[(\slashed{k}' + m_q) \gamma_\mu (\slashed{k} + m_q) \gamma_\nu \right] \frac{1}{2m_q} \delta(k'^2 - m_q^2), \qquad (2.21)$$

where $k' = k + q$ while r and r' are the quark helicities (see Fig. 2.4) and m_q is the quark mass. Equation (2.21) can be obtained from Eq. (2.13) by replacing X in it by a single particle (a quark), so that

$$\sum_{X=\text{one particle}} = \int \frac{d^3 k'}{2k'^0 (2\pi)^3} \sum_{r'=\pm 1}$$

along with $p_X \to k'$ and $P \to k$. It is then easy to show that

$$\frac{1}{4\pi m_q} \int \frac{d^3 k'}{2k'^0 (2\pi)^3} (2\pi)^4 \delta^4(k + q - k') = \frac{1}{2m_q} \delta\left((k+q)^2 - m_q^2\right), \qquad (2.22)$$

justifying the delta function factor in Eq. (2.21).

We can rewrite $\delta((k+q)^2 - m_q^2)$ as follows:

$$\delta\left((k+q)^2 - m_q^2\right) = \delta\left(2k \cdot q - Q^2\right) = \frac{1}{2k \cdot q} \delta\left(1 - \frac{Q^2}{2k \cdot q}\right), \qquad (2.23)$$

where we have used the fact that the incoming quark is on mass shell.

Calculating the trace in Eq. (2.21), comparing the result with Eq. (2.16), and using Eqs. (2.18a) and (2.18b) with P replaced by k we obtain for DIS on a point-like particle (a quark)

$$F_1^{quark}\left(x_{Bj}, Q^2\right) = m_q W_1^{quark}\left(x_{Bj}, Q^2\right) = \frac{Z_f^2}{2} \delta\left(1 - x_{Bj}\right) \qquad (2.24)$$

$$F_2^{quark}\left(x_{Bj}, Q^2\right) = \frac{Q^2}{2m_q x_{Bj}} W_2^{quark}\left(x_{Bj}, Q^2\right) = Z_f^2 \delta\left(1 - x_{Bj}\right). \qquad (2.25)$$

We have used the fact that, for DIS on a single quark, $x_{Bj} = Q^2/(2k \cdot q)$. We see that for DIS on a point-like particle the structure functions F_1 and F_2 turn out to depend only on one variable, x_{Bj}. This behavior is known as *Bjorken scaling* (Bjorken 1969).

2.2.2 Full calculation: DIS on a proton

The idea that the actual interaction in DIS occurs with the point-like constituents of a hadron (the partons) can be illustrated by studying the full DIS process. Let us consider DIS on the whole proton, as shown in Fig. 2.3. We want to calculate the diagram in Fig. 2.3 using the rules of light cone perturbation theory (LCPT) outlined in Sec. 1.3 (see also Sec. 1.4). We first rewrite all four-momenta in the light cone $(+, -, \perp)$ notation. In the IMF/Bjorken frame the proton has a very large momentum. The proton's momentum in Eq. (2.19) becomes, in light cone notation,

$$P^\mu \approx (P^+, 0, 0_\perp) \qquad (2.26)$$

with very large $P^+ \approx 2P$. Quarks and gluons in such an ultrarelativistic proton also have very large light cone plus momenta. The quark in Fig. 2.3 has four-momentum $k^\mu = (k^+, (\vec{k}_\perp^2 + m_q^2)/k^+, \vec{k}_\perp)$; we assume that it has a large k^+ component. We define the Feynman-x variable as the fraction of the light cone momentum of the proton carried by

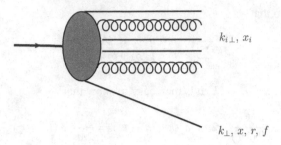

$k_{i\perp}, x_i$

$k_\perp, x, r, f$

Fig. 2.5. Light cone wave function of the proton.

this quark[2]

$$x \equiv \frac{k^+}{P^+},\tag{2.27}$$

writing $k^\mu = (xP^+, (\vec{k}_\perp^2 + m_q^2)/(xP^+), \vec{k}_\perp)$.

In LCPT every particle is on mass shell. However, we want to calculate the virtual photon–proton scattering cross section for the process shown in Fig. 2.3. By the definition of the problem the incoming photon is virtual, $q^2 = -Q^2$. Hence in LCPT we can treat this virtual photon as having an imaginary mass iQ. The virtual photon momentum (2.20) becomes, in light cone notation,

$$q^\mu = \left(q^+, \frac{\vec{q}_\perp^2 - Q^2}{q^+}, \vec{q}_\perp\right)\tag{2.28}$$

with $(q^+)^2 = \vec{q}_\perp^2 - Q^2$ in the IMF.

In the calculations below we will assume that Q^2 is very large. First, for QCD perturbation theory to be applicable Q^2 has to be much larger than the confinement scale Λ_{QCD}: $Q^2 \gg \Lambda_{QCD}^2$. Second, for the parton model (which we are about to present) to be valid, Q has to be much larger than the transverse momentum of any other particle in the problem. This applies to the quark line carrying momentum k in Fig. 2.3, for which we have $Q^2 \gg \vec{k}_\perp^2, m_q^2$. If, for a particular wave function configuration the upper boxed part of Fig. 2.3 contains n partons with transverse momenta $\vec{k}_{i\perp}$ for $i = 1, \ldots, n$, then we will assume that $Q^2 \gg \vec{k}_{i\perp}^2$ for any i. Note that $\vec{q}_\perp^2 = Q^2 + (q^+)^2 > Q^2$ is also very large.

Now let us assume that these n partons carry light cone momentum components k_i^+ or, equivalently, have Feynman-x values given by x_i for $i = 1, \ldots, n$. We can then define the light cone wave function of the $(n+1)$-parton Fock state of the proton and denote it by $\Psi_n^f(\{x_i, k_{i\perp}\}; x, k_\perp; r)$. The proton has n "spectator" partons (both quarks and gluons) and one quark carrying momentum k in Fig. 2.3 that interacts with the photon. This quark has helicity r and flavor f. The light cone wave function $\Psi_n^f(\{x_i, k_{i\perp}\}; x, k_\perp; r)$ is illustrated in Fig. 2.5. In our discussion and notation we will suppress the polarization indices of the

---

[2] The Feynman-x variable was originally defined as $x = 2k^3/\sqrt{s}$ in the center-of-mass frame with k^μ the momentum of the produced outgoing particle (Feynman 1969). Our definition here is different, but is also widely used in the community: it maps back onto the original definition at large x.

proton and the polarization, color, and flavor indices of the spectator partons: averaging over the proton polarizations and summation over the polarization, color, and flavor of the partons will always be implicitly assumed to be made after we have multiplied the wave function $\Psi_n^f(\{x_i, k_{i\perp}\}; x, k_\perp; r)$ by its complex conjugate. Note also that $k_\perp = |\vec{k}_\perp|$ (the same notation applies to the other transverse momenta).

Let us now calculate the proton's $W_{\mu\nu}$ using Eq. (2.13). Note that after the interaction the n spectator partons, along with the quark that interacts with the photon, together form what is denoted X in Eq. (2.13). Therefore, for n partons we have (see also Eq. (1.67))

$$
\sum_{X=n \text{ partons}} = \int \frac{dk'^+}{k'^+} \frac{d^2 k'_\perp}{2(2\pi)^3} \frac{1}{S_n} \sum_{r'=\pm 1} \prod_{i=1}^{n} \frac{dk_i^+}{k_i^+} \frac{d^2 k_{i\perp}}{2(2\pi)^3}
$$

$$
= \int \frac{dk'^+}{k'^+} \frac{d^2 k'_\perp}{2(2\pi)^3} \frac{1}{S_n} \sum_{r'=\pm 1} \prod_{i=1}^{n} \frac{dx_i}{x_i} \frac{d^2 k_{i\perp}}{2(2\pi)^3}, \tag{2.29}
$$

where for physical particles all integrals over the k_i^+ and k'^+ run from 0 to P^+, which translates into integrals over the x_i running from 0 to 1. Here $k'^+ = k^+ + q^+$, $\vec{k}'_\perp = \vec{k}_\perp + \vec{q}_\perp$, and r' is the helicity of the k' quark line (see Fig. 2.3). The symmetry factor S_n is defined after Eq. (1.67).

Following the definition of the hadronic tensor in Eq. (2.13) and with the help of the diagram in Fig. 2.3 we can write, using the LCPT rules presented in Secs. 1.3 and 1.4,

$$
W_{\mu\nu} = \frac{1}{4\pi m} \sum_{n, f} \int \frac{dk'^+ d^2 k'_\perp}{2k'^+ (2\pi)^3} \frac{1}{S_n} \sum_{r, r', r''} \prod_{i=1}^{n} \frac{dx_i}{x_i} \frac{d^2 k_{i\perp}}{2(2\pi)^3}
$$

$$
\times \frac{P^+}{k^+} \Psi_n^f(\{x_i, k_{i\perp}\}; x, k_\perp; r) \left[\frac{P^+}{k^+} \Psi_n^f(\{x_i, k_{i\perp}\}; x, k_\perp; r'') \right]^* Z_f^2
$$

$$
\times \bar{u}_{r'}(k') \gamma_\mu u_r(k) \left[\bar{u}_{r'}(k') \gamma_\nu u_{r''}(k) \right]^* (2\pi)^4 \delta^4\left(P + q - k' - \sum_{j=1}^{n} k_j \right). \tag{2.30}
$$

The labeling of the quark helicities r, r', and r'' is defined in Fig. 2.3. Note that, unlike in the simple case of DIS on a single quark considered above, the helicity of the quark line k in Fig. 2.3 is different on the left and on the right of the final-state cut. The factors P^+/k^+ multiplying the wave functions in Eq. (2.30) appear for two reasons. A factor $1/k^+$, which has to be included by the rules of LCPT from Sec. 1.3, is due to the internal quark line carrying momentum k and is not included in our definition of the light cone wave function outlined in Sec. 1.4. The same definition from Sec. 1.4 dictates that each light cone wave function contains a factor $1/P^+$ for each incoming line but, as the general LCPT rules in Sec. 1.3 prescribe no such factor for the full diagram for the scattering process, we need to remove this factor by multiplying the wave functions by P^+.

The delta function in Eq. (2.30) imposes the conservation of the transverse and "+" components of momenta. However, of particular importance is the conservation of the light cone energy that is also imposed by this delta function. Using Eq. (2.26) and rewriting the light cone energies of all partons in terms of the transverse and "+" components of their momenta we obtain

$$\frac{1}{k'^+}\delta\left(P^- + q^- - k'^- - \sum_{j=1}^{n} k_j^-\right)$$

$$= \frac{1}{k^+ + q^+}\delta\left(q^- - \frac{(\vec{k}_\perp + \vec{q}_\perp)^2}{k^+ + q^+} - \sum_{j=1}^{n} \frac{k_{j\perp}^2}{k_j^+}\right). \tag{2.31}$$

For simplicity we will now assume that all the partons are massless. This assumption also applies to the quark that interacts with the photon, for which we now put $m_q = 0$.

Since $Q^2, \vec{q}_\perp^2 \gg \vec{k}_\perp^2, k_{i\perp}^2$ for any i we approximate $(\vec{k}_\perp + \vec{q}_\perp)^2$ as $\vec{q}_\perp^2$ and also neglect all $k_{j\perp}^2/k_j^+$ in the argument of the delta function in Eq. (2.31). This leaves us with

$$\frac{1}{k^+ + q^+}\delta\left(q^- - \frac{(\vec{k}_\perp + \vec{q}_\perp)^2}{k^+ + q^+} - \sum_{j=1}^{n} \frac{k_{j\perp}^2}{k_j^+}\right) \approx \delta\left((k^+ + q^+)q^- - \vec{q}_\perp^2\right)$$

$$= \delta(k^+ q^- - Q^2) = \delta(x\,P^+ q^- - Q^2) \approx \delta(x\,2P\cdot q - Q^2), \tag{2.32}$$

where the last approximation was made using Eq. (2.26). Using the definition of x_{Bj} the last delta function can be rewritten as

$$\delta\left(x\,2P\cdot q - Q^2\right) = \frac{1}{2P\cdot q}\delta(x - x_{Bj}) = \frac{x_{Bj}}{Q^2}\delta(x - x_{Bj}). \tag{2.33}$$

We see that Feynman x is identical to Bjorken x. The physical meaning of x_{Bj} becomes clear: *it is the fraction of the light cone momentum of the proton carried by the struck quark!*

Since the two quantities are equal, below we will use x and x_{Bj} interchangeably, using the notation with a subscript (x_{Bj}) only in cases when we need to avoid the potential confusion of x with other quantities.

Using Eq. (2.33) in Eq. (2.30) and summing over the helicities r' yields

$$W_{\mu\nu} = \frac{1}{4m}\sum_{n,f}\int dk^+ d^2k_\perp \frac{1}{S_n}\sum_{r,r''}\prod_{i=1}^{n}\frac{dx_i}{x_i}\frac{d^2k_{i\perp}}{2(2\pi)^3}$$

$$\times \Psi_n^f\left(\{x_i, k_{i\perp}\}; \frac{k^+}{P^+}, k_\perp; r\right)\left[\Psi_n^f\left(\{x_i, k_{i\perp}\}; \frac{k^+}{P^+}, k_\perp; r''\right)\right]^* Z_f^2$$

$$\times \bar{u}_{r''}(k)\,\gamma_\nu\,(\slashed{k} + \slashed{q})\,\gamma_\mu\,u_r(k)\,\delta\left(P^+ - k^+ - \sum_{l=1}^{n} k_l^+\right)\delta^2\left(\vec{k}_\perp + \sum_{j=1}^{n} \vec{k}_{j\perp}\right)$$

$$\times \left(\frac{P^+}{k^+}\right)^2 \frac{x_{Bj}}{Q^2}\,\delta\left(x_{Bj} - \frac{k^+}{P^+}\right), \tag{2.34}$$

where we have switched from integration variables k'^+ and $\vec{k}'_\perp$ to k^+ and $\vec{k}_\perp$.

An important assumption of the parton model is that the integrals in Eq. (2.34) are convergent even if we impose no integration limit on the transverse momentum integrals.

As will be shown below, this assumption is not true in QCD, where we have to cut off the $k_\perp$-integral at Q^2 in the ultraviolet (UV), which leads to corrections to the naive parton model presented in this section; the $k_\perp$-integral converges in the UV for a theory in which partons are scalars. For now we will not address this issue and simply assume that, owing to some (perturbative or nonperturbative) physics beyond our present formalism, the $k_\perp$-integral is convergent in the UV.

We can then see that all the integrals in Eq. (2.34) "know" only about one momentum external to the integration: that momentum is P. Hence writing $\not{k} + \not{q}$ from Eq. (2.34) as $(k+q)_\alpha \gamma^\alpha$ we can argue, on the basis of Lorentz transformation properties, that after all integrations in Eq. (2.34) have been carried out the factor γ^α will have been replaced by P^α. From Eq. (2.26) we then see that only the $\alpha = +$ term will contribute to the final answer, as P^+ is larger by far than any other component of the momentum P^α. We thus can replace γ^α with γ^+ from the start, substituting $(1/2)(k+q)^- \gamma^+ \approx (Q^2/2k^+)\gamma^+$ in to Eq. (2.34) in place of $\not{k} + \not{q}$. (We have used the fact that $\vec{q}_\perp^2 \approx Q^2$ is the largest transverse momentum, while $k^+ = xP^+ \gg q^+$.) We obtain

$$
W_{\mu\nu} = \frac{1}{4m} \sum_{n,f} \int dk^+ d^2k_\perp \frac{1}{S_n} \sum_{r,r''} \prod_{i=1}^{n} \frac{dx_i}{x_i} \frac{d^2k_{i\perp}}{2(2\pi)^3}
$$

$$
\times \Psi_n^f\left(\{x_i, k_{i\perp}\}; \frac{k^+}{P^+}, k_\perp; r\right) \left[\Psi_n^f\left(\{x_i, k_{i\perp}\}; \frac{k^+}{P^+}, k_\perp; r''\right)\right]^* Z_f^2
$$

$$
\times \bar{u}_{r''}(k)\, \gamma_\nu \gamma^+ \gamma_\mu\, u_r(k)\, \delta\left(P^+ - k^+ - \sum_{l=1}^{n} k_l^+\right) \delta^2\left(\vec{k}_\perp + \sum_{j=1}^{n} \vec{k}_{j\perp}\right)
$$

$$
\times \frac{1}{2x_{Bj}k^+} \delta\left(x_{Bj} - \frac{k^+}{P^+}\right), \tag{2.35}
$$

where we have also made use of the last delta function in Eq. (2.34) to replace $(P^+/k^+)^2$ by $1/x_{Bj}^2$.

With the help of Table A.1 in appendix section A.1 and employing Eq. (2.35), it is easy to see that

$$
W^{\mu+} = W^{+\mu} \propto \bar{u}_{r''}(k)\, \gamma^\mu \gamma^+ \gamma^+\, u_r(k) = 0,
$$

$$
W^{--} \propto \bar{u}_{r''}(k)\, \gamma^- \gamma^+ \gamma^-\, u_r(k) = \bar{u}_{r''}(k)\, \gamma^-\, u_r(k) = \frac{2\delta_{r\,r''}k_\perp^2}{k^+}. \tag{2.36}
$$

To find the transverse components of $W^{\mu\nu}$ we note that, from Eq. (2.14), this tensor is symmetric. Anticipating that the final result of the integrations in Eq. (2.35) yields a symmetric tensor, we can therefore symmetrize the transverse components to get

$$
W^{ij} \propto \bar{u}_{r''}(k)\, \gamma^i \gamma^+ \gamma^i\, u_r(k) = \bar{u}_{r''}(k)\tfrac{1}{2}\left(\gamma^i \gamma^+ \gamma^j + \gamma^j \gamma^+ \gamma^i\right) u_r(k)
$$

$$
= -\tfrac{1}{2}\bar{u}_{r''}(k)\, \gamma^+ \{\gamma^j, \gamma^i\}\, u_r(k) = -g^{ij}\bar{u}_{r''}(k)\, \gamma^+\, u_r(k) = -g^{ij}\delta_{r\,r''}\, 2k^+ \tag{2.37}
$$

for $i, j = 1, 2$. As $k^+ \gg k_\perp$ we see that W^{ij} is much larger than W^{--} and is, therefore, the only nonnegligible component of the hadronic tensor $W^{\mu\nu}$. (Similarly, one can show that

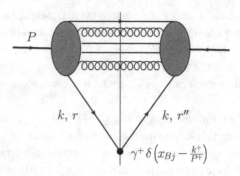

Fig. 2.6. Cut (Mueller) vertex in DIS, denoted by the solid circle.

$W^{-i} = W^{i-} \propto k_\perp^i$, which is much smaller than W^{ij} and integrates out to zero in Eq. (2.35) owing to the absence of a preferred transverse direction in the problem.)

From Eqs. (2.37) and (2.35) we see that, in the usual Feynman diagram language, the quark–photon part of the diagram in Fig. 2.3 can be replaced by a single effective vertex containing $\gamma^+\delta(x_{Bj} - k^+/P^+)$, as shown in Fig. 2.6. This effective vertex is known as a *cut vertex* or *Mueller vertex* (Mueller 1970, 1981).

From the general decomposition of $W^{\mu\nu}$ in Eq. (2.16) and using the fact that, by our frame choice, $\vec{P}_\perp = 0$ we can write

$$W^{ij} = -W_1(x_{Bj}, Q^2)\, g^{ij} + \frac{q^i q^j}{q^2} \left[W_1(x_{Bj}, Q^2) + \frac{W_2(x_{Bj}, Q^2)}{m^2} \frac{(P \cdot q)^2}{q^2} \right]. \qquad (2.38)$$

Comparing Eq. (2.38) with Eq. (2.37), for which we showed that $W^{ij} \propto g^{ij}$, we see that the hadronic tensor is given by the first term in Eq. (2.38):

$$W^{ij} = -W_1(x_{Bj}, Q^2)\, g^{ij}. \qquad (2.39)$$

Substituting Eq. (2.37) into Eq. (2.35), summing over r'', and comparing the result with Eq. (2.39) we can read off the structure function W_1:

$$W_1(x_{Bj}, Q^2) = \frac{1}{4mx_{Bj}} \sum_{n,f} Z_f^2 \int dk^+ d^2k_\perp \frac{1}{S_n} \sum_r \prod_{i=1}^n \frac{dx_i}{x_i} \frac{d^2 k_{i\perp}}{2(2\pi)^3}$$

$$\times \left| \Psi_n^f \left(\{x_i, k_{i\perp}\}; \frac{k^+}{P^+}, k_\perp; r \right) \right|^2 \delta\left(P^+ - k^+ - \sum_{l=1}^n k_l^+ \right)$$

$$\times \delta^2\left(\vec{k}_\perp + \sum_{j=1}^n \vec{k}_{j\perp} \right) \delta\left(x_{Bj} - \frac{k^+}{P^+} \right). \qquad (2.40)$$

Let us now define the *quark distribution function* by

$$q^f(x_{Bj}) = \frac{1}{2x_{Bj}} \sum_n \int d\xi\, d^2k_\perp \frac{1}{S_n} \sum_r \prod_{i=1}^n \frac{dx_i}{x_i} \frac{d^2k_{i\perp}}{2(2\pi)^3}$$

$$\times \left| \Psi_n^f \left(\{x_i, k_{i\perp}\}; \frac{k^+}{P^+}, k_\perp; r \right) \right|^2 \delta\left(1 - \xi - \sum_{l=1}^n x_l \right)$$

$$\times \delta^2\left(\vec{k}_\perp + \sum_{j=1}^n \vec{k}_{j\perp} \right) \delta\left(x_{Bj} - \xi \right), \tag{2.41}$$

where $\xi = k^+/P^+$. With the help of Eq. (2.41) we can rewrite Eq. (2.40) as

$$W_1(x_{Bj}) = \frac{1}{2m} \sum_f Z_f^2\, q^f(x_{Bj}). \tag{2.42}$$

Note that both the quark distribution function and the structure function W_1 are functions of Bjorken x only! Just as in the case of DIS on a single free quark, this is Bjorken scaling.

To find the remaining structure function, W_2, we note that, as we have just shown in Eq. (2.37), $W^{ij} \propto g^{ij}$. Therefore the term in square brackets in Eq. (2.38) must be zero. Equating it to zero, and recalling the definitions of x_{Bj} and ν from Eqs. (2.2) and (2.5), we write

$$\nu\, W_2(x_{Bj}) = 2m x_{Bj}\, W_1(x_{Bj}). \tag{2.43}$$

Using the definitions in Eqs. (2.18a) and (2.18b) we can rewrite Eq. (2.43) as

$$F_2(x_{Bj}) = 2x_{Bj}\, F_1(x_{Bj}). \tag{2.44}$$

Equation (2.44) is known as the *Callan–Gross* relation (Callan and Gross 1969). This relation is characteristic of spin-1/2 partons, such as quarks, and would be different if the proton had constituents with a different spin interacting with the virtual photon.

Combining Eqs. (2.18a), (2.42), and the Callan–Gross relation we write

$$F_1(x_{Bj}) = \frac{1}{2} \sum_f Z_f^2\, q^f(x_{Bj}), \tag{2.45}$$

$$F_2(x_{Bj}) = \sum_f Z_f^2\, x_{Bj}\, q^f(x_{Bj}). \tag{2.46}$$

We can see that both structure functions are independent of Q^2 and are functions of x_{Bj} only. Therefore, if we assume that some nonperturbative QCD effects lead to a natural UV cutoff on the transverse momenta of the partons then the DIS cross section can be described by two functions, F_1 and F_2, that are dependent on only one variable, x_{Bj}. This is a more general form of *Bjorken scaling* (Bjorken 1969). We have now shown that Bjorken scaling results from a full parton model calculation.

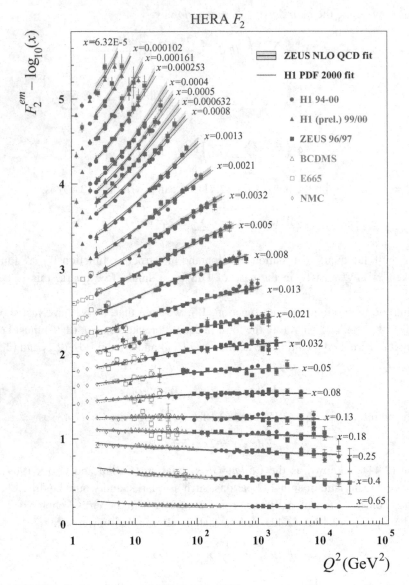

Fig. 2.7. Compilation of the world F_2 data for DIS on a proton. The proton F_2 structure function is plotted as a function of Q^2 for a range of values of x, as indicated next to the data. It can be seen that, except for very small x, F_2 is independent of Q^2, a manifestation of Bjorken scaling. (We thank Kunihiro Nagano for providing us with this figure.) A color version of this figure is available online at www.cambridge.org/9780521112574.

In Fig. 2.7 we show a summary of the world knowledge of the proton F_2 structure function. This structure function is plotted as a function of Q^2 for many different fixed values of Bjorken-x. One can clearly see that, when x is not too small, F_2 is independent of Q^2. This is the experimental manifestation of Bjorken scaling. We see that the theory we have been presenting here agrees very well with the data, at least at the qualitative level. (The curves going through the data points result from the solution of the QCD renormalization group equations in Q^2, which are presented below.)

The quark distribution function defined in Eq. (2.41) counts the number of quarks with longitudinal momentum fraction x_{Bj}. While this may not be obvious from Eq. (2.41), we may check this statement for DIS on a single quark. Comparing Eq. (2.45) for a single flavor with Eq. (2.24) yields

$$q^f_{\text{one-quark}}(x_{Bj}) = \delta(1 - x_{Bj}) \tag{2.47}$$

meaning that our target "proton" indeed consists of a single quark which carries all the "proton" momentum, i.e., the quark is at $x_{Bj} = 1$. Equation (2.47) can also be obtained from Eq. (2.41) directly by setting $n = 0$ in the latter equation and also using $|\Psi^f_0|^2 = 1$.

As one can see from Eqs. (2.45) and (2.46), the functions F_1 and F_2 have a very simple physical meaning: namely, F_1 gives the number of partons in the hadron with longitudinal momentum fraction x_{Bj} (weighted by $Z^2_f/2$) while F_2 gives the average longitudinal momentum fraction of the partons in the hadron (weighted by Z^2_f) times the number of partons.

Using Eqs. (2.45) and (2.46) we can understand the physics behind the parton model. The proton arrives with partons in its wave function, which, for the duration of the DIS interaction, can be thought of as free particles. To be specific, let us concentrate on the F_2 structure function. The interaction of each quark with the virtual photon yields a factor $Z^2_f x_{Bj}$, as seen in Eq. (2.46). The full expression for the proton structure function F_2 in Eq. (2.46) can be interpreted as the product of the number of quarks in the proton ($q^f(x_{Bj})$) and the amplitude for the interaction of each quark with the photon ($Z^2_f x_{Bj}$). We thus have a clear physical picture of a proton with noninteracting partons in its wave function scattering on a virtual photon in such a way that each parton interacts with the photon independently of the other partons. We can therefore write (for the details see Sterman (1993))

$$F_2(x_{Bj}) = \sum_f \int_0^1 d\xi \, q^f(\xi) \, C^f_2\left(\frac{x_{Bj}}{\xi}\right). \tag{2.48}$$

The distribution function q^f gives the number of quarks in the proton's wave function, while the *coefficient function* C^f_2 expresses the interaction between a quark with flavor f and the virtual photon. At the lowest order, considered here, $C^f_2 = Z^2_f \, \delta(x_{Bj}/\xi - 1)$. When used in Eq. (2.48) it leads to Eq. (2.46).

One can easily express the structure functions in terms of the photon–proton cross section $\sigma^{\gamma^* p}$ for transverse and longitudinal polarizations of the virtual photon. In particular one obtains (see Halzen and Martin (1984) along with the derivation in Sec. 4.1 below)

$$F_2(x_{Bj}, Q^2) = \frac{Q^2}{4\pi^2 \alpha_{EM}} \, \sigma^{\gamma^* p}_{tot}, \tag{2.49}$$

where $\sigma_{tot}^{\gamma^*p}$ is the total γ^*p cross section summed over all photon polarizations. With the help of Eq. (2.49), Eq. (2.48) can be rewritten directly for the cross section as

$$\sigma_{tot}^{\gamma^*p}(x_{Bj}, Q^2) = \sum_f \int \frac{d\xi}{\xi} \, \xi \, q^f(\xi) \, \hat\sigma^{\gamma^*+\text{parton}f}\left(\frac{x_{Bj}}{\xi}, Q^2\right)$$

$$= \sum_f \int dy' \, N^f(y') \, \hat\sigma^{\gamma^*+\text{parton}f}\left(e^{-(y-y')}, Q^2\right), \qquad (2.50)$$

where $N^f(y = \ln 1/x_{Bj}) = x_{Bj} \, q^f(x_{Bj})$ is the number of partons (quarks) inside the hadron having flavor f per unit *rapidity* $y = \ln(P^+/k^+) = \ln 1/x_{Bj}$. The factor $\hat\sigma^{\gamma^*+\text{parton}f}(x, Q^2)$ is the cross section for parton–virtual photon scattering. In Eq. (2.50) we have $y = \ln 1/x_{Bj}$ and $y' = \ln 1/\xi$. One can see from Eq. (2.25) that in the "naive" parton model considered here one has

$$\hat\sigma^{\gamma^*+\text{parton}f}\left(x_{Bj}, Q^2\right) = \frac{4\pi^2\alpha_{EM}}{Q^2} \, Z_f^2 \, x_{Bj} \, \delta(1 - x_{Bj}) = \frac{4\pi^2\alpha_{EM}}{Q^2} \, Z_f^2 \, \delta(y). \qquad (2.51)$$

Using Eq. (2.51) in Eq. (2.50) reduces the latter to Eq. (2.46).

Equations (2.48) and (2.50) show that, in the framework of the parton approach, finding cross sections is reduced to two separate problems: finding the light cone wave function of the hadron, which does not depend on the probe, and calculating the cross section for scattering of the parton on the probe, γ^* in the case of electron DIS. The process is illustrated in Fig. 2.10. This simple parton model with an additional obvious assumption that the partons are quarks, anti-quarks, and gluons is able to describe a striking amount of experimental data. See Feynman (1972), as well as our main textbooks Peskin and Schroeder (1995) and Halzen and Martin (1984), for more detailed comparisons of the parton model with the data.

2.3 Space–time structure of DIS processes

Equation (2.48) is very simple and intuitively sound. It would be useful to visualize it in terms of the space–time dynamics of partons. For this purpose we will rewrite Eq. (2.14) for the cross section of the virtual photon interaction as the imaginary part of the Compton scattering amplitude at zero angle. In the space–time representation it looks as follows:

$$W_{\mu\nu}\left(x_{Bj}, Q^2\right) = \frac{1}{2\pi m} \text{Im}\left\{i \int d^4x \, e^{iq\cdot x} \, \langle P | \text{T}\,[J_\mu(x) J_\nu(0)] | P \rangle\right\}, \qquad (2.52)$$

where as usual $|P\rangle$ denotes the state of the target (the proton) and T denotes time-ordering. The right-hand side of Eq. (2.52) is simply the imaginary part of the forward scattering amplitude for the photon–proton interaction. The coordinate four-vector x^μ in the forward amplitude describes the space–time separation between absorption and re-emission of the virtual photon by a quark inside the proton.

Let us first work in the rest frame of the proton. Just as in Sec. 2.1 we have $P^\mu = (m, \vec{0})$. However, now we are interested in the photon–proton interaction: we can forget about the electron in Sec. 2.1 that gave rise to the photon and choose our coordinate axis in such a way that the photon's four-momentum is $q^\mu = (q^0, \vec{0}_\perp, q^3)$. We then have $2P \cdot q = 2q^0 m = Q^2/x_{Bj}$, so that

$$q^0 = \frac{Q^2}{2mx_{Bj}} \gg Q \tag{2.53}$$

since $Q \gg m$ and $x_{Bj} \leq 1$. By the definition of Q^2 we have $0 \leq Q^2 = -q^2 = (q^3)^2 - (q^0)^2$. Hence $q^3 \geq q^0 \gg Q$. Therefore $q^0 \approx q^3 \gg Q$. We can then write

$$q^+ = q^0 + q^3 \approx 2q^0,$$

$$q^- = q^0 - q^3 = \frac{q^+ q^-}{q^+} \approx -\frac{Q^2}{2q^0} = -mx_{Bj}. \tag{2.54}$$

Writing $q \cdot x$ in the exponent in Eq. (2.52) as $\frac{1}{2}(q^+x^- + q^-x^+)$, we argue that the typical x^- range is given by $2/q^+$, while the typical x^+ range is given by $2/q^-$. Therefore

$$x^- \approx \frac{2}{q^+} \approx \frac{2mx_{Bj}}{Q^2} \ll \frac{1}{\mu},$$

$$x^+ \approx \frac{2}{|q^-|} \approx \frac{2}{mx_{Bj}} \geq \frac{1}{\mu}, \tag{2.55}$$

where $\mu \sim \Lambda_{QCD} \sim m$ is the scale of the nonperturbative (soft) QCD interactions, which gives the average transverse momenta of the partons in the parton model. From Eq. (2.55) we see that, for large Q, one has $x^- = t - z \approx 0$ and $x^+ = t + z \approx 2t \approx 2/(mx_{Bj})$. Therefore the light cone time of observation is given by

$$x^+ \approx \frac{2}{mx_{Bj}}. \tag{2.56}$$

This time is known as the *Ioffe time* (Ioffe 1969, Gribov, Ioffe, and Pomeranchuk 1966). It can be interpreted as the typical longitudinal distance of the interaction (the coherence length). We see that this longitudinal range in DIS increases with decreasing Bjorken x.[3]

We can also determine the transverse coordinate resolution of the virtual photon in DIS. Imposing the causality of the interactions in the forward scattering amplitude (2.52), i.e., $x^2 = x^+x^- - x_\perp^2 > 0$, and using Eq. (2.55) we get

$$x_\perp^2 < x^+x^- \propto \frac{4}{Q^2} \ll \frac{1}{\mu^2}. \tag{2.57}$$

We see that the typical transverse resolution of the virtual photon is of order $1/Q$. Therefore the photon can resolve very short distances, deep inside the proton: this enables it to "pick out" a quark with which to interact independently of the other "spectator" partons. This

---

[3] The careful reader will notice that for small enough x_{Bj} the light cone time, i.e., the coherence length, in Eq. (2.56) becomes larger than the size of the target proton. Therefore at least one of the electromagnetic currents J_μ in Eq. (2.52) has to be located outside the proton. How can an interaction with the proton happen outside the proton? We will explain this phenomenon in more detail later, in Chapter 4, but here we briefly note that at very small x_{Bj} the incoming current can decay into a quark–antiquark pair outside the proton, the $q\bar{q}$ pair subsequently interacting with the proton.

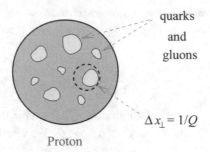

quarks
and
gluons

$\Delta x_\perp = 1/Q$

Proton

Fig. 2.8. A depiction of a proton during DIS in the transverse plane. The blobs represent partons (quarks and gluons), while the dashed circle denotes the virtual photon. A color version of this figure is available online at www.cambridge.org/9780521112574.

conclusion is illustrated in Fig. 2.8, where we show a proton in the transverse plane with the quarks and gluons in its wave function denoted by blobs with random shapes. The virtual photon is represented by a dashed circle whose size is of order $1/Q$, in agreement with Eq. (2.57). One can see explicitly now that the DIS experiment works as a microscope: varying Q^2 changes the transverse size of the photon and so changes the "resolution" of the DIS experiment, allowing the virtual photon to interact with partons of different transverse extent.

Now, let us consider DIS process in the IMF or Bjorken frame. There the proton momentum is given by Eq. (2.19) (or, equivalently, Eq. (2.26)), while the virtual photon momentum is given by Eq. (2.20) (Eq. (2.28)). We see that $2P \cdot q \approx 2Pq^0 = Q^2/x_{Bj}$, giving

$$q^0 \approx \frac{Q^2}{2x_{Bj} P}. \tag{2.58}$$

We conclude that the interaction time in the IMF is

$$t_{DIS} \approx \frac{1}{q^0} \approx \frac{2x_{Bj} P}{Q^2}. \tag{2.59}$$

This time needs to be compared with the typical time scale with which partons interact inside the proton. In the rest frame of the proton, the interparton interaction time is nonperturbatively long, of order $1/\mu$. In the IMF or Bjorken frame the time is dilated by the boost factor P/m, giving

$$t_{partons} \approx \frac{1}{\mu} \frac{P}{m}. \tag{2.60}$$

Comparing Eqs. (2.59) and (2.60) one can see clearly that since $x_{Bj}\mu m \leq \mu m \ll Q^2$ we have

$$t_{DIS} \ll t_{partons}. \tag{2.61}$$

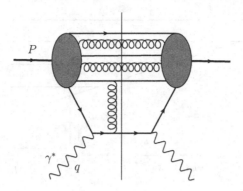

Fig. 2.9. An example of a higher-twist correction.

Therefore we have justified a main assumption in Sec. 2.2, that the typical time scale of interpartonic interactions is much longer than the typical time scale of DIS. One does not have to worry about partons interacting with each other during DIS.

The time-scale argument presented here can be supported by explicit diagrammatic calculations showing that diagrams in which the quark struck by the photon exchanges gluons with other partons, such as the graph shown in Fig. 2.9, are suppressed by powers of μ^2/Q^2 and m^2/Q^2. Such corrections are known as *higher-twist terms*. The twist of an operator is defined as its mass dimension minus its spin (Peskin and Schroeder 1995, Sterman 1993). In the operator product expansion (OPE) for the hadronic tensor $W^{\mu\nu}$ in Eq. (2.14) the contribution of higher-twist operators enters with an extra $1/Q^2$ suppression compared with the leading large-Q^2 term that we found above. In the language of LCPT the higher-twist operators correspond to a proton light cone wave function in which we tag on (i.e., detect) more than one particle. The reader particularly interested in twist expansions is referred to Sterman (1993) or Peskin and Schroeder (1995).

The transverse space dynamics is particularly simple in the IMF/Bjorken frame: from Eq. (2.58) we see that $q^0 \ll Q$, so that $Q^2 = q_\perp^2 - (q^0)^2 \approx q_\perp^2$. Hence the transverse resolution of the virtual photon is

$$x_\perp \approx \frac{1}{q_\perp} \approx \frac{1}{Q}, \tag{2.62}$$

just as in the proton's rest frame.

Equation (2.55) also has a very clear meaning in another frame, the Breit frame, where the photon momentum is equal to

$$q^\mu = (q^0 = 0, \vec{0}_\perp, q^3 = -Q) \tag{2.63}$$

and the proton's momentum is given by Eq. (2.26), as it is in the IMF. In this frame $q^+ = -q^- = -Q$; thus $x^- \propto 1/Q$ and $x^+ \propto 1/Q$, leading to $x_\perp^2 < 1/Q^2$ by a causality argument just as in the proton's rest frame. All space and time intervals between photon absorption and re-emission are short, of order $1/Q$. The photon interacts with the target during a very short time interval. The interparton interaction time in the Breit frame is the

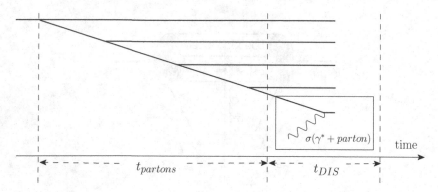

Fig. 2.10. The space–time structure of DIS in the IMF/Bjorken and Breit frames. For this illustration, it is sufficient that all partons (both quarks and gluons) are denoted by straight solid lines, for simplicity.

same as in the IMF/Bjorken frame and is given by Eq. (2.60). Then one can easily see that, with $t_{DIS} = 1/Q$ in the Breit frame and since $P Q \gg \mu m$, we still have $t_{partons} \gg t_{DIS}$ in this frame.

Having these estimates in mind we can view the deep inelastic scattering process in either the Breit or the IMF/Bjorken frame, as shown in Fig. 2.10. The fast-moving particle (the proton), long before the interaction, "produces" a system of point-like particles (partons) which can be described by a light cone wave function. At the moment of interaction, the parton with the lowest energy (the "wee" parton) interacts with the virtual photon. The virtual photon in the Breit frame is a standing wave that interacts only with partons that have the same wavelength; in other words, it interacts with the parton whose momentum is equal to $Q/2$. The last statement follows from momentum conservation for the wee parton, whose momentum is k before and k' after its interaction with photon, namely,

$$k^0 = k'^0, \quad k^3 - k'^3 = Q, \quad \vec{k}_\perp = \vec{k}'_\perp. \tag{2.64}$$

(To obtain Eq. (2.64) note that $k' = k + q$ and use Eq. (2.63).) From Eq. (2.64), and assuming that the incoming parton is on mass shell, one can show that $k^3 = -k'^3 = Q/2$. Assuming also that $Q \gg k_\perp$ and neglecting the quark mass we get $k^0 \approx k^3 = Q/2$, leading to $k^+ \approx Q$. The fraction of the proton's light cone momentum P^+ carried by the struck quark is equal to $x = k^+/P^+ \approx Q/P^+ = Q^2/(P^+q^-) = x_{Bj}$, just as in Eq. (2.33).

Therefore the DIS process happens in two stages. The first stage is the creation of many point-like partons and can be described by the light cone wave function of the fast-moving hadron. The second stage is the interaction of the slowest (wee) parton with the virtual photon, which occurs at low energies. It should be stressed that in this section we have not used the fact that the transverse momenta of partons are restricted in the UV, though we did in the previous section. This fact gives us hope that the whole structure, consisting of the wave function of the fast-moving hadron and the interaction of the parton with the photon, will remain correct in a more general approach. However, the wee parton interaction could be more complicated than in the naive parton model.

2.4 Violation of Bjorken scaling;
the Dokshitzer–Gribov–Lipatov–Altarelli–Parisi evolution equation

2.4.1 Parton distributions

Let us study the QCD corrections to the naive parton model presented above. First we rewrite the quark distribution function $q^f(x, Q^2)$ for a quark of flavor f from Eq. (2.41) as follows:

$$
q^f(x, Q^2) = \sum_n \frac{1}{x} \int \frac{d^2 k_\perp}{2(2\pi)^3} \frac{1}{S_n} \sum_{\sigma=\pm 1} \prod_{i=1}^n \frac{dx_i}{x_i} \frac{d^2 k_{i\perp}}{2(2\pi)^3}
$$

$$
\times |\Psi_n^f(\{x_i, k_{i\perp}\}; x, k_\perp; \sigma)|^2 (2\pi)^3
$$

$$
\times \delta^2\left(\vec{k}_\perp + \sum_{j=1}^n \vec{k}_{j\perp}\right) \delta\left(1 - x - \sum_{l=1}^n x_l\right). \tag{2.65}
$$

The quark carries a fraction x of the longitudinal momentum of the proton; x is identical to the Bjorken-x variable defined in Eq. (2.2). Unlike in the naive parton model the quark distribution function now depends on the momentum scale Q^2, which enters Eq. (2.65) as the renormalization scale. Roughly, this implies that the integral over the quark's transverse momentum $k_\perp$ is bounded from above by Q, so that $k_\perp \leq Q$. The same applies to the other transverse momentum integrals in (2.65) along with the virtual loop integrals within the wave function Ψ_n. (In the naive parton model, we assumed that the $k_\perp$-integrals were sufficiently convergent that one could simply replace Q in the upper limit of integration by infinity without changing the value of the integral; this is, strictly speaking, only true for super-renormalizable theories and so is not true for QCD.) The goal of this subsection is to understand this Q-dependence in more detail.

The light cone wave function $\Psi_n^f(\{x_i, k_{i\perp}\}; x, k_\perp; \sigma)$ describes a Fock state in the proton containing the quark we are measuring along with n "spectator" partons with transverse momenta $k_{i\perp}$ and longitudinal momentum fractions x_i. The sum over n runs from some small number, determined by the nonperturbative physics defining the proton, up to ∞. (If we were studying the wave function of a single quark under the assumption that it is completely perturbative, then n would run from 0 to ∞.) Note that the quark helicity, which was labeled r in Sec. 2.2 to avoid confusion with the proton polarization, will be labeled from now on by σ, since here the proton helicity does not enter our calculations explicitly.

The quark distribution function (2.65) is illustrated by the diagram in Fig. 2.11. The definition (2.65) is the LCPT analogue of the standard operator definition in the light cone gauge $A^+ = 0$ (see for instance Sterman (1993)).

In analogy with (2.65) we can define the gluon distribution function:

$$
G(x, Q^2) = \sum_n \frac{1}{x} \int \frac{d^2 k_\perp}{2(2\pi)^3} \frac{1}{S_n} \sum_{\lambda=\pm 1} \prod_{i=1}^n \frac{dx_i}{x_i} \frac{d^2 k_{i\perp}}{2(2\pi)^3}
$$

$$
\times |\Psi_n(\{x_i, k_{i\perp}\}; x, k_\perp; \lambda)|^2
$$

$$
\times (2\pi)^3 \delta^2\left(\vec{k}_\perp + \sum_{j=1}^n \vec{k}_{j\perp}\right) \delta\left(1 - x - \sum_{l=1}^n x_l\right). \tag{2.66}
$$

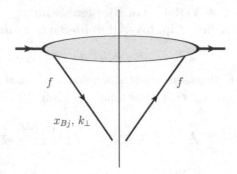

Fig. 2.11. A diagrammatic representation of the quark distribution function. The vertical solid line separates the light cone wave function from its complex conjugate.

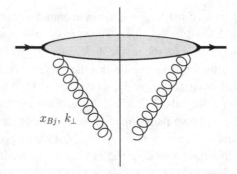

Fig. 2.12. A diagrammatic representation of the gluon distribution function.

Here $\Psi_n \left(\{x_i, k_{i\perp}\}; x, k_\perp; \lambda \right)$ is the proton light cone wave function containing n "spectators" along with a measured *gluon* having longitudinal momentum fraction x, transverse momentum $k_\perp$, and polarization λ. Again Q^2 enters (2.66) as the renormalization scale. The definition of the gluon distribution function given by Eq. (2.66) is the LCPT analogue of the operator definition in terms of gluon operators in the light cone gauge $A^+ = 0$ (Sterman 1993). It is illustrated in Fig. 2.12.

The $k_\perp$-integral in the definition of $q^f(x, Q^2)$ given in Eq. (2.65) is effectively cut off by Q, making the quark distribution function Q-dependent in general. The essential idea of Bjorken scaling is that for very large Q we can simply set the upper cutoff of the $k_\perp$-integral to infinity. In the naive parton model it is assumed that the $k_\perp$-integral is convergent in the UV, owing to some (presumed nonperturbative) universal cutoff. The resulting quark distribution becomes a function of x only, $q^f(x, Q^2 \to \infty) \approx q^f(x)$. This leads to the Bjorken scaling seen in Eqs. (2.45) and (2.46).

In reality the $k_\perp$-integral in Eq. (2.65) (and that in Eq. (2.66)) is not convergent in the UV and so needs this Q^2 cutoff: hence a Q^2-dependence remains in the quark and gluon distributions even at very high Q^2. To determine the Q^2-dependence of the distribution functions one needs to understand exactly how the proton's light cone wave function Ψ_n

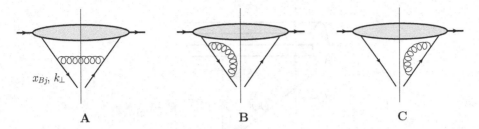

Fig. 2.13. Lowest-order QCD corrections to the quark distribution function. The virtual diagrams should be understood as including instantaneous terms.

depends on $k_\perp$. To do so we have to assume that at least part of the distribution function is perturbative. In terms of diagrams this perturbative dynamics takes place in the part of the diagram adjacent to the parton that we are describing by the distribution function. This will be justified later by the large transverse momentum of the parton. We thus need to calculate the QCD corrections to the parton distribution functions of the naive parton model pictured in Figs. 2.11 and 2.12.

2.4.2 Evolution for quark distribution

Let us start with the quark distribution function $q^f(x, Q^2)$ shown in Fig. 2.11. The lowest-order QCD corrections to $|\Psi_n^f(\{x_i, k_{i\perp}\}; x, k_\perp; \sigma)|^2$ are shown in Fig. 2.13. They consist of the "real" emission diagram A and the "virtual" diagrams B and C. (The virtual corrections in LCPT should include graphs with instantaneous terms; these are not shown explicitly.) Diagrams in which the gluon line attaches to other partons in the wave function denoted by the oval (i.e., diagrams with the gluon going into the oval) are suppressed. To see why this is so, one has to identify the resummation parameter of the calculation to be performed shortly. Indeed each diagram in Fig. 2.13 has an extra factor equal to the coupling α_s as compared with the naive parton model quark distribution in Fig. 2.11. However, we will not calculate the rest of the diagram exactly: instead we will extract the leading contribution at large Q^2. These leading contributions, after integration over $k_\perp$, will turn out to be proportional to $\ln(Q^2/\Lambda_{QCD}^2)$. Hence the diagrams in Fig. 2.13 will each give us an expression proportional to $\alpha_s \ln(Q^2/\Lambda_{QCD}^2)$. This will be the resummation parameter of our approximation: for each power of α_s we will pick up one power of $\ln(Q^2/\Lambda_{QCD}^2)$. Owing to asymptotic freedom at large Q^2 we have $\alpha_s(Q^2) \ll 1$ while $\ln(Q^2/\Lambda_{QCD}^2) \gg 1$. Our resummation parameter is thus the product of a small quantity (the coupling) and a large quantity (the logarithm), and therefore

$$\alpha_s \ln \frac{Q^2}{\Lambda_{QCD}^2} \sim 1. \tag{2.67}$$

The resummation of the parameter in Eq. (2.67) is called the leading logarithmic approximation (LLA).

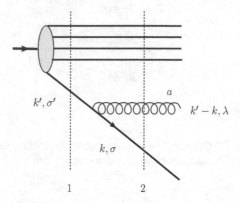

Fig. 2.14. Lowest-order correction to the proton light cone wave function contributing to the quark distribution. The vertical dotted lines denote intermediate states.

As one can show explicitly using the techniques we will develop below, the diagrams with an extra gluon connecting to the oval are, in fact, outside the leading logarithmic approximation. That is, they would generate powers of the coupling α_s not enhanced by powers of the logarithm of Q^2. This is why such diagrams are neglected in our analysis.

We will begin by calculating the diagram in Fig. 2.13A. Instead of calculating the diagram for the wave function squared it is better to start by calculating corrections to the wave function itself. The correction to the light cone wave function corresponding to Fig. 2.13A is shown in Fig. 2.14. There the intermediate states are denoted by the dotted vertical lines and are labeled 1 and 2.

Denoting by $\Psi_{n-1}^f(\{x_i, k_{i\perp}\}; x', k'_\perp; \sigma')$ the wave function for a proton with $n-1$ spectator partons (i.e., without the gluon emitted in Fig. 2.14), we note that the energy denominator corresponding to intermediate state 1 (denoted by the left-hand vertical dotted line) is already included in Ψ_{n-1}. Using the rules of LCPT outlined in Sec. 1.3 and their modification for the calculation of wave functions in Sec. 1.4, we can write down the contribution to the proton's wave function from the diagram in Fig. 2.14 as

$$\Psi_n^f\left(\{k_i^+, k_{i\perp}\}; x, k_\perp; \sigma\right) = \frac{g t^a\, \theta(k^+)\, \theta(k'^+ - k^+)}{(k'-k)^- + k^- + \sum\limits_{j=1}^{n-1} k_j^- - P^-}$$

$$\times\; \frac{\bar{u}_\sigma(k)\,\gamma\cdot\epsilon_\lambda^*(k'-k)\,u_{\sigma'}(k')}{k'^+}\; \Psi_{n-1}^f\left(\{k_i^+, k_{i\perp}\}; x', k'_\perp; \sigma'\right). \tag{2.68}$$

Here g is the QCD coupling, t^a is the color matrix (the gluon carries color a), and $x' = k'^+/P^+$. The quark line carrying momentum k' is internal and therefore contributes a factor $1/k'^+$ which is not included in the definition of the light cone wave function Ψ_{n-1}^f and so has to be included explicitly in Eq. (2.68). The intermediate state 2 from Fig. 2.14 gives the light cone energy denominator in Eq. (2.68):

$$\frac{1}{(k'-k)^- + k^- + \sum_{j=1}^{n-1} k_j^- - P^-} \equiv \frac{1}{\frac{(\vec{k}'_\perp - \vec{k}_\perp)^2}{k'^+ - k^+} + \frac{\vec{k}_\perp^2}{k^+} + \sum_{j=1}^{n-1} \frac{\vec{k}_j^2}{k_j^+} - P^-}. \tag{2.69}$$

Here P^- is the light cone energy of the incoming proton state. It is negligibly small, as it is inversely proportional to the large light cone plus momentum of the proton, $P^- \sim 1/P^+$, not enhanced by a large transverse momentum. (Indeed, for a "proton" consisting of a single valence quark of mass m_q one has $P^- = (\vec{P}_\perp^2 + m_q^2)/P^+$ with $\vec{P}_\perp$ the transverse momentum of the "proton".)

We are working in the $A^+ = 0$ light cone gauge. The gluon polarization vector is

$$\epsilon_\lambda^\mu(k' - k) = \left(0, \frac{2\vec{\epsilon}_\perp^\lambda \cdot k'^+ - k^+}{\vec{k}'_\perp - \vec{k}_\perp}, \vec{\epsilon}_\perp^\lambda \right)$$

in the $(+, -, \perp)$ notation with $\vec{\epsilon}_\perp^\lambda = -(1/\sqrt{2})(\lambda, i)$. One can thus write

$$\bar{u}_\sigma(k) \gamma \cdot \epsilon_\lambda^*(k' - k) u_{\sigma'}(k') = \bar{u}_\sigma(k) \gamma^+ u_{\sigma'}(k') \frac{\vec{\epsilon}_\perp^{\lambda*} \cdot (\vec{k}'_\perp - \vec{k}_\perp)}{k'^+ - k^+}$$
$$- \bar{u}_\sigma(k) \vec{\gamma}_\perp u_{\sigma'}(k') \cdot \vec{\epsilon}_\perp^{\lambda*}. \tag{2.70}$$

Using the tables for Dirac matrix elements from appendix section A.1 one obtains after some algebra

$$\bar{u}_\sigma(k) \gamma \cdot \epsilon_\lambda^*(k' - k) u_{\sigma'}(k') = -\frac{\delta_{\sigma\sigma'}}{\sqrt{z}(1-z)} \vec{\epsilon}_\perp^{\lambda*} \cdot (\vec{k}_\perp - z\vec{k}'_\perp)$$
$$\times [1 + z + \sigma\lambda(1-z)], \tag{2.71}$$

where $z = k^+/k'^+$ and we have assumed that the quarks are massless for simplicity. In arriving at Eq. (2.71) we have used $\vec{\epsilon}_\perp^{\lambda*} \times \vec{k}_\perp = i\lambda \vec{\epsilon}_\perp^{\lambda*} \cdot \vec{k}_\perp$, which is valid in two dimensions.

We will be working in the approximation where all transverse momenta are ordered:

$$Q^2 \gg k_\perp^2 \gg k_\perp'^2 \gg k_{n-1,\perp}^2 \gg \cdots \gg k_{1,\perp}^2 \sim \Lambda_{QCD}^2. \tag{2.72}$$

Such a regime corresponds to the LLA discussed above. One also assumes that all relevant large transverse momentum scales are much larger than the quark masses, which justifies the massless quark approximation we have just used. In the regime defined by Eq. (2.72) the light cone energy denominator becomes (see Eq. (2.69))

$$\frac{1}{(k'-k)^- + k^- + \sum_{j=1}^{n-1} k_j^- - P^-} \approx \frac{k'^+ z(1-z)}{\vec{k}_\perp^2}. \tag{2.73}$$

Substituting Eqs. (2.71) and (2.73) into Eq. (2.68) and assuming that $k_\perp^2 \gg k_\perp'^2$ yields

$$\Psi_n^f(\{x_i, k_{i\perp}\}; x, k_\perp; \sigma) = -gt^a\,\theta(z)\theta(1-z)\,\delta_{\sigma\sigma'}\,\sqrt{z}\,\frac{\vec{\epsilon}_\perp^{\lambda*}\cdot\vec{k}_\perp}{k_\perp^2}$$

$$\times\,[1+z+\sigma\lambda(1-z)]\,\Psi_{n-1}^f\left(\{x_i, k_{i\perp}\}; x', k_\perp'; \sigma'\right).\quad(2.74)$$

Multiplying the wave function (2.74) by its complex conjugate and summing over the quark and gluon polarizations and colors we get

$$\sum_{\sigma,\sigma',\lambda,a} |\Psi_n^f(\{x_i, k_{i\perp}\}; x, k_\perp; \sigma)|^2 = 8\pi\alpha_s\,C_F\,\theta(z)\theta(1-z)\,z(1+z^2)\,\frac{1}{k_\perp^2}$$

$$\times\sum_{\sigma'=\pm1} |\Psi_{n-1}^f\left(\{x_i, k_{i\perp}\}; x', k_\perp'; \sigma'\right)|^2;\quad(2.75)$$

in arriving at Eq. (2.75) we have used the fact that $\vec{\epsilon}_\perp^\lambda = -(1/\sqrt{2})(\lambda, i)$ and $\sum_{a=1}^{N_c^2-1} t^a\,t^a = C_F$, where

$$C_F = \frac{N_c^2 - 1}{2N_c}\quad(2.76)$$

is the Casimir operator in the fundamental representation of SU(N_c).

Substituting Eq. (2.75) into the definition of the quark distribution function (2.65) yields the contribution of the diagram in Fig. 2.13A:

$$q_A^f(x, Q^2) = \sum_n \frac{1}{x}\int \prod_{i=1}^{n-1} \frac{dx_i}{x_i}\,\frac{d^2k_{i\perp}}{2(2\pi)^3}\,\frac{d^2k_\perp}{2(2\pi)^3}\,\frac{d(k'^+ - k^+)}{k'^+ - k^+}\,\frac{d^2(\vec{k}_\perp' - \vec{k}_\perp)}{2(2\pi)^3}$$

$$\times\,8\pi\alpha_s C_F\,\theta(z)\theta(1-z)\,\frac{z(1+z^2)}{k_\perp^2}\sum_{\sigma'=\pm1} |\Psi_{n-1}^f\left(\{x_i, k_{i\perp}\}; x', k_\perp'; \sigma'\right)|^2$$

$$\times\,(2\pi)^3\,\delta^2\left(\vec{k}_\perp' + \sum_{j=1}^{n-1}\vec{k}_{j\perp}\right)\delta\left(1 - x' - \sum_{l=1}^{n-1}x_l\right).\quad(2.77)$$

Note that the symmetry factor S_n from Eq. (2.65) is eliminated by the momentum ordering (2.72), which makes the particles in the wave function distinct. Since we are keeping k^+ fixed, the integral over $k'^+ - k^+$ can be rewritten as follows:

$$\int_0^{P^+ - k^+} \frac{d(k'^+ - k^+)}{k'^+ - k^+} = \int_{k^+}^{P^+} \frac{dk'^+}{k'^+ - k^+} = \int_x^1 \frac{dz}{z(1-z)}.\quad(2.78)$$

Then we can rewrite Eq. (2.77):

$$q_A^f(x, Q^2) = \frac{\alpha_s C_F}{2\pi} \frac{1}{x} \int\limits^{Q^2} \frac{dk_\perp^2}{k_\perp^2} \int\limits_x^1 dz \frac{1+z^2}{1-z}$$

$$\times \sum_n \int \prod_{i=1}^{n-1} \frac{dx_i}{x_i} \frac{d^2 k_{i\perp}}{2(2\pi)^3} \frac{d^2 k_\perp'}{2(2\pi)^3} \sum_{\sigma'=\pm 1} |\Psi_{n-1}^f(\{x_i, k_{i\perp}\}; x', k_\perp'; \sigma')|^2$$

$$\times (2\pi)^3 \delta^2\left(\vec{k}_\perp' + \sum_{j=1}^{n-1} \vec{k}_{j\perp}\right) \delta\left(1 - x' - \sum_{l=1}^{n-1} x_l\right), \qquad (2.79)$$

where the integral over $k_\perp'$ is cut off by $k_\perp$ from above owing to our momentum ordering $Q^2 \gg k_\perp^2 \gg k_\perp'^2 \gg \Lambda_{QCD}^2$.

Comparing with Eq. (2.65) we recognize the last two lines of Eq. (2.79) as $x' q^f(x', k_\perp^2)$. Equation (2.79) thus gives

$$q_A^f(x, Q^2) = \frac{\alpha_s C_F}{2\pi} \frac{1}{x} \int\limits^{Q^2} \frac{dk_\perp^2}{k_\perp^2} \int\limits_x^1 dz \frac{1+z^2}{1-z} x' q^f(x', k_\perp^2). \qquad (2.80)$$

Remembering that $z = k^+/k'^+ = x/x'$, we write

$$q_A^f(x, Q^2) = \frac{\alpha_s C_F}{2\pi} \int\limits^{Q^2} \frac{dk_\perp^2}{k_\perp^2} \int\limits_x^1 \frac{dz}{z} \frac{1+z^2}{1-z} q^f\left(\frac{x}{z}, k_\perp^2\right). \qquad (2.81)$$

This is the contribution of diagram A in Fig. 2.13 to the quark distribution function. As promised above, it contains the coupling α_s as a factor and a logarithmic integral $dk_\perp^2/k_\perp^2$ cut off by Q^2 in the UV and by some nonperturbative scale $\sim \Lambda_{QCD}^2$ in the infrared (IR). We have thus shown that the leading large Q^2 contribution of diagram A in Fig. 2.13 to the quark distribution function is proportional to $\alpha_s \ln Q^2/\Lambda_{QCD}^2$.

Now imagine that we slowly increase Q^2. As Q^2 gets larger, the phase space for the emitted gluons increases, generating larger and larger $\ln(Q^2/\Lambda_{QCD}^2)$ values and thus increasing the probability of gluon emission. The modification $\delta q_A^f(x, Q^2)$ of the quark distribution with increasing Q^2 due to the gluon emission in Fig. 2.13A can be obtained by differentiating Eq. (2.81) with respect to Q^2:

$$Q^2 \frac{\partial q_A^f(x, Q^2)}{\partial Q^2} = \frac{\alpha_s C_F}{2\pi} \int\limits_x^1 \frac{dz}{z} \frac{1+z^2}{1-z} q^f\left(\frac{x}{z}, Q^2\right). \qquad (2.82)$$

An example of a diagram that does not give a leading logarithmic contribution is shown in Fig. 2.15, where the oval of Fig. 2.13 is reduced to a gluon line for simplicity. The dotted vertical lines in Fig. 2.15 represent the intermediate states contributing light cone energy denominators. The diagram is of order α_s^2. Let us show that it does not give an LLA contribution, by using the results obtained above. We will work in the transverse

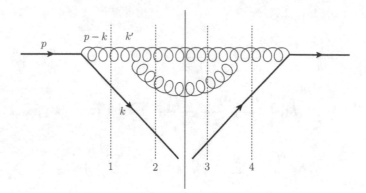

Fig. 2.15. An example of a diagram outside the leading-logarithmic approximation.

momentum ordering approximation of Eq. (2.72): $k_\perp^2 \gg k_\perp'^2$ in terms of the momentum labeling in Fig. 2.15. Keeping track of the transverse momenta we see that the energy denominators of the intermediate states 1 through 4 in Fig. 2.15 each give $1/k_\perp^2$, since they are dominated by the large light cone energy of the k-quark line (cf. Eq. (2.73)). In the same large-$k_\perp$ approximation each quark–gluon splitting gives a factor $\vec{k}_\perp$ in the amplitude (cf. Eq. (2.71)) for the net contribution of $k_\perp^2$. Assuming that gluon–gluon splitting gives a similar factor $k_\perp^2$ (this will be demonstrated explicitly in Sec. 2.4.4 below), we conclude that the contribution of the graph in Fig. 2.15 is proportional to $(1/k_\perp^2)^4(k_\perp^2)^2 = 1/k_\perp^4$. Performing the integrals over $k_\perp^2$ and $k_\perp'^2$ with the $k_\perp^2 \gg k_\perp'^2 \gg \Lambda_{QCD}^2$ ordering, we find that the diagram in Fig. 2.15 is proportional to

$$\alpha_s^2 \int\limits_{\Lambda_{QCD}^2}^{Q^2} dk_\perp^2 \int\limits_{\Lambda_{QCD}^2}^{k_\perp^2} \frac{dk_\perp'^2}{k_\perp^4} \approx \alpha_s^2 \int\limits_{\Lambda_{QCD}^2}^{Q^2} \frac{dk_\perp^2}{k_\perp^2} = \alpha_s^2 \ln \frac{Q^2}{\Lambda_{QCD}^2}. \tag{2.83}$$

We observe that this diagram is certainly beyond the LLA, as it brings in two powers of α_s with only one power of $\ln(Q^2/\Lambda_{QCD}^2)$, whereas an LLA diagram at the same order in α_s would bring in two powers of $\ln(Q^2/\Lambda_{QCD}^2)$. Therefore, it (and other graphs not included in Fig. 2.13) is subleading and can be neglected in the LLA.

The contributions of diagrams B and C in Fig. 2.13 to the change in the quark distribution can be calculated directly, similarly to that of diagram A. However, instead of embarking upon another possibly tedious calculation we will derive these contributions using a unitarity argument.

Unitarity argument Let us start with a proton state $|\Psi\rangle$, normalized for simplicity to 1, $\langle\Psi|\Psi\rangle = 1$. Single-gluon corrections of the diagrams in Fig. 2.13 modify the state $|\Psi\rangle$ as follows:

$$|\Psi\rangle \to |\Psi'\rangle = |\Psi\rangle + R|\Psi\rangle + V|\Psi\rangle. \tag{2.84}$$

Here the new state $|\Psi'\rangle$ consists of a sum of the following terms: (i) the "old" state $|\Psi\rangle$ corresponding to no gluon corrections at all; (ii) the "real" emission shown in Fig. 2.13A,

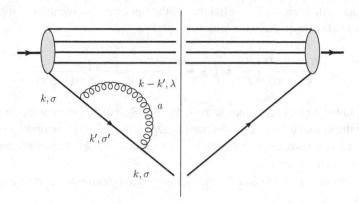

Fig. 2.16. Virtual correction to the quark–quark splitting function.

which turns $|\Psi\rangle$ into $R|\Psi\rangle$ where R denotes the factor relating Ψ_n and Ψ_{n-1} in Eq. (2.74); (iii) the "virtual" emission shown in Figs. 2.13B, C, where the gluon is reabsorbed back into the wave function from which it was emitted, thus leaving the number of partons unchanged and generating a contribution $V|\Psi\rangle$.

Requiring unitarity, i.e., probability conservation $\langle\Psi'|\Psi'\rangle = \langle\Psi|\Psi\rangle = 1$, in Eq. (2.84) leads to

$$R^\dagger R + V + V^* = 0 \qquad (2.85)$$

to order g^2. Therefore, the sum of the contributions of diagrams B and C in Fig. 2.13 is

$$V + V^* = -R^\dagger R. \qquad (2.86)$$

We see that instead of calculating diagrams B and C in Fig. 2.13 we can simply multiply the contribution of diagram A by its conjugate, integrate over the phase space, and sum over the quantum numbers of the produced and measured partons, multiplying the result by -1.

Using the unitarity prescription of Eq. (2.86) along with Eq. (2.77) for the contribution of diagram A, we write the contribution of diagrams B and C from Fig. 2.13 (along with instantaneous terms) as

$$q_{B+C}^f(x, Q^2) = -\sum_n \int \prod_{i=1}^n \frac{dx_i}{x_i} \frac{d^2k_{i\perp}}{2(2\pi)^3} \frac{1}{x'} \frac{d^2k'_\perp}{2(2\pi)^3} \frac{d(k^+ - k'^+)}{k^+ - k'^+} \frac{d^2k_\perp}{2(2\pi)^3}$$

$$\times 8\pi\alpha_s C_F \theta(z)\theta(1-z) \frac{z(1+z^2)}{k_\perp^2} \sum_{\sigma=\pm1} |\Psi_n^f(\{x_i, k_{i\perp}\}; x, k_\perp; \sigma)|^2$$

$$\times (2\pi)^3 \delta^2\left(\vec{k}_\perp + \sum_{j=1}^n \vec{k}_{j\perp}\right) \delta\left(1 - x - \sum_l x_l\right). \qquad (2.87)$$

Equation (2.87) is illustrated in Fig. 2.16. In arriving at Eq. (2.87) we have swapped k and k' as compared with the real emission diagram shown in Fig. 2.14. After emitting a gluon, the quark now carries momentum k'; the fraction of the light cone momentum of the incoming quark k^+ carried by the quark in the loop is $z = k'^+/k^+$. As above,

$$\int\limits_0^{P^+} \frac{d(k^+ - k'^+)}{k^+ - k'^+} \frac{1}{x'} = \int\limits_0^{k^+} \frac{dk'^+ P^+}{k'^+(k^+ - k'^+)} = \frac{1}{x} \int\limits_0^1 \frac{dz}{z(1 - z)}. \tag{2.88}$$

Note that the lower limit of the z-integration in Eq. (2.88) is different from that in Eq. (2.78): this is due to the virtual nature of the diagram in Fig. 2.16. It is also important to remember that now the large transverse momentum is $k'_\perp$, so that $k'^2_\perp \gg k^2_\perp$, which accounts for the factor $k'^2_\perp$ in the denominator in Eq. (2.87).

With the help of Eq. (2.88) and the quark distribution definition (2.65) we can rewrite Eq. (2.87) as

$$q_{B+C}^f(x, Q^2) = -\frac{\alpha_s C_F}{2\pi} \int\limits^{Q^2} \frac{dk'^2_\perp}{k'^2_\perp} \int\limits_0^1 dz \frac{1 + z^2}{1 - z} q^f(x, k'^2_\perp), \tag{2.89}$$

where, owing to the constraint $k'^2_\perp \gg k^2_\perp$, we may cut off the $k_\perp$-integral in Eq. (2.87) by $k'^2_\perp$ in the UV. The result is that $k'^2_\perp$ is the scale of the quark distribution function on the right-hand side of Eq. (2.89). Note that the $k'_\perp$-integral is a loop integral and is, in general, divergent: it has to be regularized, and so a graph with a counterterm should be added to the diagram in Fig. 2.16. Since Q is the renormalization scale, to leading-logarithmic accuracy we simply cut off the $k'^2_\perp$-integral in Eq. (2.89) by Q^2 in the UV.

Equation (2.89) is the contribution of the virtual diagrams B and C in Fig. 2.13. Just as for the real diagram A, we now imagine that we slowly increase Q^2: the contribution of graphs B and C to the variation in the quark distribution function is

$$Q^2 \frac{\partial q_{B+C}^f(x, Q^2)}{\partial Q^2} = -\frac{\alpha_s C_F}{2\pi} \int\limits_0^1 dz \frac{1 + z^2}{1 - z} q^f(x, Q^2). \tag{2.90}$$

The total modification of the quark distribution, $\delta q^f(x, Q^2) = \delta q_A^f(x, Q^2) + \delta q_{B+C}^f(x, Q^2)$, is obtained by summing Eqs. (2.82) and (2.90). This yields

$$Q^2 \frac{\partial q^f(x, Q^2)}{\partial Q^2} = \frac{\alpha_s C_F}{2\pi} \left[\int\limits_x^1 \frac{dz}{z} \frac{1 + z^2}{1 - z} q^f\left(\frac{x}{z}, Q^2\right) \right.$$

$$\left. - \int\limits_0^1 dz \frac{1 + z^2}{1 - z} q^f(x, Q^2) \right]. \tag{2.91}$$

To write Eq. (2.91) in the standard notation, we define the quark–quark *splitting function* $P_{qq}(z)$ by

$$P_{qq}(z) \equiv C_F \left[\frac{1 + z^2}{(1 - z)_+} + \frac{3}{2} \delta(1 - z) \right] \tag{2.92}$$

with the "plus" notation defined in Sterman (1993),

$$\int\limits_{x}^{1} dz\, \frac{1}{(1-z)_+}\, f(z) = \int\limits_{x}^{1} dz\, \frac{1}{1-z}\, [f(z) - f(1)] + f(1)\ln(1-x), \qquad (2.93)$$

for an arbitrary function $f(z)$ defined for $0 \le x \le 1$. With the help of $P_{qq}(z)$ we rewrite Eq. (2.91) in the more compact form

$$Q^2\, \frac{\partial q^f(x, Q^2)}{\partial Q^2} = \frac{\alpha_s}{2\pi} \int\limits_{x}^{1} \frac{dz}{z}\, P_{qq}(z)\, q^f\!\left(\frac{x}{z}, Q^2\right). \qquad (2.94)$$

We have thus obtained a differential equation for the quark distribution function. The initial condition for this equation is usually given by the quark distribution $q^f(x, Q_0^2)$ at some initial virtuality Q_0^2. At low Q_0^2 such an initial condition is likely to be due to some nonperturbative (large-α_s) physics: it cannot be calculated using perturbative techniques and is usually inferred from the data. Given the initial condition $q^f(x, Q_0^2)$, Eq. (2.94) allows one to uniquely construct the quark distribution function at all $Q^2 > Q_0^2$ (with leading-logarithmic accuracy). Therefore Eq. (2.94) *evolves* the quark distribution function in Q^2 from some initial value at Q_0^2 to its value at another scale Q^2: equations like (2.94) are usually referred to as *evolution equations*. The variation of a distribution function with Q^2 is known as the Q^2-*evolution* of the distribution function.

The physical meaning of the splitting function $P_{qq}(z)$ is clear from our derivation of Eq. (2.94): $P_{qq}(z)$ is proportional to the probability of finding one quark in another quark's wave function, with the "measured" quark carrying a fraction z of the original quark's light cone momentum.

Another important question concerns the scale of the coupling constant α_s in Eq. (2.94). Without going into details of the calculation of the running coupling corrections, we simply note that, up to a z-dependent factor, the scale is simply Q^2, so that $\alpha_s = \alpha_s(Q^2)$. Thus the coupling runs with the perturbative (hard) scale of the problem, justifying the use of perturbation theory.

2.4.3 The DGLAP evolution equations

Equation (2.94) is not complete yet: so far we have ignored gluons. Indeed, a quark in the proton's wave function may also result from the splitting of a gluon into a $q\bar{q}$ pair! Thus the gluon distribution $G(x, Q^2)$ also contributes to the modification of the quark distribution. Conversely, the gluon distribution also gets modified owing to the splitting of gluons into gluon pairs, or the emission of gluons from quarks as in Fig. 2.13A.

Including the gluon contribution requires additional calculations, similar to those carried out above. Before outlining these calculations let us first present the result.

We define the *flavor nonsinglet distribution function* by

$$\Delta^{f\bar{f}}(x, Q^2) = q^f(x, Q^2) - q^{\bar{f}}(x, Q^2), \qquad (2.95)$$

where $\bar{f}$ denotes the antiquark of flavor f. Since the splitting of a gluon into $q\bar{q}$ pairs contributes *equally* to the creation of quarks and anti-quarks in the proton's wave function, it should not contribute to the nonsinglet distribution $\Delta^{f\bar{f}}(x, Q^2)$. Hence the evolution of $\Delta^{f\bar{f}}(x, Q^2)$ is driven only by the quark evolution from Eq. (2.94). We thus write

$$Q^2 \frac{\partial \Delta^{f\bar{f}}(x, Q^2)}{\partial Q^2} = \frac{\alpha_s(Q^2)}{2\pi} \int_x^1 \frac{dz}{z} P_{qq}(z) \Delta^{f\bar{f}}\left(\frac{x}{z}, Q^2\right). \tag{2.96}$$

To take the gluon contribution into account we define the *flavor singlet distribution function*

$$\Sigma(x, Q^2) = \sum_f \left[q^f(x, Q^2) + q^{\bar{f}}(x, Q^2)\right]. \tag{2.97}$$

The evolution equations for $\Sigma(x, Q^2)$ and $G(x, Q^2)$ read

$$Q^2 \frac{\partial}{\partial Q^2} \begin{pmatrix} \Sigma(x, Q^2) \\ G(x, Q^2) \end{pmatrix} = \frac{\alpha_s(Q^2)}{2\pi} \int_x^1 \frac{dz}{z} \begin{pmatrix} P_{qq}(z) & P_{qG}(z) \\ P_{Gq}(z) & P_{GG}(z) \end{pmatrix}$$

$$\times \begin{pmatrix} \Sigma(x/z, Q^2) \\ G(x/z, Q^2) \end{pmatrix}. \tag{2.98}$$

Equations (2.96) and (2.98) are known as the Dokshitzer–Gribov–Lipatov–Altarelli–Parisi (DGLAP) evolution equations. The QED version of these equations (involving electrons and photons) in (x, Q^2)-space was originally derived by Gribov and Lipatov (1972), while the QCD version was obtained independently by Altarelli and Parisi (1977) and by Dokshitzer (1977). In the Mellin moment space (to be defined shortly) the QED equations were derived by Christ, Hasslacher, and Mueller (1972) and the QCD equations were derived by Georgi and Politzer (1974) and by Gross and Wilczek (1974).

Equations (2.96) and (2.98) contain the splitting function, $P_{qq}(z)$ from Eq. (2.92), along with three other splitting functions, $P_{qG}(z)$, $P_{Gq}(z)$, and $P_{GG}(z)$. For reference purposes, let us first list all the splitting functions, even though we have already found $P_{qq}(z)$ above. They are

$$P_{qq}(z) = C_F \left[\frac{1 + z^2}{(1 - z)_+} + \frac{3}{2}\delta(1 - z)\right], \tag{2.99a}$$

$$P_{Gq}(z) = C_F \frac{1 + (1 - z)^2}{z}, \tag{2.99b}$$

$$P_{qG}(z) = N_f \left[z^2 + (1 - z)^2\right], \tag{2.99c}$$

$$P_{GG}(z) = 2N_c \left[\frac{z}{(1 - z)_+} + \frac{1 - z}{z} + z(1 - z)\right] + \frac{11N_c - 2N_f}{6}\delta(1 - z). \tag{2.99d}$$

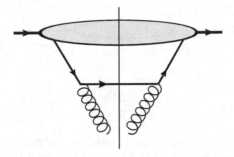

Fig. 2.17. The diagram contributing to the splitting function $P_{Gq}(z)$.

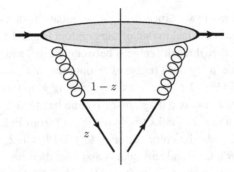

Fig. 2.18. The diagram contributing to the splitting function $P_{qG}(z)$.

The "plus" notation is defined above in Eq. (2.93).

The splitting function $P_{Gq}(z)$ is easy to find knowing $P_{qq}(z)$: $P_{Gq}(z)$ represents the probability of finding a gluon in a quark's light cone wave function. Its contribution consists of one diagram, pictured in Fig. 2.17. One can see that the calculation of $P_{Gq}(z)$ would be similar to that of diagram A in Fig. 2.13. The main difference would be in the fact that now it is the gluon that one wants to "measure", and therefore it is the gluon line that carries the longitudinal momentum fraction z of the quark. Since in the calculation of Fig. 2.13A the gluon line carried the momentum fraction $1 - z$, all we have to do to find $P_{Gq}(z)$ is to replace z by $1 - z$ in the contribution of graph A. To single out the contribution of diagram A we need to remove the contributions of the virtual diagrams B and C in Fig. 2.13 from Eq. (2.99a), which is easily accomplished by removing the plus sign in the subscript on the right-hand side and dropping the delta function term, yielding

$$P_{qq}^{real}(z) = C_F \frac{1 + z^2}{1 - z}. \tag{2.100}$$

Replacing z by $1 - z$ in $P_{qq}^{real}(z)$ yields $P_{Gq}(z)$, Eq. (2.99b), which is the correct result for the gluon–quark splitting function.

Finding the quark–gluon splitting function $P_{qG}(z)$ is a little more subtle. The only diagram contributing to the splitting function $P_{qG}(z)$ is shown in Fig. 2.18. (One also

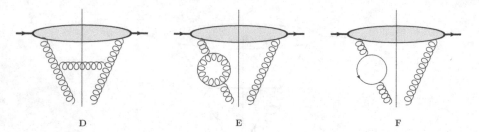

Fig. 2.19. The diagrams contributing to the splitting function $P_{GG}(z)$. The complex conjugates of the last two diagrams (E and F) have to be included in the calculation, along with the instantaneous terms in the quark and gluon propagators in the loops.

has to add in a diagram where we "measure" the antiquark instead of the quark, but the contribution of this diagram is equal to that of the graph in Fig. 2.18.) Comparing this with Fig. 2.14 we see that there are three differences between $P_{qG}(z)$ and the real part of $P_{qq}(z)$: (i) the incoming quark line in Fig. 2.14 becomes an outgoing antiquark line in Fig. 2.18 and the outgoing gluon line in Fig. 2.14 becomes an incoming gluon line in Fig. 2.18; (ii) the color factors are different in the two diagrams; (iii) one has to sum over all quark flavors f and over both quarks and anti-quarks to obtain $P_{qG}(z)$ from Fig. 2.18. Differences (ii) and (iii) are easily addressed. The color factor in Fig. 2.18 is $1/2$, which replaces C_F in Eq. (2.99a). The sum over quarks and anti-quarks and over their flavors trivially gives $2N_f$. Hence in the end one has to replace C_F in Eq. (2.99a) by $(1/2) \times 2N_f = N_f$. Difference (i) can be taken into account by applying the crossing symmetry. In the end the prescription is

$$P_{qG}(z) = \frac{N_f}{C_F} z \, P_{qq}^{real} \left(1 - \frac{1}{z} \right), \qquad (2.101)$$

which, with the help of $P_{qq}^{real}(z) = C_F(1 + z^2)/(1 - z)$, gives Eq. (2.99c). Indeed, the heuristic derivation of $P_{qG}(z)$ given here needs to be verified by explicit diagrammatic calculations. We leave the explicit calculation of $P_{qG}(z)$ using the diagram in Fig. 2.18 as an exercise for the reader; see Exercise 2.2 at the end of the chapter.

Finding the remaining splitting function $P_{GG}(z)$ requires some explicit diagrammatic calculations as well. We will present them in the next (special-topic) chapter.

2.4.4 Gluon–gluon splitting function*

Our goal here is to derive the gluon–gluon splitting function $P_{GG}(z)$. To calculate $P_{GG}(z)$ one has to sum the graphs shown in Fig. 2.19. There we show only half the diagrams with virtual corrections; the complex conjugates of graphs E and F need to be calculated too. As in the case of the quark–quark splitting function $P_{qq}(z)$, we will calculate only the real emission diagram D in Fig. 2.19 and derive the contributions of the remaining virtual diagrams E and F (and their conjugates) by using unitarity.

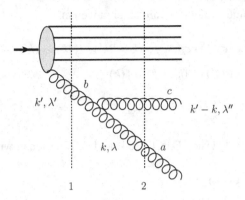

Fig. 2.20. A gluon splitting into two gluons in the proton light cone wave function. As usual, the vertical dotted lines denote intermediate states.

Just as in Sec. 2.4.2, to calculate the graph in Fig. 2.19D for the light cone wave function squared we first need to find the wave function itself. To that end we start with the diagram pictured in Fig. 2.20. Again, the intermediate state 1 is included in the wave function Ψ_{n-1} at the previous step of the evolution. Using the rules of LCPT outlined in Secs. 1.3 and 1.4, we can write the contribution of the graph in Fig. 2.20 as follows:

$$\Psi_n\left(\{x_i, k_{i\perp}\}; x, k_\perp; \lambda\right) = \frac{ig f^{abc}\theta(k^+)\theta(k'^+ - k^+)}{(k'-k)^- + k^- + \sum_{j=1}^{n-1} k_j^- - P^-} \frac{1}{k'^+}$$

$$\times \left[(k'+k)\cdot\epsilon_{\lambda''}^*(k'-k)\,\epsilon_\lambda^*(k)\cdot\epsilon_{\lambda'}(k') + (k-2k')\cdot\epsilon_\lambda^*(k)\right.$$

$$\times \epsilon_{\lambda'}(k')\cdot\epsilon_{\lambda''}^*(k'-k) + (k'-2k)\cdot\epsilon_{\lambda'}(k')\,\epsilon_{\lambda''}^*(k'-k)\cdot\epsilon_\lambda^*(k)\bigg]$$

$$\times \Psi_{n-1}\left(\{x_i, k_{i\perp}\}; x', k'_\perp; \lambda'\right), \qquad (2.102)$$

where now $\Psi_n\left(\{x_i, k_{i\perp}\}; x, k_\perp; \lambda\right)$ is the light cone wave function of the proton containing n "spectator" partons and the *gluon* being tagged. As usual $x = k^+/P^+$ and $x' = k'^+/P^+$ are the fractions of the proton's light cone momentum P^+ carried by the gluons.

Using the gluon polarizations in the $A^+ = 0$ light cone gauge,

$$\epsilon_\mu^\lambda(k) = \left(0, \frac{2\vec{\epsilon}_\perp^\lambda \cdot \vec{k}_\perp}{k^+}, \vec{\epsilon}_\perp^\lambda\right), \qquad (2.103a)$$

$$\epsilon_\mu^{\lambda'}(k') = \left(0, \frac{2\vec{\epsilon}_\perp^{\lambda'} \cdot \vec{k}'_\perp}{k'^+}, \vec{\epsilon}_\perp^{\lambda'}\right), \qquad (2.103b)$$

$$\epsilon_\mu^{\lambda''}(k'-k) = \left(0, \frac{2\vec{\epsilon}_\perp^{\lambda''} \cdot (\vec{k}'_\perp - \vec{k}_\perp)}{k'^+ - k^+}, \vec{\epsilon}_\perp^{\lambda''}\right) \qquad (2.103c)$$

with $\vec{\epsilon}_\perp^\lambda = -(1/\sqrt{2})(\lambda, i)$ (and similar expressions involving λ' and λ''), and imposing the transverse momentum ordering $|\vec{k}_\perp| \gg |\vec{k}'_\perp|$ (and, therefore, simply neglecting all terms containing $\vec{k}'_\perp$), after some straightforward algebra we get

$$(k' + k) \cdot \epsilon^*_{\lambda''}(k' - k) \, \epsilon^*_\lambda(k) \cdot \epsilon_{\lambda'}(k') + (k - 2k') \cdot \epsilon^*_\lambda(k) \, \epsilon_{\lambda'}(k') \cdot \epsilon^*_{\lambda''}(k' - k)$$

$$+ (k' - 2k) \cdot \epsilon_{\lambda'}(k') \, \epsilon^*_{\lambda''}(k' - k) \cdot \epsilon^*_\lambda(k)$$

$$\approx \frac{2}{1 - z} \vec{k}_\perp \cdot \vec{\epsilon}^{\lambda''*}_\perp \, \vec{\epsilon}^{\lambda*}_\perp \cdot \vec{\epsilon}^{\lambda'}_\perp + \frac{2}{z} \vec{k}_\perp \cdot \vec{\epsilon}^{\lambda*}_\perp \, \vec{\epsilon}^{\lambda'}_\perp \cdot \vec{\epsilon}^{\lambda''*}_\perp - 2\vec{k}_\perp \cdot \vec{\epsilon}^{\lambda'}_\perp \, \vec{\epsilon}^{\lambda*}_\perp \cdot \vec{\epsilon}^{\lambda''*}_\perp. \quad (2.104)$$

Here, as usual, $z = k^+/k'^+$. Using Eqs. (2.104) and (2.73) we can write Eq. (2.102) as

$$\Psi_n \left(\{x_i, k_{i\perp}\}; x, k_\perp; \lambda \right)$$

$$= igf^{abc}\theta(z)\theta(1 - z)\frac{z(1 - z)}{\vec{k}^2_\perp}$$

$$\times \left(\frac{2}{1 - z}\vec{k}_\perp \cdot \vec{\epsilon}^{\lambda''*}_\perp \vec{\epsilon}^{\lambda*}_\perp \cdot \vec{\epsilon}^{\lambda'}_\perp + \frac{2}{z}\vec{k}_\perp \cdot \vec{\epsilon}^{\lambda*}_\perp \vec{\epsilon}^{\lambda'}_\perp \cdot \vec{\epsilon}^{\lambda''*}_\perp - 2\vec{k}_\perp \cdot \vec{\epsilon}^{\lambda'}_\perp \vec{\epsilon}^{\lambda*}_\perp \cdot \vec{\epsilon}^{\lambda''*}_\perp \right)$$

$$\times \Psi_{n-1} \left(\{x_i, k_{i\perp}\}; x', k'_\perp; \lambda' \right). \quad (2.105)$$

(We can use Eq. (2.73) since the approximations used in calculating the splitting function $P_{qq}(z)$ are the same as those that we are assuming here for the splitting function $P_{GG}(z)$.)

Multiplying the wave function in Eq. (2.105) by its complex conjugate and summing over polarizations and colors yields

$$\sum_{\lambda,\lambda',\lambda'',a,b,b',c} |\Psi_n \left(\{x_i, k_{i\perp}\}; x, k_\perp; \lambda \right)|^2$$

$$= 16\pi\alpha_s N_c \theta(z)\theta(1 - z)\frac{1}{\vec{k}^2_\perp}$$

$$\times \left[z^2 + (1 - z)^2 + z^2(1 - z)^2 \right] \sum_{\lambda',b} |\Psi_{n-1} \left(\{x_i, k_{i\perp}\}; x', k'_\perp; \lambda' \right)|^2.$$

$$(2.106)$$

Note that the definition of the gluon distribution corresponding to Fig. 2.12 implies a summation over the colors of the two gluon lines. The color of the gluon line to the left of the cut is equal to the color of the gluon line to the right of the cut. We have made this color summation explicit in Eq. (2.106) to facilitate the calculation of the color factor: the color of the internal gluon line, which is labeled b in Fig. 2.20, is denoted b' in the complex conjugate wave function. In arriving at Eq. (2.106) we have used $f^{abc}f^{ab'c} = N_c\delta^{bb'}$ and $|\vec{k}_\perp \cdot \vec{\epsilon}^\lambda_\perp|^2 = \vec{k}^2_\perp/2$.

Following the steps outlined in Sec. 2.4.2 for the quark distribution function, which led to Eq. (2.80), we infer from Eq. (2.106) that the contribution of diagram D in Fig. 2.19 to the gluon distribution function is

$$G_D(x, Q^2) = \frac{\alpha_s N_c}{\pi} \frac{1}{x} \int\limits^{Q^2} \frac{dk_\perp^2}{k_\perp^2} \int\limits_x^1 \frac{dz}{z(1-z)} \left[z^2 + (1-z)^2 + z^2(1-z)^2 \right]$$

$$\times x' G(x', k_\perp^2), \tag{2.107}$$

with $z = x/x'$. Again assuming that we are varying Q^2, Eq. (2.107) can be trivially rewritten as

$$Q^2 \frac{\partial G_D(x, Q^2)}{\partial Q^2} = \frac{\alpha_s N_c}{\pi} \int\limits_x^1 \frac{dz}{z} \left[\frac{z}{1-z} + \frac{1-z}{z} + z(1-z) \right] G\left(\frac{x}{z}, Q^2 \right). \tag{2.108}$$

Using the unitarity argument of Sec. 2.4.2 we can calculate the contribution of diagram E in Fig. 2.19 along with its complex conjugate and all the virtual gluon graphs with instantaneous terms (cf. Eq. (2.90)), obtaining

$$Q^2 \frac{\partial G_E(x, Q^2)}{\partial Q^2} = -\frac{\alpha_s N_c}{\pi} \frac{1}{2} \int\limits_0^1 dz \left[\frac{z}{1-z} + \frac{1-z}{z} + z(1-z) \right] G(x, Q^2). \tag{2.109}$$

The factor $1/2$ in Eq. (2.109) is simply a symmetry factor, as the two propagators in the loop of graph E are identical gluons. The z-integration in Eq. (2.109) has two singularities: one at $z = 1$ and the other at $z = 0$. The singularities correspond to either one or the other gluon in the loop of diagram E having a small longitudinal momentum. The two singularities have therefore identical physical origins. We rewrite them as one singularity at $z = 1$:

$$\int\limits_0^1 dz \left[\frac{z}{1-z} + \frac{1-z}{z} + z(1-z) \right] = \int\limits_0^1 dz \left[-1 + \frac{1}{1-z} + \frac{1}{z} - 1 + z(1-z) \right]$$

$$= \int\limits_0^1 dz \left[\frac{2}{1-z} - 2 + z(1-z) \right] = \int\limits_0^1 dz \frac{2}{1-z} - \frac{11}{6}. \tag{2.110}$$

With the help of this rearrangement the sum of diagrams D and E is (cf. Eq. (2.91))

$$Q^2 \frac{\partial G_{D+E}(x, Q^2)}{\partial Q^2} = \frac{\alpha_s N_c}{\pi} \left\{ \int\limits_x^1 \frac{dz}{z} \left[\frac{z}{1-z} + \frac{1-z}{z} + z(1-z) \right] G\left(\frac{x}{z}, Q^2 \right) \right.$$

$$\left. - \int\limits_0^1 dz \frac{1}{1-z} G(x, Q^2) + \frac{11}{12} G(x, Q^2) \right\}. \tag{2.111}$$

Here we are not going to calculate the contribution of diagram F in Fig. 2.19 explicitly. Instead we will use the splitting function $P_{qG}(z)$ illustrated in Fig. 2.18 and given in

Eq. (2.99c). As one can see from Figs. 2.18 and 2.19, the contribution of graph F in the latter is simply a virtual correction to the diagram in Fig. 2.18. With the help of $P_{qG}(z)$ from Eq. (2.99c) and the unitarity argument of Sec. 2.4.2 we obtain the contribution of diagram F:

$$Q^2 \frac{\partial G_F(x, Q^2)}{\partial Q^2} = -\frac{\alpha_s N_f}{2\pi} \frac{1}{2} \int\limits_0^1 dz \left[z^2 + (1-z)^2 \right] G(x, Q^2).$$ (2.112)

The factor $1/2$ is inserted to remove the double-counting associated with tagging on both the quark and the antiquark in the calculation of $P_{qG}(z)$. Equation (2.112) trivially gives

$$Q^2 \frac{\partial G_F(x, Q^2)}{\partial Q^2} = -\frac{\alpha_s}{2\pi} \frac{N_f}{3} G(x, Q^2).$$ (2.113)

Combining Eqs. (2.111) and (2.113) we arrive at the contribution of all three diagrams in Fig. 2.19:

$$Q^2 \frac{\partial G(x, Q^2)}{\partial Q^2} = \frac{\alpha_s}{2\pi} \left\{ 2N_c \int\limits_x^1 \frac{dz}{z} \left[\frac{z}{1-z} + \frac{1-z}{z} + z(1-z) \right] G\left(\frac{x}{z}, Q^2 \right) \right.$$

$$\left. - 2N_c \int\limits_0^1 dz \frac{1}{1-z} G(x, Q^2) + \frac{11N_c - 2N_f}{6} G(x, Q^2) \right\}.$$ (2.114)

(Even though Eq. (2.114) looks like a closed integro-differential equation, one has to remember that the quark distribution's contribution is not included in its right-hand side and that the full DGLAP evolution for the gluon distribution is given in Eq. (2.98).) Rewriting Eq. (2.114) in the compact form

$$Q^2 \frac{\partial G(x, Q^2)}{\partial Q^2} = \frac{\alpha_s}{2\pi} \int\limits_x^1 \frac{dz}{z} P_{GG}(z) G\left(\frac{x}{z}, Q^2 \right),$$ (2.115)

we immediately see that

$$P_{GG}(z) = 2N_c \left[\frac{z}{(1-z)_+} + \frac{1-z}{z} + z(1-z) \right] + \frac{11N_c - 2N_f}{6} \delta(1-z),$$

which is exactly Eq. (2.99d)! We have thus derived the gluon–gluon splitting function.

2.4.5 *General solution of the DGLAP equations*

To solve the DGLAP equations (2.96) and (2.98), one usually writes them first in *moment space*. The *moment* $f_\omega(Q^2)$ of a distribution function $f(x, Q^2)$ is defined by the Mellin transform

$$f_\omega(Q^2) \equiv \int\limits_0^1 dx\, x^\omega f(x, Q^2), \tag{2.116}$$

where $f = \Delta^{f\bar{f}}$ or Σ for the nonsinglet or singlet quark distribution functions respectively and $f = G$ for the gluon distribution. Inverting Eq. (2.116), we write the distribution function as

$$f(x, Q^2) = \int\limits_{a-i\infty}^{a+i\infty} \frac{d\omega}{2\pi i} x^{-\omega-1} f_\omega(Q^2), \tag{2.117}$$

where the integral in ω-space runs along a contour parallel to the imaginary axis and to the right of all the singularities of the moment $f_\omega(Q^2)$ (which can be chosen by adjusting the arbitrary real number a).

As one can show (see Exercise 2.5), in the moment space the DGLAP equations (2.96) and (2.98) become

$$Q^2 \frac{\partial \Delta_\omega^{f\bar{f}}(Q^2)}{\partial Q^2} = \frac{\alpha_s(Q^2)}{2\pi} \gamma_{qq}(\omega) \Delta_\omega^{f\bar{f}}(Q^2) \tag{2.118}$$

and

$$Q^2 \frac{\partial}{\partial Q^2} \begin{pmatrix} \Sigma_\omega(Q^2) \\ G_\omega(Q^2) \end{pmatrix} = \frac{\alpha_s(Q^2)}{2\pi} \begin{pmatrix} \gamma_{qq}(\omega) & \gamma_{qG}(\omega) \\ \gamma_{Gq}(\omega) & \gamma_{GG}(\omega) \end{pmatrix} \begin{pmatrix} \Sigma_\omega(Q^2) \\ G_\omega(Q^2) \end{pmatrix}. \tag{2.119}$$

In arriving at Eqs. (2.118) and (2.119) we have defined *anomalous dimensions* $\gamma_{ij}(\omega)$ by

$$\gamma_{ij}(\omega) = \int\limits_0^1 dz\, z^\omega P_{ij}(z), \tag{2.120}$$

where i, j can each be equal to either q or G. With the help of Eqs. (2.99) and (2.120) one can show that the DGLAP anomalous dimensions are (Georgi and Politzer 1974, Gross and Wilczek 1974)

$$\gamma_{qq}(\omega) = C_F \left[\frac{3}{2} + \frac{1}{(1+\omega)(2+\omega)} - 2\psi(\omega+2) + 2\psi(1) \right], \tag{2.121a}$$

$$\gamma_{Gq}(\omega) = C_F \left[\frac{1}{2+\omega} + \frac{2}{\omega(1+\omega)} \right], \tag{2.121b}$$

$$\gamma_{qG}(\omega) = N_f \left[\frac{1}{1+\omega} - \frac{2}{(2+\omega)(3+\omega)} \right], \tag{2.121c}$$

$$\gamma_{GG}(\omega) = \frac{11N_c - 2N_f}{6}$$

$$+ 2N_c \left[\frac{1}{\omega(1+\omega)} + \frac{1}{(2+\omega)(3+\omega)} - \psi(\omega+2) + \psi(1) \right], \tag{2.121d}$$

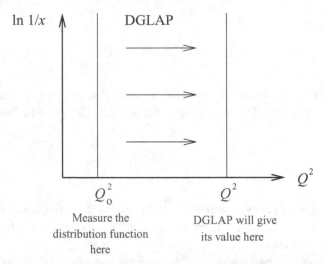

Fig. 2.21. The DGLAP equations in the $(\ln 1/x,\, Q^2)$-plane.

where $\psi(w) = \Gamma'(w)/\Gamma(w)$ is the digamma function. Note that $\psi(1) = -\gamma_E$, with γ_E Euler's constant. We leave the derivation of the anomalous dimensions (2.121) as an exercise; see Exercise 2.5.

Equations (2.118) and (2.119) are easy to solve. Suppose that the (usually nonperturbative) initial conditions for the equations are given at some initial scale Q_0^2. That is, we know $\Delta^{f\bar{f}}(x, Q_0^2)$, $\Sigma(x, Q_0^2)$, and $G(x, Q_0^2)$. Using Eq. (2.116) we can find the initial conditions for the moments, obtaining $\Delta_\omega^{f\bar{f}}(Q_0^2)$, $\Sigma_\omega(Q_0^2)$, and $G_\omega(Q_0^2)$. Solving Eqs. (2.118) and (2.119) we can now find the moments of the distribution functions at all Q^2:

$$\Delta_\omega^{f\bar{f}}(Q^2) = \exp\left\{\int_{Q_0^2}^{Q^2} \frac{dQ'^2}{Q'^2} \frac{\alpha_s(Q'^2)}{2\pi} \gamma_{qq}(\omega)\right\} \Delta_\omega^{f\bar{f}}(Q_0^2), \tag{2.122}$$

$$\begin{pmatrix} \Sigma_\omega(Q^2) \\ G_\omega(Q^2) \end{pmatrix} = \exp\left\{\int_{Q_0^2}^{Q^2} \frac{dQ'^2}{Q'^2} \frac{\alpha_s(Q'^2)}{2\pi} \begin{pmatrix} \gamma_{qq}(\omega) & \gamma_{qG}(\omega) \\ \gamma_{Gq}(\omega) & \gamma_{GG}(\omega) \end{pmatrix}\right\} \begin{pmatrix} \Sigma_\omega(Q_0^2) \\ G_\omega(Q_0^2) \end{pmatrix}. \tag{2.123}$$

Equations (2.122) and (2.123) allow one to find the distribution functions in moment space. With the help of Eq. (2.117) one then can transform the moments of the distribution functions back into x-space, thus obtaining the distribution functions solving the DGLAP equations at all Q^2.

The way in which the DGLAP equations work is depicted in Fig. 2.21 in the $(\ln 1/x,\ Q^2)$-plane, which we will often use to demonstrate our results. The initial values of the distribution functions for DGLAP evolution are set at some initial scale Q_0^2 for all the relevant values of x: thus the initial conditions are given along the vertical line on the left in Fig. 2.21. Given the initial conditions, the DGLAP equations then give the distribution functions at other values of Q^2. For instance, using the DGLAP equations one may obtain distribution functions along the vertical line on the right in Fig. 2.21. Thus the DGLAP equations *evolve* the distribution functions in Q^2 from some initial conditions at Q_0^2 to their values at some other Q^2, as indicated by the arrows in Fig. 2.21. Note that the curves shown in Fig. 2.7 resulted from using the DGLAP equations, having adjusted the initial conditions to fit the DIS data.

Indeed the DGLAP equations (2.98) and (2.96) presented above are valid only at the leading-logarithmic level. They are often referred to as the leading-order (LO) DGLAP equations, since the integral kernel on the right-hand side is given at the lowest order in α_s (i.e., at order α_s). Higher-order corrections to the splitting functions would generate terms with higher powers of α_s on the right-hand sides of Eqs. (2.98) and (2.96). For instance Eqs. (2.98) and (2.96), with right-hand sides calculated up to $O(\alpha_s^2)$, are referred to as next-to-leading-order DGLAP or simply NLO DGLAP. The next order after that is called next-to-next-to-leading-order DGLAP (NNLO DGLAP), etc. Note that at such higher orders the naive factorization relations (2.45) and (2.46) (see also Eq. (2.48)) between the structure functions and the quark distribution function would be modified. Even the LO DGLAP evolution of Eqs. (2.98) and (2.96) obviously violates Bjorken scaling. It also generates corrections to the Callan–Gross relation (2.44).

2.4.6 Double logarithmic approximation

Let us now study structure functions and parton distributions at small Bjorken x using the DGLAP equations. This limit is interesting and important for our discussion, since small x corresponds to high energy $\hat{s}$ of virtual photon–proton scattering, as one can see from Eqs. (2.6). A brief inspection of Fig. 2.7 shows that the structure function F_2 clearly rises at small x. The question that we would like to address is whether DGLAP evolution can provide a theoretical explanation for such a rise.

To answer this question we need to analyze Eqs. (2.96) and (2.98) at small x. At small x the z-integral in Eqs. (2.96) and (2.98) may get extra enhancement from the small-z region. To see this let us study the small-z asymptotics of the splitting functions. Using Eqs. (2.99) one can show that only two of the splitting functions are singular at small z:

$$P_{Gq}(z)\Big|_{z\ll 1} \approx \frac{2C_F}{z}, \quad P_{GG}(z)\Big|_{z\ll 1} \approx \frac{2N_c}{z}. \tag{2.124}$$

Thus, in Eqs. (2.96) and (2.98) only the second line of Eq. (2.98) is enhanced at small x. We conclude that the evolution of the gluon distribution $G(x, Q^2)$ runs much faster than that of the quark distributions (both singlet and nonsinglet), at small x. Therefore we can neglect the evolution of the quark distribution functions compared with that of the gluon. Also, the quark contribution to the gluon evolution, which enters via $P_{Gq}(z)$ into Eq. (2.98), is negligible as well: as $\Sigma(x, Q^2)$ is small owing to the lack of small-x enhancement to its own evolution, it would not contribute much to the gluon evolution.

Neglecting the quark distribution in the DGLAP equation (2.98) and using the approximation for the gluon–gluon splitting function from Eq. (2.124), we can write down an evolution equation for the gluon distribution only,

$$Q^2 \frac{\partial G(x, Q^2)}{\partial Q^2} = \frac{\alpha_s(Q^2)}{2\pi} \int_x^1 \frac{dz}{z} \frac{2N_c}{z} G\left(\frac{x}{z}, Q^2\right), \qquad (2.125)$$

which of course is valid only at small x.

Before we solve Eq. (2.125), let us clarify the approximation that we have made in arriving at this equation. To see this more clearly, let us redefine z as x/x' and write Eq. (2.125) as

$$Q^2 \frac{\partial x G(x, Q^2)}{\partial Q^2} = \frac{\alpha_s(Q^2) N_c}{\pi} \int_x^1 \frac{dx'}{x'} x' G(x', Q^2). \qquad (2.126)$$

Differentiating Eq. (2.126) with respect to $\ln(1/x)$, we can write it as

$$\frac{\partial^2 x G(x, Q^2)}{\partial \ln(1/x) \partial \ln(Q^2/Q_0^2)} = \frac{\alpha_s(Q^2) N_c}{\pi} x G(x, Q^2) \qquad (2.127)$$

with Q_0 a constant initial-virtuality scale.

For simplicity let us imagine that the coupling constant is fixed, $\alpha_s(Q^2) = \alpha_s$. We can then see clearly from Eq. (2.127) that its solution iterates powers of α_s multiplied not just by one logarithm, $\ln(Q^2/Q_0^2)$, as in the DGLAP equations, but by two logarithms, $\ln(1/x) \ln(Q^2/Q_0^2)$. Thus the resummation parameter of Eq. (2.127) is

$$\alpha_s \ln \frac{1}{x} \ln \frac{Q^2}{Q_0^2}. \qquad (2.128)$$

Thus at small coupling $\alpha_s \ll 1$, large $Q^2 \gg Q_0^2$, and small x such that $\ln(1/x) \gg 1$, we see that the small coupling α_s is multiplied by two large logarithms, which makes the resummation parameter (2.128) large and important to resum. Resummation of a series in powers of the parameter (2.128) is called the double logarithmic approximation (DLA).

With the DLA parameter (2.128) the approximations we made in obtaining Eq. (2.125) become clear. The absence of $1/z$ singularities in $P_{qq}(z)$ and $P_{qG}(z)$ insures that no $\ln(1/x)$ factor is generated in each step of the DGLAP evolution for the singlet and nonsinglet quark structure functions. Hence the evolution of $\Sigma(x, Q_0^2)$ and of $\Delta^{f\bar{f}}(x, Q_0^2)$ is subleading in the DLA parameter (2.128) and can be neglected in the approximation that resums only powers of the logarithms of both Q^2 and $1/x$ in Eq. (2.128).

Now let us solve Eq. (2.125). Substituting the approximate gluon–gluon splitting function from Eq. (2.124) into Eq. (2.120) we obtain

$$\gamma_{GG}(\omega) \approx \frac{2N_c}{\omega}. \qquad (2.129)$$

One can see that the small-z singularity in $P_{GG}(z)$ translates into a singularity at $\omega = 0$ in $\gamma_{GG}(\omega)$. This is an important result, which we will use below.

With the help of Eq. (2.129) we can write Eq. (2.125) in moment space:

$$Q^2 \frac{\partial G_\omega(Q^2)}{\partial Q^2} = \frac{\alpha_s(Q^2) N_c}{\pi} \frac{1}{\omega} G_\omega(Q^2). \qquad (2.130)$$

From Eqs. (2.121) one can see that only $\gamma_{GG}(\omega)$ and $\gamma_{Gq}(\omega)$ have singularities at $\omega = 0$: using this observation we could have derived the DLA DGLAP evolution equation in moment space (2.130) directly from Eq. (2.119).

The solution of Eq. (2.130) is easily found and reads

$$
G_\omega(Q^2) = \exp\left\{ \int_{Q_0^2}^{Q^2} \frac{dQ'^2}{Q'^2} \frac{\alpha_s(Q'^2)N_c}{\pi\omega} \right\} G_\omega(Q_0^2). \tag{2.131}
$$

Inverting the Mellin transform (2.116) with the help of Eq. (2.117), we obtain the gluon distribution function in the DLA:

$$
xG(x, Q^2) = \int_{a-i\infty}^{a+i\infty} \frac{d\omega}{2\pi i} \exp\left\{ \omega \ln\frac{1}{x} + \int_{Q_0^2}^{Q^2} \frac{dQ'^2}{Q'^2} \frac{\alpha_s(Q'^2)N_c}{\pi\omega} \right\} G_\omega(Q_0^2). \tag{2.132}
$$

The Q'^2-integral is easy to carry out. Taking the one-loop running coupling constant

$$
\alpha_s(Q^2) = \frac{1}{\beta_2 \ln(Q^2/\Lambda_{QCD}^2)}
$$

and assuming that $Q_0^2 > \Lambda_{QCD}^2$, we can write Eq. (2.132) as

$$
xG(x, Q^2) = \int_{a-i\infty}^{a+i\infty} \frac{d\omega}{2\pi i} \exp\left\{ \omega \ln\frac{1}{x} + \frac{N_c}{\pi\beta_2\omega} \ln\frac{\ln(Q^2/\Lambda_{QCD}^2)}{\ln(Q_0^2/\Lambda_{QCD}^2)} \right\} G_\omega(Q_0^2). \tag{2.133}
$$

Note that, with the inclusion of the running coupling corrections, the transverse logarithm $\ln(Q^2/Q_0^2)$ in Eq. (2.128) turns into the logarithm of the ratio of logarithms seen in the exponent of Eq. (2.133).

The integral in Eq. (2.133) cannot be calculated exactly without explicit knowledge of the initial conditions, which give $G_\omega(Q_0^2)$. However, it can be evaluated approximately for very small x and very large Q^2 using the saddle point (steepest descent) approximation. To do so we rewrite Eq. (2.133) as

$$
xG(x, Q^2) = \int_{a-i\infty}^{a+i\infty} \frac{d\omega}{2\pi i} e^{P(\omega)} G_\omega(Q_0^2) \tag{2.134}
$$

with all the x- and Q^2-dependent terms assembled in the exponent:

$$
P(\omega) = \omega \ln\frac{1}{x} + \frac{N_c}{\pi\beta_2\omega}\rho(Q^2), \tag{2.135}
$$

where we have defined an abbreviated notation

$$
\rho(Q^2) \equiv \ln\frac{\ln(Q^2/\Lambda_{QCD}^2)}{\ln(Q_0^2/\Lambda_{QCD}^2)} = \ln\frac{\alpha_s(Q_0^2)}{\alpha_s(Q^2)}. \tag{2.136}
$$

Indeed, $P(\omega)$ is also a function of x and Q^2: we have suppressed these arguments for brevity.

First we need to find the saddle points of the exponent $P(\omega)$, which are defined by the condition

$$P'(\omega = \omega_{sp}) = 0, \tag{2.137}$$

where the prime denotes a (partial) derivative with respect to ω. For $P(\omega)$ from Eq. (2.135) we get the saddle points

$$\omega_{sp} = \pm\sqrt{\frac{N_c}{\pi\beta_2}\frac{\rho(Q^2)}{\ln(1/x)}}. \tag{2.138}$$

One can easily argue that at small x the saddle point with the plus sign in Eq. (2.138) dominates. From here on we will label by ω_{sp} the expression in Eq. (2.138) with the plus sign.

Our next step is to approximate the exponent $P(\omega)$ by its Taylor expansion around the saddle point up to the quadratic term:

$$P(\omega) \approx P(\omega_{sp}) + \tfrac{1}{2}P''(\omega_{sp})(\omega - \omega_{sp})^2, \tag{2.139}$$

where the term linear in $\omega - \omega_{sp}$ is zero owing to the condition (2.137). Since $P''(\omega_{sp})$ is real and positive, distorting the integration contour in Eq. (2.134) so that it goes through ω_{sp} when crossing the real axis in the complex ω-plane (i.e., setting $a = \omega_{sp}$), we can define a new integration variable w by

$$\omega - \omega_{sp} \equiv iw. \tag{2.140}$$

Note that w is real along the new integration contour.

With this contour distortion and variable redefinition, Eq. (2.134) becomes

$$xG(x, Q^2) \approx e^{P(\omega_{sp})}G_{\omega_{sp}}(Q_0^2)\int\limits_{-\infty}^{\infty}\frac{dw}{2\pi}e^{-P''(\omega_{sp})w^2/2}, \tag{2.141}$$

where we also assume that G_ω is a slowly varying function of ω such that, owing to saddle point dominance, $G_\omega(Q_0^2) \approx G_{\omega_{sp}}(Q_0^2)$. Performing w-integration yields

$$xG(x, Q^2) \approx \frac{G_{\omega_{sp}}(Q_0^2)}{\sqrt{2\pi P''(\omega_{sp})}}e^{P(\omega_{sp})}. \tag{2.142}$$

With the help of Eqs. (2.138) (with the plus sign), (2.135), and (4.176) we obtain the DLA gluon distribution function in the saddle point approximation,

$$xG(x, Q^2) \approx \frac{G_{\omega_{sp}}(Q_0^2)}{\sqrt{4\pi}}\left\{\frac{N_c}{\pi\beta_2}\ln\frac{\ln(Q^2/\Lambda_{QCD}^2)}{\ln(Q_0^2/\Lambda_{QCD}^2)}\right\}^{1/4}\left(\ln\frac{1}{x}\right)^{-3/4}$$

$$\times \exp\left\{2\sqrt{\frac{N_c}{\pi\beta_2}\ln\frac{\ln(Q^2/\Lambda_{QCD}^2)}{\ln(Q_0^2/\Lambda_{QCD}^2)}\ln\frac{1}{x}}\right\}. \tag{2.143}$$

To justify the expansion (2.139) that led ultimately to Eq. (2.143) we need to estimate the next (cubic) term in the expansion, which previously we neglected:

$$P'''(\omega_{sp})(\omega - \omega_{sp})^3. \tag{2.144}$$

Since in the integral in Eq. (2.141) the typical width is

$$\omega - \omega_{sp} \sim \frac{1}{\sqrt{P''(\omega_{sp})}}, \tag{2.145}$$

we see that

$$P'''(\omega_{sp})(\omega - \omega_{sp})^3 \sim P'''(\omega_{sp})[P''(\omega_{sp})]^{-3/2}. \tag{2.146}$$

Using Eqs. (2.138) and (2.135) one can readily show that

$$P'''(\omega_{sp})(\omega - \omega_{sp})^3 \sim \left[\rho(Q^2)\right]^{-1/4} \left(\ln \frac{1}{x}\right)^{-1/4}, \tag{2.147}$$

which is negligibly small at small x and large Q^2, justifying our approximation.

Our main result from Eq. (2.143) is that

$$xG(x, Q^2) \sim \exp\left\{2\sqrt{\frac{N_c}{\pi\beta_2} \ln \frac{\ln(Q^2/\Lambda_{QCD}^2)}{\ln(Q_0^2/\Lambda_{QCD}^2)} \ln \frac{1}{x}}\right\}. \tag{2.148}$$

That is, $xG(x, Q^2)$ increases as x decreases and/or Q^2 increases. The rise in $xG(x, Q^2)$ with decreasing x is therefore a prediction of the DGLAP evolution. As we can see from Eq. (2.148), the DGLAP equation predicts a rise in $xG(x, Q^2)$ with decreasing x that is faster than any power of $\ln(1/x)$ but is slower than a power of $1/x$. A rising gluon distribution would translate into a rising (but smaller) quark distribution; both would lead to an increase in the structure function $F_2(x, Q^2)$ at small x, which is in (at least) qualitative agreement with the data in Fig. 2.7. A detailed analysis of the DIS data shows that DGLAP-based fits are able to describe most data (after a suitable choice of initial conditions is made), as demonstrated by the curves in Fig. 2.7.

A physical picture of DGLAP evolution is shown in Fig. 2.22 using the transverse plane representation of the proton from Fig. 2.8. On the left of Fig. 2.22 we show a proton with partons in it, as seen by a virtual photon with virtuality Q_0 corresponding to the resolution scale $1/Q_0$ in the transverse plane. On the right we show what happens when the same proton is probed by a virtual photon with higher virtuality, $Q > Q_0$, which is able to resolve shorter transverse distances $1/Q$. When probing the partons (quarks) at shorter distances the photon is able to distinguish that each quark may fluctuate into itself along with, say, several gluons and/or quark–antiquark pairs, as we see from the DGLAP splitting functions. The net number of partons at the higher scale Q is thus larger than at the scale Q_0, in agreement with the prediction from Eq. (2.148). To illustrate how the DGLAP equation works in practice, we will present some distribution functions extracted from DIS experiments on protons. One usually distinguishes contributions to the quark distribution function coming from the valence quarks (the two u quarks and the d quark in the proton) and from the sea quarks (all the other quarks in the proton).

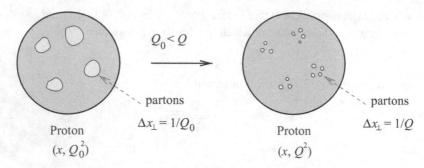

Fig. 2.22. A graphical illustration of the DGLAP evolution equations. The blobs indicate partons (quarks and gluons). A color version of this figure is available online at www.cambridge.org/9780521112574.

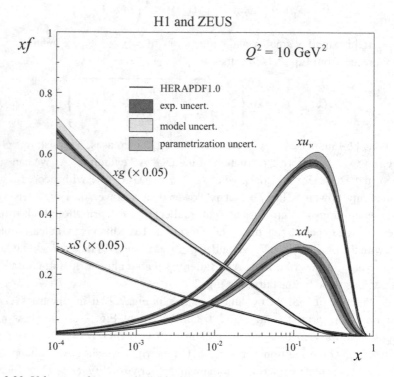

Fig. 2.23. Valence and sea quark distributions in the proton, plotted along with the gluon distribution, as functions of Bjorken x for fixed $Q^2 = 10$ GeV^2. (Reprinted with kind permission from Springer Science +Business Media: H1 and ZEUS collaboration (2010).) A color version of this figure is available online at www.cambridge.org/9780521112574.

Figure 2.23 shows the valence quark distributions $xu_v(x, Q^2)$ and $xd_v(x, Q^2)$, along with the sea quark distribution xS and the gluon distribution xg. All distributions are plotted as functions of x for fixed $Q^2 = 10$ GeV^2. The curves in Fig. 2.23 are the result of a combined NLO DGLAP-based fit of the data from the H1 and ZEUS collaborations at DESY (H1

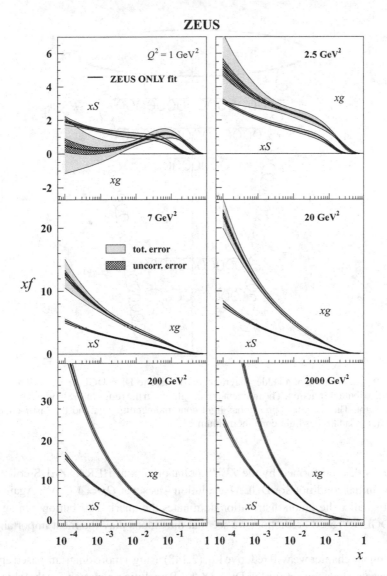

Fig. 2.24. Gluon and sea quark distributions in the proton plotted as functions of Bjorken x for six different values of Q^2. (Reprinted with permission from ZEUS collaboration (2003). Copyright 2003 by the American Physical Society.) A color version of this figure is available online at www.cambridge.org/9780521112574.

and ZEUS collaboration 2010). Note that the sea quark and gluon distributions were scaled down by a factor 0.05 to fit into the same plot as the valence quark distributions. One can see clearly that the gluon and sea quark distributions dominate at small x, in qualitative agreement with the DLA DGLAP predictions.

In Fig. 2.24 we give the sea quark and gluon distributions as functions of x for six different values of Q^2. The curves in Fig. 2.24 are the results of an NLO DGLAP-based

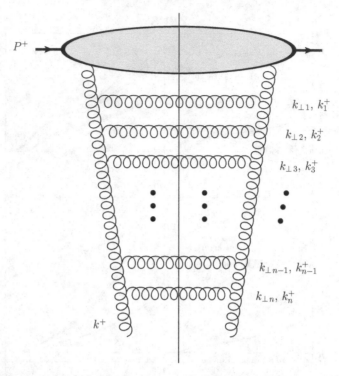

Fig. 2.25. An example of a ladder diagram contributing to DLA DGLAP evolution for the gluon distribution function. The momenta of the gluons in the rungs of the ladder are labeled on the right. The incoming proton has light cone momentum P^+, and the last t-channel gluon in the ladder has light cone momentum k^+.

fit to the DIS data performed by the ZEUS collaboration at HERA (ZEUS collaboration 2003). The initial condition for DGLAP evolution was set at $Q_0^2 = 1 \text{ GeV}^2$. Again one can see that at small x the gluon distribution dominates the quark distributions. In agreement with the DGLAP-based prediction, we see that the gluons play the most important role at small x.

In closing the chapter we will rederive Eq. (2.148) using a more diagram-based approach. Let us construct the solution for the DLA DGLAP evolution equation for the gluon distribution (2.125) by summing diagrams that iterate the kernel of the evolution equation given by the real part of the gluon–gluon splitting function in Fig. 2.19D. (Note that the parts of the splitting functions that are singular at small z, given by Eq. (2.124), are entirely due to the real emission diagrams). Diagrams iterating the gluon emission kernel have a "ladder" structure, as shown in Fig. 2.25. There the transverse momenta of the gluons in the rungs of the ladder, when ordered as

$$k_{\perp n}^2 \gg k_{\perp n-1}^2 \gg \cdots \gg k_{\perp 2}^2 \gg k_{\perp 1}^2 \tag{2.149}$$

give the transverse logarithms of DGLAP evolution. As we are interested in the DLA limit, the longitudinal momenta of the gluons in Fig. 2.25 can be ordered too, as

$$k_1^+ \gg k_2^+ \gg \cdots \gg k_{n-1}^+ \gg k_n^+ \gg k^+, \qquad (2.150)$$

to generate the logarithms of x. Each rung of the ladder generates a logarithmic integral over longitudinal momenta dk^+/k^+, a logarithmic integral over transverse momenta $dk_\perp^2/k_\perp^2$, and a power of the coupling constant $\alpha_s(k_\perp^2)$. Each rung also brings in a color factor N_c and a factor $1/\pi$ coming from more careful diagram evaluation, which we will not perform here. Ordering all the integrations, we get

$$xG(x, Q^2) \sim \sum_{n=0}^{\infty} \int_{Q_0^2}^{Q^2} \frac{dk_{\perp n}^2}{k_{\perp n}^2} \, \bar{\alpha}_s(k_{\perp n}^2) \int_{Q_0^2}^{k_{\perp n}^2} \frac{dk_{\perp n-1}^2}{k_{\perp n-1}^2} \bar{\alpha}_s(k_{\perp n-1}^2) \cdots$$

$$\times \int_{Q_0^2}^{k_{\perp 2}^2} \frac{dk_{\perp 1}^2}{k_{\perp 1}^2} \, \bar{\alpha}_s(k_{\perp 1}^2) \int_{k^+}^{P^+} \frac{dk_1^+}{k_1^+} \int_{k^+}^{k_1^+} \frac{dk_2^+}{k_2^+} \cdots \int_{k^+}^{k_{n-1}^+} \frac{dk_n^+}{k_n^+}, \qquad (2.151)$$

where

$$\bar{\alpha}_s(Q^2) \equiv \frac{\alpha_s(Q^2) N_c}{\pi}. \qquad (2.152)$$

Performing the integrals yields (as $x = k^+/P^+$)

$$xG(x, Q^2) \sim \sum_{n=0}^{\infty} \frac{1}{(n!)^2} \left[\int_{Q_0^2}^{Q^2} \frac{dk_\perp^2}{k_\perp^2} \, \bar{\alpha}_s(k_\perp^2) \ln \frac{1}{x} \right]^n \qquad (2.153)$$

or, equivalently,

$$xG(x, Q^2) \sim \sum_{n=0}^{\infty} \frac{1}{(n!)^2} \left[\frac{N_c}{\pi \beta_2} \rho(Q^2) \ln \frac{1}{x} \right]^n, \qquad (2.154)$$

which after summation gives a modified Bessel function:

$$xG(x, Q^2) \sim I_0 \left(2 \sqrt{\frac{N_c}{\pi \beta_2} \rho(Q^2) \ln \frac{1}{x}} \right). \qquad (2.155)$$

The exact index of the modified Bessel function depends on the initial conditions for the evolution and is not always 0 (Gorshkov *et al.* 1968). Using the large-argument asymptotics of the modified Bessel function, $I_\nu(z) \sim e^z$, we obtain Eq. (2.148). The prefactor in front of the exponent, shown in Eq. (2.143), can be obtained similarly, by keeping the prefactor in the asymptotics of the modified Bessel function and matching the initial conditions to those used in obtaining Eq. (2.143).

The derivation we have presented shows the diagrammatic origin of the result (2.148). Diagrams also allow one to understand the space–time structure of the parton emissions.

Consider the proton in Fig. 2.25, which, as throughout this chapter, is moving in the light cone plus direction. The light cone times of gluon emissions, which we label x_i^+ for the ith gluon shown in the ladder in Fig. 2.25, owing to the uncertainty principle are given by $x_i^+ \approx 1/k_i^-$. As the gluons in the rungs of the ladder are on mass shell, $k_i^- = k_{\perp i}^2/k_i^+$ and $x_i^+ \approx k_i^+/k_{\perp i}^2$. The DGLAP ordering of transverse momenta (2.149) of itself insures that

$$x_1^+ \gg x_2^+ \gg \cdots \gg x_n^+. \tag{2.156}$$

The ordering of longitudinal momenta (2.150) merely reinforces the ordering of gluon lifetimes (2.156). We see that the gluons with the lowest transverse momentum and/or largest longitudinal momentum are emitted earliest and have the longest lifetimes. Conversely the gluons with the largest transverse momenta and/or smallest longitudinal momenta are emitted last and exist over the shortest lifetimes. This time-ordering of gluon emissions is not only important for our understanding of DGLAP evolution, but will be useful when we start talking about the small-x evolution equations, as it applies there too.

Further reading

A detailed pedagogical discussion of DIS and the DGLAP evolution equations covering topics omitted in this chapter can be found in Halzen and Martin (1984), Sterman (1993), Peskin and Schroeder (1995), Ellis, Stirling, and Webber (1996), and Weinberg (1996).

The reader can find NLO splitting functions for DGLAP evolution in Ellis, Stirling, and Webber (1996). For further discussion of the running coupling scale in DGLAP evolution we refer the reader to Dokshitzer and Shirkov (1995).

Exercises

2.1 Show that, in general, the hadronic tensor $W^{\mu\nu}(p, q)$ can be written in the form (2.16). Do this by observing that it is a function of two four-vectors p^μ and q^μ only, demanding that $W^{\mu\nu}$ is symmetric ($W^{\mu\nu} = W^{\nu\mu}$), and imposing the conditions (2.15).

2.2* Calculate the splitting function $P_{qG}(z)$ in light cone perturbation theory using the diagram in Fig. 2.18. You should get Eq. (2.99c).

2.3 Show that the DGLAP equations conserve the longitudinal momentum of the partons. Starting from Eq. (2.98), and using Eqs. (2.99), show that

$$\int_0^1 dx \, x \left[\Sigma(x, Q^2) + G(x, Q^2) \right] \tag{2.157}$$

is independent of Q^2. With the help of Eq. (2.119) argue that this momentum conservation requires that all the anomalous dimensions are zero at $\omega = 1$, i.e., $\gamma_{ij}(\omega = 1) = 0$.

2.4 Show that the DGLAP equations conserve baryon number. Starting from Eq. (2.96), and using Eq. (2.99a), show that

$$\int\limits_0^1 dx \, \Delta^{f\bar{f}}(x, Q^2) \tag{2.158}$$

is independent of Q^2.

2.5 **(a)** Starting from Eqs. (2.96) and (2.98), and with the help of Eq. (2.116), derive the DGLAP equations in moment space, obtaining Eqs. (2.118) and (2.119) with the anomalous dimensions defined in Eq. (2.120).

(b) Explicitly derive the DGLAP anomalous dimensions shown above in Eqs. (2.121): that is, use Eq. (2.120) to integrate the splitting functions given by Eqs. (2.99).

2.6 Using the methods in Sec. 2.4.6, solve the DGLAP equation for the gluon distribution,

$$Q^2 \frac{\partial}{\partial Q^2} G(x, Q^2) = \frac{\alpha_s}{2\pi} \int_x^1 \frac{dz}{z} P_{GG}(z) \, G\left(\frac{x}{z}, Q^2\right),$$

with

$$P_{GG}(z) = \frac{2N_c}{z}$$

in the small-x asymptotics but now with *fixed* coupling constant α_s (i.e., for α_s independent of Q^2). In particular show that, in the saddle point approximation, the small-x asymptotics for the gluon distribution is given by

$$xG(x, Q^2) \sim \exp\left(2\sqrt{\frac{\alpha_s N_c}{\pi} \ln\frac{1}{x} \ln\frac{Q^2}{Q_0^2}}\right). \tag{2.159}$$

3

Energy evolution and leading logarithm-$1/x$ approximation in QCD

We now begin the presentation of our main subject: high energy QCD, also known as small-x physics. We argue that at small Bjorken x it is natural to try to resum leading logarithms of $1/x$, that is, powers of $\alpha_s \ln 1/x$. Resummation of this parameter in the linear approximation corresponding to low parton density is accomplished by the Balitsky–Fadin–Kuraev–Lipatov (BFKL) evolution equation, which we describe in this chapter using the standard approach based on Feynman diagrams. Note that our derivation of the BFKL equation in this chapter is rather introductory in nature; a more rigorous re-derivation employing LCPT is left until for the next chapter. We point out some problems with the linear BFKL evolution; in particular we argue that it violates unitarity constraints for the scattering cross section. We describe initial attempts to solve the BFKL unitarity problem by introducing nonlinear corrections to the BFKL evolution, resulting in the Gribov–Levin–Ryskin and Mueller–Qiu (GLR–MQ) evolution equation. We discuss properties of the GLR–MQ evolution equation and, for the first time, introduce the saturation scale Q_s.

3.1 Paradigm shift

Our goal in this book is to study the high energy behavior of QCD. In the context of DIS the high energy asymptotics can be explored by fixing the photon virtuality Q^2 and taking the photon–proton center-of-mass energy squared $\hat{s}$ to be large. In this limit the Bjorken-x variable becomes small, as follows from Eq. (2.6). The small-x asymptotics is therefore synonymous with the high energy limit of QCD:

$$\text{small } x \iff \text{high energy } s. \tag{3.1}$$

The small-x asymptotics of the gluon distribution function $xG(x, Q^2)$ in the framework of DGLAP evolution was discussed in Section 2.4.6. For the LLA DGLAP, the small-x asymptotics corresponds to summation of the parameter

$$\alpha_s \ln \frac{1}{x} \ln \frac{Q^2}{Q_0^2}, \tag{3.2}$$

which constitutes the double logarithm approximation (DLA). While in Sec. 2.4.6 we worked out the running coupling case, the small-x asymptotics of the gluon distribution function for fixed coupling can be shown to be that in Eq. (2.159). The resulting gluon

Table 3.1. *The transverse and longitudinal leading logarithmic approximations (LLAs)* *and the double logarithmic approximation (DLA)*

Approximation	Coupling	Transverse logarithm	Longitudinal logarithm
LLA in Q^2	$\alpha_s(Q^2) \ll 1$	$\alpha_s \ln(Q^2/Q_0^2) \approx 1$	$\alpha_s \ln 1/x \ll 1$
LLA in $1/x$	$\alpha_s \ll 1$	$\alpha_s \ln(Q^2/Q_0^2) \ll 1$	$\alpha_s \ln 1/x \approx 1$
DLA	$\alpha_s(Q^2) \ll 1$	$\alpha_s \ln(Q^2/Q_0^2) \ll 1$	$\alpha_s \ln 1/x \ll 1$
		but $\alpha_s \ln(Q^2/Q_0^2) \ln 1/x \approx 1$	

distribution grows with decreasing x in such a way that

$$\left(\frac{1}{x}\right)^a \gg xG(x, Q^2) \propto \exp\left(2\sqrt{\frac{\alpha_s N_c}{\pi} \ln \frac{1}{x} \ln \frac{Q^2}{Q_0^2}}\right) \gg \ln^n \frac{1}{x}, \qquad (3.3)$$

which is faster than any positive power n of $\ln 1/x$ but slower than any positive power a of $1/x$.

The asymptotics of the gluon distribution (3.3) is valid in the double logarithmic limit of small x and large Q^2. However, if one is interested in studying the high energy (Regge) limit of QCD, one simply needs to fix Q^2 at some, not necessarily large, value and study the small-x asymptotics. As there is no need to take the large-Q^2 limit, $\ln(Q^2/Q_0^2)$ is now neither a large nor a small parameter. We therefore drop it from Eq. (3.2) and aim to resum the parameter

$$\alpha_s \ln \frac{1}{x}. \qquad (3.4)$$

Resummation of a series in powers of the parameter (3.4) is referred to as the leading-logarithmic approximation (LLA) in $1/x$. As with previous logarithmic approximations we assume that the relevant transverse momentum scales are large enough that $\alpha_s \ll 1$. At small x we have $\ln 1/x \gg 1$, so that $\alpha_s \ln 1/x \sim 1$ and is an important parameter to resum. (Indeed, as we have seen from Sec. 2.4.6 already, and as will be clear from the calculations below, for gluon distribution functions and for total hadronic scattering cross sections one can have at most one power of $\ln 1/x$ per power of the coupling α_s, i.e., there is no resummation parameter like $\alpha_s \ln^2 1/x$ in xG though there are other observables, such as $\Delta^{f\bar{f}}$, which depend on this parameter: however, these are suppressed at high energy and the presentation of their low-x asymptotics is beyond the scope of this book.) As we will see in the next chapter, the resummation of gluon emissions in the light cone wave function presented in Sec. 2.4.2 can also be done in the LLA in $1/x$ instead of the LLA in Q^2, as used for DGLAP evolution.

Table 3.1 gives for comparison the two leading logarithmic approximations, that in Q^2 of Eq. (2.67) leading to the DGLAP equations and that in $1/x$ from Eq. (3.4), which we will study below. As discussed in Sec. 2.3, the photon virtuality Q determines the transverse size resolution of a DIS experiment, while Bjorken x determines the longitudinal (Ioffe) lifetime of the partonic fluctuation: we therefore refer to $\ln Q^2$ as the transverse logarithm and to $\ln 1/x$ as the longitudinal logarithm. As one can see from Table 3.1, the two LLA regimes should give identical results when they overlap in the double logarithmic approximation (DLA).

As discussed in the previous chapter, the LLA in Q^2 leads to the evolution described by the DGLAP equations, which allows us to determine the number of partons with transverse size larger than $1/Q$ if we know the number of partons with size larger than $1/Q_0$. Formally speaking, Q_0 is chosen to be large enough that $\alpha_s(Q_0^2) \ll 1$. In x-evolution we hope to find the number of partons of roughly the same transverse size at low x if we know this number at some $x = x_0$. Therefore Fig. 2.22 would have to be modified for small-x evolution. We will return to this subject later, after deriving the linear small-x evolution equation.

Resummation of the leading logarithms of $1/x$ instead of those of Q^2 is the essential paradigm shift needed in studying the small-x asymptotics. The equation resumming leading logarithms of $1/x$ will be, unlike the DGLAP equation, an evolution equation in x not an evolution equation in Q^2. A main goal of this chapter is to develop the technique of summing such longitudinal logarithmic contributions. We will show that the summation of powers of $\alpha_s \ln 1/x$ leads to gluon distributions increasing as a power of $1/x$ at small x, namely as $(1/x)^{1+\text{const}\,\alpha_s}$. For hadron–hadron scattering cross sections, $\ln 1/x$ is replaced by $\ln s$ (cut off by some dimensionful scale), so that the resummation of longitudinal logarithms gives cross sections growing as a power of the center-of-mass energy: $\sigma_{tot} \sim s^{1+\text{const}\,\alpha_s}$.

3.2 Two-gluon exchange: the Low–Nussinov pomeron

We start our analysis of high energy scattering with the lowest-order diagrams. As mentioned earlier, in this chapter we will be using the usual Feynman diagram technique. For simplicity let us consider the high energy scattering of two quark–antiquark bound states (quarkonia) on each other. We assume that the quarkonia either resulted from a splitting of virtual photons of high virtuality Q ($\gamma^* + \gamma^*$ scattering) or consist of quarks sufficiently heavy to insure the applicability of perturbative QCD methods.

Before we start the calculation let us formulate a general rule for high energy scattering, which will be confirmed by explicit calculations below, albeit for the particular case of gluons. Consider a high energy scattering event in which a particle of spin j is exchanged in the t-channel between some scatterers, as shown in Fig. 3.1. The rule is simple: if one wants to count the powers of the center-of-mass energy squared s in the total scattering cross section then the contribution of each t-channel exchange of particle with spin j to the scattering cross section is (Regge 1959, 1960)[1]

$$s^{j-1}. \tag{3.5}$$

To avoid confusion between contributions to the scattering amplitude and to the cross section we note that in our (standard) normalization the cross section is $\sigma \sim |M|^2/s^2$, where

---

[1] This simple rule applies only to counting powers of s and cannot be used to count the powers of $\ln s$, which is a much slower function of s than a power and is therefore neglected by the rule.

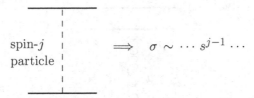

Fig. 3.1. A t-channel exchange of a particle with spin j between two particles scattering at high energy. The exchange shown is assumed to be part of some amplitude squared contributing to the scattering cross section. The contribution of each particle exchange to the resulting scattering cross section is s^{j-1}.

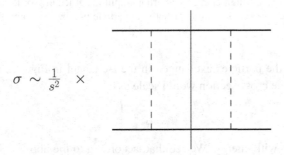

Fig. 3.2. The scattering cross section as the amplitude squared of the t-channel exchange diagram from Fig. 3.1 divided by the appropriate kinematic factors, including s^2. The vertical solid line denotes the final-state cut.

M is the scattering amplitude (see e.g. Amsler *et al.* (2008)). An exchange of k particles of spin j in the amplitude and k particles in the complex conjugate amplitude leads to a cross section scaling as $\sigma \sim s^{(j-1)2k}$, while the amplitude with k exchanged particles would then scale as $M \sim s^{1+(j-1)k}$. Hence one-particle exchange contributes s^j to the amplitude ($k = 1$), while the exchange of two particles ($k = 2$) gives a factor s^{2j-1} in the amplitude, etc.

As an example, consider the contribution of the squared amplitude in Fig. 3.1 to the total scattering cross section, as shown in Fig. 3.2. According to the above rule the cross section receives contributions from the exchanges of two t-channel particles of spin j, each contributing s^{j-1}. The resulting scattering cross section scales as

$$\sigma \sim s^{2(j-1)}. \tag{3.6}$$

Thus, if the particles exchanged in the t-channel were gluons with spin $j = 1$, the cross section would scale as

$$\sigma_{gluons} \sim s^0. \tag{3.7}$$

On the basis of rule (3.6) we would expect the cross section due to a two-gluon exchange to be constant with energy. This is an important observation, which we will soon verify by explicit calculations.

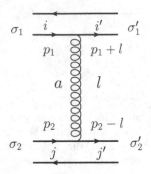

Fig. 3.3. Onium–onium high energy scattering amplitude at leading order. The arrows on the quark lines denote the directions of both the particle number flow and the momentum flow.

Alternatively, if the particles exchanged in the t-channel in Fig. 3.2 were quarks with spin $j = 1/2$ then the cross section would scale as

$$\sigma_{quarks} \sim \frac{1}{s} \qquad (3.8)$$

and would decrease with energy. We see that, according to the above rule, the gluon contribution to the scattering cross section dominates the quark contribution. This conclusion is certainly in line with our earlier observation in Sec. 2.4.6 that the gluon distribution dominates in DIS at small x. We see that in high energy processes gluons play a more important role than quarks.

Let us consider the case when scalar particles are exchanged in the t-channel of Fig. 3.2 (we are now going beyond QCD and are considering a scalar theory). The cross section would scale as

$$\sigma_{scalars} \sim \frac{1}{s^2} \qquad (3.9)$$

and is also, like the cross section for quark exchanges, small at high energy.

Finally, if spin-2 particles, such as gravitons, are exchanged in the t-channel of Fig. 3.2 then one gets

$$\sigma_{gravitons} \sim s^2 \qquad (3.10)$$

and the cross section would grow rather fast with energy. Luckily, despite this energy enhancement, gravity is rather weakly coupled at the energies of modern-day accelerators and does not contribute significantly to the total cross sections.

Let us now return to QCD and to the high energy scattering of two quarkonia (to which we will often simply refer to as "onia"). In view of the above rule, and as can be shown by a simple calculation, at high energy the dominant lowest-order contribution to the QCD scattering amplitude is due to a t-channel gluon exchange, as shown in Fig. 3.3.

We are working in the center-of-mass frame, where the top onium (along with its quark and antiquark) in Fig. 3.3 has a large plus light cone component of momentum, while the lower onium has a large minus momentum component. Specifically, for simplicity neglecting the quark masses one may choose the incoming quarks in Fig. 3.3 to be light-like:

$$p_1^\mu = (p_1^+ \equiv P^+ = \sqrt{s}, 0, 0_\perp) \quad \text{and} \quad p_2^\mu = (0, p_2^- \equiv P^- = \sqrt{s}, 0_\perp), \tag{3.11}$$

using the $(+, -, \perp)$ notation. Note that, in our high energy kinematics, P^+ and P^- are the two largest momentum scales in the problem; all other momenta are assumed to be much smaller than P^+ and P^-. This is known as the *eikonal approximation*.

A simple calculation in the covariant (Feynman) gauge yields the amplitude for the diagram in Fig. 3.3:

$$iM_{qq \to qq}^0 = -ig^2 (t^a)_{i'i} (t^a)_{j'j} \frac{1}{l_\perp^2} \, \bar{u}_{\sigma_1'}(p_1 + l)\gamma^\mu u_{\sigma_1}(p_1) \, \bar{u}_{\sigma_2'}(p_2 - l)\gamma_\mu u_{\sigma_2}(p_2). \tag{3.12}$$

In arriving at Eq. (3.12) we have used the fact that the outgoing quarks are on mass shell, so that

$$0 = (p_1 + l)^2 = p_1^+ l^- + l^2, \tag{3.13}$$

giving

$$l^- = -\frac{l^2}{p_1^+} = -\frac{l^2}{P^+} \approx 0. \tag{3.14}$$

Similarly

$$l^+ = \frac{l^2}{p_2^-} = \frac{l^2}{P^-} \approx 0 \tag{3.15}$$

and, therefore,

$$l^2 \approx -l_\perp^2. \tag{3.16}$$

We see that in the high energy approximation the exchanged gluon has no longitudinal momentum: we will refer to it as an instantaneous or Coulomb gluon.

To keep only leading powers of P^+ and P^- we use the following trick: we consider that the spinors of the quark line with the large plus momentum (the upper line in Fig. 3.3) are chosen in the Lepage and Brodsky (1980) convention while the spinors in the quark line with the large minus momentum (the lower line in Fig. 3.3) are also chosen in the Lepage and Brodsky (1980) convention but with the P^- and P^+ momenta interchanged (see Eqs. (1.50) and (1.51)). Using Table A.1 in Appendix A we see that γ^+ dominates in the upper quark line of Fig. 3.3 since it carries a large P^+ momentum while γ^- dominates in the lower quark line, which carries a large P^- momentum. With the help of Table A.1 we then obtain[2]

$$M_{qq \to qq}^0(\vec{l}_\perp) = -2g^2 (t^a)_{i'i} (t^a)_{j'j} \delta_{\sigma_1 \sigma_1'} \delta_{\sigma_2 \sigma_2'} \frac{s}{l_\perp^2}. \tag{3.17}$$

---

[2] One may also use standard notation for Dirac spinors (see e.g Peskin and Schroeder (1995)). In this case, neglecting l compared to p_1 and p_2, one should use the relation $\bar{u}_{\sigma'}(p)\gamma^\mu u_\sigma(p) = 2p^\mu \delta_{\sigma\sigma'}$, which follows from the Gordon identity, to simplify Eq. (3.12).

The square of the amplitude in Eq. (3.17) leads to the following high energy cross section:

$$\sigma^0_{qq \to qq} = \frac{2\alpha_s^2 C_F}{N_c} \int \frac{d^2 l_\perp}{(l_\perp^2)^2}. \tag{3.18}$$

We see that, in agreement with the rule in Eq. (3.6), the cross section due to two t-channel gluon exchanges is independent of energy at high energy. This feature of QCD was first noticed by Low (1975) and Nussinov (1976). The two t-channel gluon exchange cross section is sometimes called the *Low–Nussinov pomeron*, since this result was the first successful attempt to describe hadronic cross sections in the framework of perturbative QCD: in pre-QCD language hadronic cross sections were described as being due to the t-channel exchange of a hypothetical particle with the quantum numbers of the vacuum called *the pomeron*, named after I. Y. Pomeranchuk (1958). The contribution of the pomeron to the scattering amplitude is

$$M \sim s^{\alpha(t)}, \tag{3.19}$$

where s and t are Mandelstam variables and $\alpha(t)$ is the "angular momentum" of the pomeron, usually referred to as the *pomeron trajectory*. The contribution of a single pomeron exchange to the total cross section is

$$\sigma_{tot} \sim s^{\alpha(0)-1}. \tag{3.20}$$

Here $\alpha(0)$ is the value of the pomeron trajectory at $t = 0$, which is the point where it intercepts the angular momentum axis in the (t, α)-plane. Therefore $\alpha(0)$ is referred to as the *pomeron intercept* and is sometimes denoted by α_P. As one can see from Eq. (3.20), the pomeron intercept always comes in the combination $\alpha(0) - 1$: according to a common notation, we will often refer to $\alpha(0) - 1 = \alpha_P - 1$ as itself the pomeron intercept. Frequently one uses a linear expansion of the pomeron trajectory near $t = 0$:

$$\alpha(t) \approx \alpha(0) + \alpha' t. \tag{3.21}$$

The parameter α' is called the *slope* of the pomeron trajectory. A tantalizing feature of strong interactions is that the linear approximation (3.21) actually describes the pomeron trajectory $\alpha(t)$ rather well at all values of t. This observation gave rise to the development of string theory, which started out as a candidate theory for strong interactions (see e.g. Green, Schwarz, and Witten (1987)).

From Eq. (3.18) it is clear that the Low–Nussinov pomeron has intercept $\alpha(0) - 1 = 0$. In high energy proton–proton (pp) (and proton–antiproton, $p\bar{p}$) collisions, analysis of the experimental data showed that the total cross section grows approximately as follows (Donnachie and Landshoff 1992):

$$\sigma^{pp}_{tot} \sim s^{0.08}. \tag{3.22}$$

That is, using pre-QCD language, the pomeron intercept $\alpha_P - 1 = 0.08$. Since soft non-perturbative QCD physics is probably responsible for much of the total pp cross section observed at many modern-day accelerators, the pomeron with intercept $\alpha_P - 1 = 0.08$ is usually called the "soft pomeron".

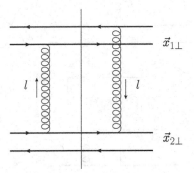

Fig. 3.4. A diagram contributing to the onium–onium high energy scattering cross section at leading order. The arrows next to the gluon lines indicate the direction of momentum flow and the vertical straight line denotes the final state cut.

We see that the prediction of Low and Nussinov that $\alpha_P - 1 = 0$, while it does not give the correct pomeron intercept, is not far from it, in the sense of giving a cross section that at least does not decrease with energy. (Of course there is no *a priori* reason to expect a perturbative calculation to describe the total pp scattering cross section, but it is good to have at least qualitative agreement between the two.) As we will see below, higher-order perturbative corrections to the cross section (3.18) generate a positive order-α_s contribution to the $\alpha_P - 1 = 0$ result. Note that the fact that experimental measurement of the total pp scattering cross section (3.22) gives a result that does not fall off with energy but instead rises slowly with s, when combined with the above rule for counting powers of s (see (3.6)), demonstrates that there must exist a spin-1 particle responsible for strong interactions – the gluon. This is exactly the argument for the existence of gluons mentioned in Sec. 1.1.

The $l_\perp$-integral in Eq. (3.18) has an infrared (IR) divergence. This is natural since we are calculating a cross section for the scattering of free color charges (quarks). To make the cross section IR-finite we need to remember that the scattering quarks are part of the onium wave functions. Suppose that the $q\bar{q}$ pairs have separations $\vec{x}_{1\perp}$ and $\vec{x}_{2\perp}$ in transverse coordinate space, though the impact parameter between the two onia has been integrated out. By summing diagrams with all possible gluon connections to quarks and antiquarks, one of which is shown in Fig. 3.4, one can then show that the total onium–onium scattering cross section is

$$\sigma_{tot}^{onium+onium} = \int d^2x_{1\perp} d^2x_{2\perp} \int_0^1 dz_1 dz_2 \, |\Psi(\vec{x}_{1\perp}, z_1)|^2 \, |\Psi(\vec{x}_{2\perp}, z_2)|^2 \hat{\sigma}_{tot}^{onium+onium} \quad (3.23)$$

with

$$\hat{\sigma}_{tot}^{onium+onium} = \frac{2\alpha_s^2 C_F}{N_c} \int \frac{d^2 l_\perp}{(l_\perp^2)^2} \left(2 - e^{-i\vec{l}_\perp \cdot \vec{x}_{1\perp}} - e^{i\vec{l}_\perp \cdot \vec{x}_{1\perp}}\right) \left(2 - e^{-i\vec{l}_\perp \cdot \vec{x}_{2\perp}} - e^{i\vec{l}_\perp \cdot \vec{x}_{2\perp}}\right), \quad (3.24)$$

at the lowest order in α_s. Here $\Psi(\vec{x}_\perp, z)$ is the onium light cone wave function with quark light cone momentum fraction z. The exact form of the wave function is not important

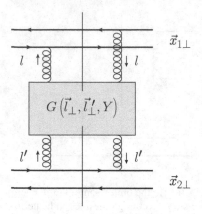

Fig. 3.5. A general representation of the onium–onium scattering cross section at high energy. The rectangle denotes all leading-ln s corrections to the two-gluon exchange cross section from Fig. 3.5.

for the moment. The summation and averaging over all appropriate quantum numbers is implicit in the $|\Psi|^2$ factors in Eq. (3.23).

The $l_\perp$-integral in Eq. (3.24) is now finite; if we average over the directions of $\vec{x}_{1\perp}$ and $\vec{x}_{2\perp}$ then it can be easily carried out, giving

$$\langle \hat{\sigma}_{tot}^{onium+onium} \rangle = \frac{4\pi \alpha_s^2 C_F}{N_c} x_<^2 \left(\ln \frac{x_>}{x_<} + 1 \right), \tag{3.25}$$

where $x_{>(<)} = \max (\min)\{|\vec{x}_{1\perp}|, |\vec{x}_{2\perp}|\}$ and $\langle \cdots \rangle$ denotes angular averaging.

We will now look for corrections to this lowest-order result.

3.3 The Balitsky–Fadin–Kuraev–Lipatov evolution equation

As discussed in Sec. 3.1, in high energy scattering (or at small Bjorken x) one would like to sum the longitudinal logarithms, i.e., the powers of $\alpha_s \ln s$ (or $\alpha_s \ln 1/x$). We will denote the sum of all such corrections to the Born-level onium–onium scattering cross section found above in Sec. 3.2 by the shaded rectangle in Fig. 3.5.

Generalizing the cross section in Eq. (3.24) we write

$$\hat{\sigma}_{tot}^{onium+onium} = \frac{2\alpha_s^2 C_F}{N_c} \int \frac{d^2 l_\perp d^2 l'_\perp}{l_\perp^2 l'^2_\perp} \left(2 - e^{-i\vec{l}_\perp \cdot \vec{x}_{1\perp}} - e^{i\vec{l}_\perp \cdot \vec{x}_{1\perp}} \right)$$

$$\times \left(2 - e^{-i\vec{l}'_\perp \cdot \vec{x}_{2\perp}} - e^{i\vec{l}'_\perp \cdot \vec{x}_{2\perp}} \right) G \left(\vec{l}_\perp, \vec{l}'_\perp, Y \right), \tag{3.26}$$

where l and l' are the momenta of the gluon lines on each side of the shaded rectangle, as illustrated in Fig. 3.5. We also define the *rapidity* variable $Y = \ln(s|\vec{x}_{1\perp}||\vec{x}_{2,\perp}|)$; it is important that $Y \sim \ln s$, though the exact cutoff under the logarithm of the energy is not important in the leading-logarithmic approximation that we would like to apply here. The shaded rectangle in Fig. 3.5 brings in a factor $G(\vec{l}_\perp, \vec{l}'_\perp, Y)$. The lowest-order expression

(3.24) is recovered by substituting

$$G\left(\vec{l}_\perp, \vec{l}_\perp', Y = 0\right) = G_0\left(\vec{l}_\perp, \vec{l}_\perp'\right) = \delta^2\left(\vec{l}_\perp - \vec{l}_\perp'\right) \tag{3.27}$$

in Eq. (3.26).

Below we will construct an equation for $G\left(\vec{l}_\perp, \vec{l}_\perp', Y\right)$ by analyzing the one-gluon order-α_s corrections to Eq. (3.24) and to Fig. 3.4.

3.3.1 Effective emission vertex

Let us start with the real one-gluon corrections to Fig. 3.4, i.e., corrections where the extra gluon is present in the final state (the gluon is cut). The difference between the quark–quark scattering cross section (3.18) and the onium–onium cross section (3.24) is only in the so-called impact factors, which at the lowest order, considered here, are simply factors like $2 - e^{-i\vec{l}_\perp \cdot \vec{x}_{1\perp}} - e^{i\vec{l}_\perp \cdot \vec{x}_{1\perp}}$; see the large parentheses in Eq. (3.26). Thus we will first consider corrections to the quark–quark high energy scattering amplitude.

All possible real-gluon emission corrections to the quark–quark scattering amplitude of Fig. 3.3 are shown in Fig. 3.6. In order to extract the leading-ln s contribution we assume that $k^+ \ll P^+$, $k^- \ll P^-$, where k^μ is the momentum of the produced (i.e., final-state) gluon. If one performs the calculation in the covariant (Feynman) gauge then all the diagrams in Fig. 3.6 will contribute (Fadin, Kuraev, and Lipatov 1975). Here we will perform the calculation in the $\eta \cdot A = A^+ = 0$ light cone gauge. The advantage of this gauge is that in it diagrams D and E in Fig. 3.6 do not contribute (at high energy). To see this we again use the same trick and choose the Lepage and Brodsky (1980) convention for spinors for the upper quark line and the same convention with P^- and P^+ interchanged for the lower line. Again with the help of Table A.1, one can see that the dominant contribution of each diagram at high energy comes from the γ^+'s in the quark–gluon vertices in the upper quark line and from the γ^-'s in the the quark–gluon vertices in the lower quark line. (For instance, the numerator of the quark propagator corresponding to the lower line in diagram D gives $(p_2 - q) \cdot \gamma \approx (1/2)p_2^- \gamma^+$, so that, since $(\gamma^+)^2 = 0$, the adjacent vertices can only give either γ^- or $\gamma^\perp$; the $\gamma^\perp$ contribution is suppressed by powers of P^-, though, leaving only the γ^- vertices.) The polarization vector of the outgoing gluon in the $A^+ = 0$ light cone gauge can be parametrized as

$$\epsilon_\lambda^\mu(k) = \left(0, \frac{2\vec{\epsilon}_\perp^\lambda \cdot \vec{k}_\perp}{k^+}, \vec{\epsilon}_\perp^\lambda\right) \tag{3.28}$$

with transverse vector $\vec{\epsilon}_\perp^\lambda = -(1/\sqrt{2})(\lambda, i)$. We see that the γ^- from the emission vertex of the gluon carrying momentum k in graphs D and E is multiplied by $\epsilon_\lambda^{+*} = 0$ and therefore gives zero. We are left with diagrams A, B, and C to calculate.

Let us start by calculating diagrams B and C. Using the Feynman rules we write for diagram B (note that $\eta^\mu = (0, 2, 0_\perp)$)

$$iM^{\mathrm{B}}_{qq \to qqG} = g^3 (t^a t^c)_{i'i} (t^c)_{j'j} \bar{u}_{\sigma'_1}(p_1 - k + q) \epsilon^*_\lambda(k) \frac{\slashed{p}_1 + \slashed{q}}{(p_1 + q)^2} \gamma^\mu u_{\sigma_1}(p_1)$$

$$\times \bar{u}_{\sigma'_2}(p_2 - q) \gamma^\nu u_{\sigma_2}(p_2) \frac{-i}{q^2} \left(g_{\mu\nu} - \frac{\eta_\mu q_\nu + \eta_\nu q_\mu}{\eta \cdot q} \right). \tag{3.29}$$

(The colors and polarizations of the incoming and outgoing quarks are labeled in the same way as in Fig. 3.3.) As in the lowest-order case we note that the outgoing quarks are on mass shell, so that $(p_1 - k + q)^2 = 0$ and $(p_2 - q)^2 = 0$. These conditions give

$$q^- = k^- + O\left(\frac{1}{P^+} \right), \quad q^+ = O\left(\frac{1}{P^-} \right). \tag{3.30}$$

We see that $q^2 \approx -\vec{q}_\perp^{\,2}$ and, therefore, the gluon with momentum q is an instantaneous (Coulomb) gluon.

Using the Dirac equation we see that the q_ν-term in the large parentheses in Eq. (3.29), which postmultiplies the matrix element $\bar{u}_{\sigma'_2}(p_2 - q) \gamma^\nu u_{\sigma_2}(p_2)$, is zero. The η_ν-term in the gluon propagator of Eq. (3.29) gives $\eta_\nu \bar{u}_{\sigma'_2}(p_2 - q) \gamma^\nu u_{\sigma_2}(p_2) = \bar{u}_{\sigma'_2}(p_2 - q) \gamma^+ u_{\sigma_2}(p_2)$, which, if one uses the Lepage and Brodsky (1980) convention for spinors with the $+$ and $-$ momenta interchanged, is suppressed by $1/P^{-2}$ in comparison with the leading-order term arising from $\bar{u}_{\sigma'_2}(p_2 - q) \gamma^- u_{\sigma_2}(p_2)$. This leaves us with only the $g_{\mu\nu}$-term within the large parentheses. Therefore $\nu = -$ and $\mu = +$ gives the dominant contribution in the matrix elements in Eq. (3.29). Making the approximations $(p_1 + q) \cdot \gamma \approx (1/2) P^+ \gamma^-$ and $(p_1 + q)^2 \approx P^+ q^-$ in the quark propagator, and using Table A.1 along with Eq. (3.30), yields

$$iM^{\mathrm{B}}_{qq \to qqG} = 4ig^3 (t^a t^c)_{i'i} (t^c)_{j'j} \delta_{\sigma_1 \sigma'_1} \delta_{\sigma_2 \sigma'_2} \frac{s}{q_\perp^2} \frac{\vec{\epsilon}_\perp^{\lambda*} \cdot \vec{k}_\perp}{k_\perp^2}. \tag{3.31}$$

A similar calculation for diagram C in Fig. 3.6 would readily show that its contribution $M^{\mathrm{C}}_{qq \to qqG}$ is different from $M^{\mathrm{B}}_{qq \to qqG}$ in Eq. (3.31) by an overall minus sign and by a change in the order of the color matrices: $t^a t^c \to t^c t^a$. The sum of the two graphs is

$$M^{\mathrm{B}}_{qq \to qqG} + M^{\mathrm{C}}_{qq \to qqG} = -4ig^3 f^{abc} (t^b)_{i'i} (t^c)_{j'j} \delta_{\sigma_1 \sigma'_1} \delta_{\sigma_2 \sigma'_2} \frac{s}{q_\perp^2} \frac{\vec{\epsilon}_\perp^{\lambda*} \cdot \vec{k}_\perp}{k_\perp^2}. \tag{3.32}$$

It is important to note that, for an Abelian theory such as QED, the sum of diagrams B and C would be zero, owing to the absence of color matrices. This makes physical sense: as we will see below in more detail, the high energy approximation used above implies that each quark moves without recoil along its light cone throughout the scattering process. For an electron this would mean that high energy scattering does not affect it at all: it does not acquire any acceleration. Therefore, without acceleration or deceleration the electron will not radiate; this statement is equivalent to the cancellation of graphs B and C in QED. In non-Abelian theories such as QCD, radiation is caused not only by acceleration but also by color rotation. Thus the recoilless motion of the quarks does not mean the absence of radiation: what happens in diagrams B and C in Fig. 3.6 is that the color of the upper quark line is rotated by the t-channel gluon exchange interaction with the lower quark line.

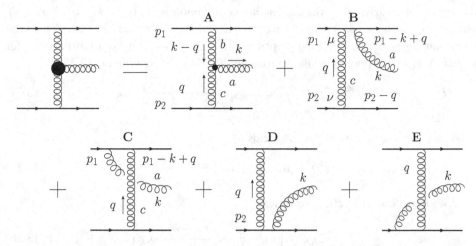

Fig. 3.6. The effective real-gluon emission vertex (the Lipatov vertex), defined as the sum of all gluon emission diagrams. The triple gluon vertex is denoted by the smaller solid circle while the Lipatov vertex is shown by the larger solid circle.

Because of this color flip, diagrams B and C no longer cancel, which means that, unlike in QED, the radiations from the upper quark before (C) and after (B) the interaction do not cancel. Thus, we see that in QCD the gluon radiation can be entirely due to color rotation: this is an essentially non-Abelian feature of the theory.

We now need to calculate the amplitude represented by diagram A in Fig. 3.6. Applying the Feynman rules we write

$$
iM^A_{qq \to qqG} = g^3 f^{abc} (t^b)_{i'i} (t^c)_{j'j}\, \epsilon^*_{\lambda \rho}(k)\, \bar{u}_{\sigma'_1}(p_1 - k + q) \gamma^{\mu'} u_{\sigma_1}(p_1)
$$
$$
\times\, \bar{u}_{\sigma'_2}(p_2 - q) \gamma^{\nu'} u_{\sigma_2}(p_2) \left[(k - 2q)^\rho g^{\mu\nu} + (q - 2k)^\nu g^{\mu\rho} + (q + k)^\mu g^{\nu\rho} \right]
$$
$$
\times\, \frac{1}{q^2} \left(g_{\nu\nu'} - \frac{\eta_\nu q_{\nu'} + \eta_{\nu'} q_\nu}{\eta \cdot q} \right) \frac{1}{(k-q)^2} \left(g_{\mu\mu'} - \frac{\eta_\mu (k-q)_{\mu'} + \eta_{\mu'} (k-q)_\mu}{\eta \cdot (k-q)} \right).
$$
$$(3.33)$$

(The Lorentz indices, while not explicitly shown in Fig. 3.6A, should be self-evident.) The evaluation proceeds along lines similar to the calculation of graphs B and C. We first note that in the q-line propagator the $q_{\nu'}$-term is zero by the Dirac equation (current conservation), while the $\eta_{\nu'}$-term is strongly suppressed at high energy (see Table A.1): this leaves only the $g_{\nu\nu'}$-term to contribute. As a side observation, we note that the $A^+ = 0$ light cone gauge is equivalent to the covariant (Feynman) gauge for the lower part of the diagrams in Fig. 3.6. This is a useful tool, which we will employ again later.

Again, using Table A.1 and the Lepage and Brodsky (1980) convention for spinors (with $+$ and $-$ momenta interchanged), we see that γ^- gives the leading contribution in the matrix element $\bar{u}_{\sigma'_2}(p_2 - q) \gamma^{\nu'} u_{\sigma_2}(p_2)$. Since in the $A^+ = 0$ light cone gauge we have $\epsilon^+_\lambda = 0$ and since the propagator of the $k - q$ line is also zero when either of its (upper) indices is $+$, we conclude that only the $g^{\mu\rho}$ term in the triple-gluon vertex contributes at high energy. Similarly, γ^+ gives the leading contribution in $\bar{u}_{\sigma'_1}(p_1 - k + q)\gamma^{\mu'}u_{\sigma_1}(p_1)$. Finally, since we have taken care to have the external lines carry the same momenta (and other quantum numbers) in all the graphs in Fig. 3.6, we see that Eq. (3.30) applies for diagram A as well. Applying all the above simplifications, we rewrite Eq. (3.33) as

$$M^A_{qq \to qqG} = 4ig^3 f^{abc}(t^b)_{i'i}(t^c)_{j'j}\delta_{\sigma_1\sigma'_1}\delta_{\sigma_2\sigma'_2}\frac{s}{q^2_\perp(\vec{k}_\perp - \vec{q}_\perp)^2}\vec{\epsilon}^{\lambda*}_\perp \cdot (\vec{k}_\perp - \vec{q}_\perp). \quad (3.34)$$

Adding all the diagrams in Fig. 3.6 yields

$$M_{qq \to qqG} = 2ig^2(t^b)_{i'i}(t^c)_{j'j}\delta_{\sigma_1\sigma'_1}\delta_{\sigma_2\sigma'_2}\frac{s}{q^2_\perp(\vec{k}_\perp - \vec{q}_\perp)^2}\vec{\epsilon}^{\lambda*}_\perp \cdot \vec{\Gamma}^{abc}_\perp, \quad (3.35)$$

where we have defined an effective vertex

$$\vec{\Gamma}^{abc}_\perp = 2gf^{abc}\left[\vec{k}_\perp - \vec{q}_\perp - \frac{(\vec{k}_\perp - \vec{q}_\perp)^2}{k^2_\perp}\vec{k}_\perp\right]. \quad (3.36)$$

The vertex $\vec{\Gamma}^{abc}_\perp$ was first derived in Fadin, Kuraev, and Lipatov (1975, 1977). In the literature it is usually referred to as the *Lipatov vertex*. It is pictured on the left-hand side of Fig. 3.6, where it is denoted by the large solid circle. The origin of this notation can be seen in Eq. (3.35), which one can regard as containing the propagators of the two t-channel gluon lines ($\vec{q}^2_\perp$ and $(\vec{k}_\perp - \vec{q}_\perp)^2$), the color factors and Kronecker deltas coming from the quark lines, an overall factor of s characteristic of the leading high energy amplitudes, and the vertex $\vec{\Gamma}^{abc}_\perp$. We see that all five diagrams A–E in Fig. 3.6 can be thought of as one diagram with an effective Lipatov vertex $\vec{\Gamma}^{abc}_\perp$ instead of the triple-gluon vertex.

We have to add here that there are gauges in which diagrams B through E in Fig. 3.6 do not contribute in the high energy limit, so that the amplitude is simply given by diagram A. An example would be the $A^0 = 0$ gauge.

Squaring the amplitude in Eq. (3.35), we write the corresponding cross section as

$$\sigma_{qq \to qqG} = \frac{2\alpha^3_s C_F}{\pi^2} \int \frac{d^2k_\perp d^2q_\perp}{k^2_\perp q^2_\perp(\vec{k}_\perp - \vec{q}_\perp)^2} \int\limits_{k^2_\perp/P^-}^{P^+} \frac{dk^+}{k^+}. \quad (3.37)$$

As in the derivation of the DGLAP evolution equations in Chapter 2, we obtain a logarithmic longitudinal integral – the integral over k^+ in the above expression. Since Eq. (3.37) was derived in the high energy approximation with $P^+ \gg k^+$, in order to obtain the leading logarithmic ($\ln s$) contribution we put P^+ as the upper limit of the k^+ integral (the same applies to the lower limit of this integral): defining the rapidity of the gluon by[3]

$$y = \ln\frac{P^-}{k^-}, \quad (3.38)$$

---

[3] The standard rapidity definition is $y = (1/2)\ln(k^+/k^-)$ in the center-of-mass frame. Our definition here is different by an overall shift, making the rapidity equal to zero in the direction of one of the onia and equal to $Y = \ln(s/k^2_\perp)$ in the direction of the other.

we rewrite the gluon production cross section as

$$\sigma_{qq\to qqG} = \frac{2\alpha_s^3 C_F}{\pi^2} \int \frac{d^2k_\perp d^2q_\perp}{k_\perp^2 q_\perp^2 (\vec{k}_\perp - \vec{q}_\perp)^2} \int_0^Y dy. \tag{3.39}$$

Here $Y = \ln s/k_\perp^2$ is the total rapidity interval between the colliding quarks; since our goal is to track the leading-$\ln s$ contribution to the cross section, this Y is for us not different from the $Y = \ln(s|\vec{x}_{1,\perp}||\vec{x}_{2\perp}|)$ defined earlier in the chapter. The difference between the two does not contain $\ln s$ and can be disregarded at degree of our precision. For simplicity, every time we discuss leading-$\ln s$ asymptotics we will assume that $Y = \ln(s/m_\perp^2)$, with $m_\perp$ some transverse momentum scale the exact value of which is irrelevant in the leading-$\ln s$ approximation.

Just as for the quark–quark scattering in Eq. (3.18), the cross section in Eq. (3.37) has IR divergences at $\vec{q}_\perp = 0$ and $\vec{q}_\perp = \vec{k}_\perp$ since we are considering free quark scattering. Generalizing this result to onium–onium scattering by summing over all interactions of quarks and antiquarks, we include the impact factors that regulate these divergences,[4] obtaining, after relabeling the momenta to match those in Eq. (3.26),

$$\hat{\sigma}_{1,real}^{onium+onium} = \frac{2\alpha_s^2 C_F}{N_c} \int \frac{d^2l_\perp d^2l'_\perp}{l_\perp^2 l_\perp'^2} \left(2 - e^{-i\vec{l}_\perp \cdot \vec{x}_{1\perp}} - e^{i\vec{l}_\perp \cdot \vec{x}_{1\perp}}\right)$$
$$\times \left(2 - e^{-i\vec{l}'_\perp \cdot \vec{x}_{2\perp}} - e^{i\vec{l}'_\perp \cdot \vec{x}_{2\perp}}\right) G_1^{real}\left(\vec{l}_\perp, \vec{l}'_\perp, Y\right), \tag{3.40}$$

where

$$G_1^{real}\left(\vec{l}_\perp, \vec{l}'_\perp, Y\right) = \frac{\alpha_s N_c}{\pi^2} Y \frac{1}{(\vec{l}_\perp - \vec{l}'_\perp)^2}. \tag{3.41}$$

A contribution to Eq. (3.40) is shown in Fig. 3.7; the circles denote Lipatov vertices. (Indeed, one has to sum the diagram in Fig. 3.7 over all connections of t-channel gluons to all quark lines to obtain Eq. (3.40).)

We have made the first step in understanding the structure of the shaded rectangle in Fig. 3.5 by calculating the lowest-order real-emission correction to Fig. 3.4. Our results so far can be written as (see Eqs. (3.26) and (3.27))

$$G\left(\vec{l}_\perp, \vec{l}'_\perp, Y\right) = G_0\left(\vec{l}_\perp, \vec{l}'_\perp\right) + G_1^{real}\left(\vec{l}_\perp, \vec{l}'_\perp, Y\right) + \cdots$$

$$= G_0\left(\vec{l}_\perp, \vec{l}'_\perp\right) + \frac{\alpha_s N_c}{\pi^2} \int_0^Y dy \int \frac{d^2q_\perp}{(\vec{l}_\perp - \vec{q}_\perp)^2} G_0\left(\vec{q}_\perp, \vec{l}'_\perp\right) + \cdots \tag{3.42}$$

While at the moment Eq. (3.42) may seem like a trivial rewriting of Eq. (3.41), it will be useful later.

---

[4] The divergence at $k_\perp = 0$ still remains in the real-gluon contribution to the total cross section; it will be discussed further in Chapter 8 when we consider gluon production.

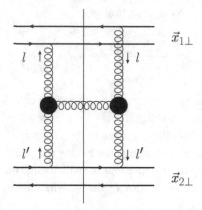

Fig. 3.7. The sum of the real one-gluon corrections to the two-gluon exchange cross section of Fig. 3.4 represented with the help of the effective Lipatov vertices from Fig. 3.6, denoted by the solid circles.

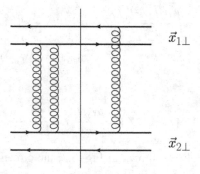

Fig. 3.8. An example of a leading-ln s order-α_s virtual correction to the Born amplitude.

3.3.2 Virtual corrections and reggeized gluons

To complete our calculation of the order-α_s corrections to the Born-level onium–onium scattering cross section in Eq. (3.24) we need to include the virtual corrections, i.e., diagrams where the extra gluon is not present in the final state. In the amplitude squared we are interested in interference terms between the leading-order single-gluon exchange amplitude of Fig. 3.3 and order-α_s^2 diagrams including one-gluon virtual corrections to it: an example of such a diagram is shown in Fig. 3.8.

Diagrams representing the main types of virtual correction to quark–quark scattering are shown in Fig. 3.9. (All other virtual corrections may be obtained by mirror reflections of the graphs in Fig. 3.9.) As before we assume that all momenta are much smaller than P^+ and P^-.

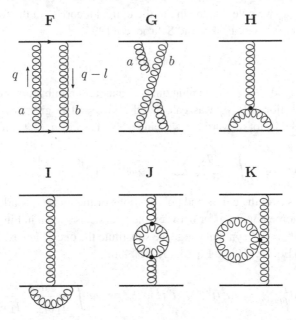

Fig. 3.9. The main classes of the leading-order virtual corrections to the quark–quark scattering amplitude from Fig. 3.3. The small solid circles denote three- and four-gluon vertices.

Our goal is to find the order-α_s^2 amplitude in Fig. 3.9. Instead of a lengthy direct calculation we use the following double-subtracted dispersion relation for scattering amplitudes:[5]

$$M(s, t) = M(s = 0, t) + s\partial_s M(s = 0, t)$$

$$+ \frac{s^2}{\pi} \left[\int\limits_{4m^2}^{\infty} ds' \frac{\mathrm{Im}_s M(s', t)}{s'^2(s' - s)} + \int\limits_{4m^2-t}^{\infty} du' \frac{\mathrm{Im}_u M(u', t)}{\left(4m^2 - t - u'\right)^2 (u' - u)} \right] \quad (3.43)$$

where ∂_s denotes a partial derivative with respect to s. The first term in the brackets in Eq. (3.43) contains a discontinuity in the s-channel, while the second term has a discontinuity in the u-channel. (To underscore this, we have relabeled the argument of the amplitude M in the second term to show explicitly its dependence on u and have replaced s' by $4m^2 - t - u'$ in the denominator of the second term.) Above, having in mind the $qq \rightarrow qq$ scattering amplitude, we have assumed here that quarks have a small mass m, such that $4m^2$ is the particle production threshold in both the s- and the u-channel.

We see from Eq. (3.43) that in order to find the order-α_s^2 amplitude in Fig. 3.9 we need the diagrams that have an imaginary part. Therefore we do not need diagrams H–K in Fig. 3.9 since those lead to amplitudes that are purely real (they cannot be cut). The amplitudes given by diagrams F and G have s- and u-channel discontinuities correspondingly. Denoting them

---

[5] The dispersion relations used here are derived in Appendix B. Their derivation can also be found in Forshaw and Ross (1997), in Collins, P. D. B. (1977), and, in a slightly different form, in Weinberg (1996).

$M^F_{qq \to qq}$ and $M^G_{qq \to qq}$ we note that, owing to the optical theorem, in the forward scattering case we would have (see e.g. Peskin and Schroeder (1995))

$$\text{Im } M^F_{qq \to qq}(\text{forward}) = s\sigma^0_{qq \to qq}; \qquad (3.44)$$

here we have averaged over the incoming quarks' quantum numbers in the forward amplitude. The cross section $\sigma^0_{qq \to qq}$ was calculated above and is given by Eq. (3.18). The scattering in Fig. 3.9 is not forward but we can easily correct Eq. (3.44) for that and write

$$\text{Im } M^F_{qq \to qq} = \int \frac{d^2 q_\perp}{(4\pi)^2 s} \sum M^0_{qq \to qq}(\vec{q}_\perp) \left[M^0_{qq \to qq}(\vec{q}_\perp - \vec{l}_\perp) \right]^*, \qquad (3.45)$$

where the sum runs over the colors and polarizations of the internal quark lines in graph F and $\vec{l}_\perp$ is the net momentum transfer between the quarks, as shown in Fig. 3.9. Working in the covariant ($\partial_\mu A^\mu = 0$) Feynman gauge we substitute the one-gluon exchange amplitude $M^0_{qq \to qq}(\vec{q}_\perp)$ from Eq. (3.17) into Eq. (3.45), obtaining

$$\text{Im } M^F_{qq \to qq} = 4\alpha_s^2 (t^b t^a)_{i'i} (t^b t^a)_{j'j} \delta_{\sigma_1 \sigma_1'} \delta_{\sigma_2 \sigma_2'} \int \frac{d^2 q_\perp}{q_\perp^2 (\vec{q}_\perp - \vec{l}_\perp)^2} s. \qquad (3.46)$$

(Here the colors and helicities of the external quark lines are labeled in the same way as in Eq. (3.17) and as shown in Fig. 3.3.)

The imaginary part of the amplitude $M^G_{qq \to qq}$ is obtained from Im $M^F_{qq \to qq}$ by interchanging s and u and the color indices a and b along one quark line:

$$\text{Im } M^G_{qq \to qq} = 4\alpha_s^2 (t^b t^a)_{i'i} (t^a t^b)_{j'j} \delta_{\sigma_1 \sigma_1'} \delta_{\sigma_2 \sigma_2'} \int \frac{d^2 q_\perp}{q_\perp^2 (\vec{q}_\perp - \vec{l}_\perp)^2} u. \qquad (3.47)$$

Substituting Eqs. (3.46) and (3.47) into Eq. (3.43), we find the total order-α_s^2 amplitude for $qq \to qq$ scattering:

$$M^1_{qq \to qq}(s, t = -l_\perp^2)$$

$$= M^1_{qq \to qq}(s = 0, t) + s\partial_s M^1_{qq \to qq}(s = 0, t) + \frac{4\alpha_s^2 s^2}{\pi} \delta_{\sigma_1 \sigma_1'} \delta_{\sigma_2 \sigma_2'} \int \frac{d^2 q_\perp}{q_\perp^2 (\vec{q}_\perp - \vec{l}_\perp)^2} (t^b t^a)_{i'i}$$

$$\times \left[(t^b t^a)_{j'j} \int_{4m^2}^{\infty} ds' \frac{1}{s'(s' - s)} + (t^a t^b)_{j'j} \int_{4m^2 - t}^{\infty} du' \frac{u'}{(4m^2 - t - u')^2 (u' - u)} \right]. \qquad (3.48)$$

Note that the s'-integral in Eq. (3.48) is divergent for $s > 4m^2$ owing to a singularity at $s' = s$ and should be understood as the $\epsilon \to 0$ limit of the regulated expression obtained by replacing s by $s + i\epsilon$ in it and integrating. The u'-integral is regulated in a similar way.

We require the high energy asymptotics of the amplitude $M^1_{qq \to qq}(s, t)$. Moreover, we need the leading-ln s contribution; noting that at high energy $u \approx -s$ and neglecting $s' \ll s$ and $4m^2 - t \ll u' \ll |u|$, while cutting off the s'-integral by s and the u'-integral by $|u| \approx s$

in the UV to obtain the leading logarithms of s, yields

$$M^1_{qq \to qq}(s, t = -l^2_\perp) = -\frac{4\alpha^2_s s}{\pi} \delta_{\sigma_1 \sigma'_1} \delta_{\sigma_2 \sigma'_2} \int \frac{d^2 q_\perp}{q^2_\perp (\vec{q}_\perp - \vec{l}_\perp)^2}$$

$$\times (t^b t^a)_{i'i} \left[(t^b t^a)_{j'j} \ln \left(\frac{-s}{4m^2} \right) - (t^a t^b)_{j'j} \ln \left(\frac{s}{4m^2 - t} \right) \right]. \quad (3.49)$$

The factor $\ln(-s)$ arises from the exact integration over s' in (3.48); it reflects the fact that this amplitude has a branch cut at $s > 0$. (Note that integration over u' does not lead to such singularities.) In arriving at Eq. (3.49) we have dropped $M^1_{qq \to qq}(s = 0, t) + s \partial_s M^1_{qq \to qq}(s = 0, t)$, since these terms grow with energy as at most s, which is subleading compared with the $s \ln s$ scaling of the leading part of the term that we have kept on the right-hand side of Eq. (3.49).

We see from Eq. (3.49) that, as in real-gluon emission, the only reason why the diagrams F and G do not cancel is the presence of color factors: in QED the leading logarithms of energy would vanish for graphs F and G taken together. Keeping $\ln s \sim Y$ terms only we obtain

$$M^1_{qq \to qq} = \frac{2\alpha^2_s N_c s}{\pi} (t^a)_{i'i} (t^a)_{j'j} \delta_{\sigma_1 \sigma'_1} \delta_{\sigma_2 \sigma'_2} \int \frac{d^2 q_\perp}{q^2_\perp (\vec{q}_\perp - \vec{l}_\perp)^2} Y. \quad (3.50)$$

Note that the color factors have become the same as for the single-gluon exchange amplitude (3.17): we see that the two exchanged gluons in diagrams F and G are in the *color octet* state. We also see that when the two t-channel gluons are in the color singlet state, as in Fig. 3.4 and Eq. (3.24), no logarithms of the energy are generated, whereas when the two gluons are in the color octet state one gets a $\ln s$ contribution, as we have just seen.

Comparing Eq. (3.50) and Eq. (3.17) we can rewrite the former as

$$M^1_{qq \to qq}(\vec{l}_\perp) = M^0_{qq \to qq}(\vec{l}_\perp) \omega_G(l_\perp) Y, \quad (3.51)$$

where we have defined the *gluon Regge trajectory*

$$\omega_G(l_\perp) = -\frac{\alpha_s N_c}{4\pi^2} \int d^2 q_\perp \frac{l^2_\perp}{q^2_\perp (\vec{q}_\perp - \vec{l}_\perp)^2}. \quad (3.52)$$

One can show further that virtual corrections to the amplitude in Eq. (3.17) that are of higher order in α_s, bringing in leading logarithms of s, lead to a simple exponentiation of the result (3.51), so that one can replace the gluon propagator (3.17) by (Fadin, Kuraev, and Lipatov 1975, 1976, 1977; see Sec. 3.3.5 below)

$$\frac{ig_{\mu\nu}}{l^2_\perp} \to \frac{ig_{\mu\nu}}{l^2_\perp} e^{\omega_G(l_\perp)Y} \sim \frac{ig_{\mu\nu}}{l^2_\perp} s^{\omega_G(l_\perp)}. \quad (3.53)$$

As discussed in Section 3.2, the propagator (3.53) can be viewed as describing the exchange of a particle with spin $j = 1 + \omega_G(l_\perp)$. We will refer to this "quasi-particle" as a *reggeized gluon*. It is given by the sum of all leading-$\ln s$ virtual corrections to a single-gluon exchange and is illustrated in Fig. 3.10. We will denote reggeized gluons by a thick corkscrew line, as shown in Fig. 3.10.

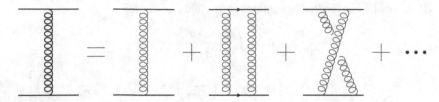

Fig. 3.10. Reggeized gluon (bold corkscrew line) represented as the sum of all leading-ln s corrections to the single-gluon exchange amplitude for $qq \rightarrow qq$ scattering.

To find the order-α_s virtual correction to the two-gluon exchange cross section, we have to consider the interference between the lowest-order amplitude in Fig. 3.3 and the amplitude in Fig. 3.9:

$$M^1_{qq \rightarrow qq} \left(M^0_{qq \rightarrow qq} \right)^* + M^0_{qq \rightarrow qq} \left(M^1_{qq \rightarrow qq} \right)^* = |M^0_{qq \rightarrow qq}|^2 2\omega_G(l_\perp)Y. \tag{3.54}$$

The correction contributing to the shaded rectangle in Fig. 3.5 is then

$$G^{virtual}_1 \left(\vec{l}_\perp, \vec{l}_\perp', Y \right) = G_0 \left(\vec{l}_\perp, \vec{l}_\perp' \right) 2\omega_G(l_\perp)Y. \tag{3.55}$$

Equation (3.55) resums diagrams like that shown in Fig. 3.8. Using (3.55) we can now include all order-α_s corrections in Eq. (3.42), turning it into

$$G \left(\vec{l}_\perp, \vec{l}_\perp', Y \right) = G_0 \left(\vec{l}_\perp, \vec{l}_\perp' \right) + \frac{\alpha_s N_c}{\pi^2} \int_0^Y dy \int \frac{d^2 q_\perp}{(\vec{l}_\perp - \vec{q}_\perp)^2}$$

$$\times \left[G_0 \left(\vec{q}_\perp, \vec{l}_\perp' \right) - \frac{l_\perp^2}{2q_\perp^2} G_0 \left(\vec{l}_\perp, \vec{l}_\perp' \right) \right] + O(\alpha_s^2). \tag{3.56}$$

We will now discuss how to generalize this result to all orders in α_s.

3.3.3 The BFKL equation

The shaded rectangle in Fig. 3.5 can be written as a sum of gluon corrections order by order in α_s:

$$G \left(\vec{l}_\perp, \vec{l}_\perp', Y \right) = \sum_{m=0}^\infty G_m \left(\vec{l}_\perp, \vec{l}_\perp', Y \right), \tag{3.57}$$

where each G_m is of order α_s^m, where α_s is the coupling constant. One can readily see that Eq. (3.56) represents the first (order-α_s) iteration for the solution of the following equation:

$$\frac{\partial G \left(\vec{l}_\perp, \vec{l}_\perp', Y \right)}{\partial Y} = \frac{\alpha_s N_c}{\pi^2} \int \frac{d^2 q_\perp}{(\vec{l}_\perp - \vec{q}_\perp)^2} \left[G \left(\vec{q}_\perp, \vec{l}_\perp', Y \right) - \frac{l_\perp^2}{2q_\perp^2} G \left(\vec{l}_\perp, \vec{l}_\perp', Y \right) \right] \tag{3.58}$$

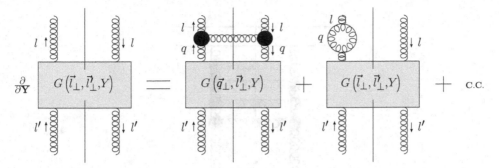

Fig. 3.11. Diagrammatic representation of the BFKL evolution equation (3.58). The second diagram on the right-hand side represents virtual corrections; c.c. denotes the complex conjugate contribution.

with initial condition

$$G\left(\vec{l}_{\perp}, \vec{l}'_{\perp}, Y = 0\right) = \delta^2\left(\vec{l}_{\perp} - \vec{l}'_{\perp}\right). \tag{3.59}$$

In fact, as was shown by Fadin, Kuraev, and Lipatov (1977) and by Balitsky and Lipatov (1978), Eq. (3.58) correctly resums all leading-ln s corrections to the Born-level onium–onium scattering amplitude of Fig. 3.4. This is the Balitsky–Fadin–Kuraev–Lipatov (BFKL) evolution equation (Fadin, Kuraev, and Lipatov 1977, Balitsky and Lipatov 1978). The object $G(\vec{l}_{\perp}, \vec{l}'_{\perp}, Y)$ is called the Green function of the BFKL equation: it describes the propagation of two t-channel gluons over the rapidity interval Y.

The BFKL equation for the Green function $G(\vec{l}_{\perp}, \vec{l}'_{\perp}, Y)$ is represented graphically in Fig. 3.11. It shows that in one step of BFKL evolution the Green function gets corrected either by real-gluon emissions (summarized in the first diagram on the right-hand side by the square of a Lipatov vertex) or by virtual corrections on either of the two t-channel gluon lines (represented by the second diagram on the right-hand side of Fig. 3.11 together with its complex conjugate). Iterations of the corrections shown in Fig. 3.11 lead to the representation of BFKL evolution by "ladder" diagrams such as those shown in Fig. 3.12.

By performing the calculation at order α_s^3 of the onium–onium scattering amplitude we have obtained two main ingredients that describe high energy scattering in the leading-ln s approximation: the reggeized gluon and and the new effective (Lipatov) vertex of Eq. (3.36). Fadin, Kuraev, and Lipatov (1975, 1976, 1977) proved that the general diagram contributing to the high energy amplitude at the leading-ln s level can be written as a sum over the produced gluons of the simple ladder-type diagram shown in Fig. 3.12. In this diagram, each vertex is of the type (3.36), each t-channel gluon is a reggeized gluon with propagator (3.53), while all the produced (s channel) gluons are the regular gluons of the QCD Lagrangian. We are not going to reproduce here the original proof of the BFKL equation. Instead we will rederive this equation in the next chapter in a more rigorous way using LCPT: for a complete derivation using the conventional techniques outlined here we refer the reader to Lipatov (1997, 1999), Del Duca (1995), or Forshaw and Ross (1997).

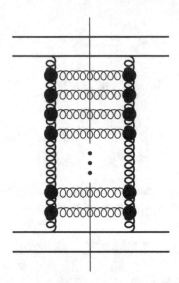

Fig. 3.12. Representation of BFKL evolution as a ladder diagram with effective Lipatov vertices (large solid circles) and reggeized gluons (bold corkscrew lines).

Comparing Fig. 3.12 with Fig. 2.25 we see that both the DGLAP and the BFKL equations effectively resum ladder diagrams, though the "rungs" of the two ladders and the vertices should be understood differently. (Of course, Fig. 2.25 was used to illustrate DLA DGLAP, which is contained in BFKL evolution, as we will shortly see: however, a general DGLAP evolution can also be represented by a ladder diagram resulting from iteration of the DGLAP splitting functions. Unlike the BFKL ladder, the DGLAP ladder would include quarks.) Another major difference between the DGLAP and BFKL ladders is the kinematics of the produced partons. As we saw in Eq. (2.72), the partons in DGLAP evolution are strongly ordered in their transverse momenta, while there is no ordering in their longitudinal momenta. In the BFKL case the kinematics is opposite: it is the longitudinal momenta that are strongly ordered, while there are no constraints on the transverse momenta. One can see this from our derivation of one iteration of the BFKL equation, presented above: we assumed that the plus components of the momenta of all gluons in Figs. 3.6 and 3.9 are much smaller than P^+, while the minus components are much smaller than P^-. At the same time we did not impose any constraints on the transverse momenta of the gluons in Figs. 3.6 and 3.9. If the momenta of the gluons in the ladder are labeled $k_1^\mu, k_2^\mu, \ldots, k_n^\mu$ (as shown in Fig. 3.13), the BFKL kinematics corresponds to

$$P^+ \gg k_1^+ \gg k_2^+ \gg \cdots \gg k_n^+, \tag{3.60a}$$

$$k_1^- \ll k_2^- \ll \cdots \ll k_n^- \ll P^-, \tag{3.60b}$$

$$k_{1\perp} \sim k_{2\perp} \sim \cdots \sim k_{n\perp}. \tag{3.60c}$$

The kinematics in Eqs. (3.60) is known as the *multi-Regge kinematics* and is also sometimes referred to as the *multi-peripheral model*. It is illustrated by Fig. 3.13. Since the first two

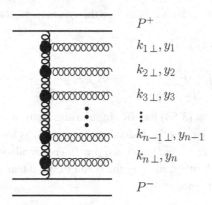

P^+

$k_{1\perp}, y_1$

$k_{2\perp}, y_2$

$k_{3\perp}, y_3$

$k_{n-1\perp}, y_{n-1}$

$k_{n\perp}, y_n$

P^-

Fig. 3.13. A scattering amplitude the square of which gives the BFKL ladder diagram of Fig. 3.12. The produced gluons have multi-Regge kinematics.

of Eqs. (3.60) imply in terms of rapidities that

$$Y \gg y_1 \gg y_2 \gg \cdots \gg y_n \gg 0, \tag{3.61}$$

where y_i, the rapidity of the ith produced gluon, is defined in Eq. (3.38), we see that the multi-Regge kinematics corresponds to the situation where the produced gluons uniformly cover the whole available rapidity interval. Note that owing to this property, the BFKL approach gives us the possibility to calculate the exclusive production cross section for any given number of gluons in the multi-Regge kinematics.

To better understand the dynamics resulting from the BFKL evolution, let us now find the solution of the BFKL equation.

3.3.4 Solution of the BFKL equation

To find the general solution of Eq. (3.58) we need to find eigenfunctions of its integral kernel K_{BFKL}, defined by

$$\int d^2 q_\perp K_{BFKL}(l, q) f(\vec{q}_\perp) \equiv \frac{1}{\pi} \int \frac{d^2 q_\perp}{(\vec{l}_\perp - \vec{q}_\perp)^2} \left[f(\vec{q}_\perp) - \frac{l_\perp^2}{2q_\perp^2} f(\vec{l}_\perp) \right] \tag{3.62}$$

for an arbitrary function $f(\vec{q}_\perp)$. The BFKL kernel (3.62) is conformally invariant. Therefore, one would expect that its set of eigenfunctions consists of powers of the transverse momentum times the complex exponentials of an integer number multiplying the azimuthal angle:

$$l_\perp^{2(\gamma-1)} e^{in\phi_l} \tag{3.63}$$

with γ an arbitrary complex number (analogous to the DGLAP anomalous dimension). Here ϕ_l is the angle between the vector $\vec{l}_\perp$ and some chosen axis in the transverse plane, and n is an integer. To see that the functions in Eq. (3.63) are indeed BFKL eigenfunctions we need to find the action of the BFKL kernel on these functions, that is, we need to evaluate

$$
\int d^2 q_\perp K_{BFKL}(l, q) \, q_\perp^{2(\gamma-1)} e^{in\phi_q}
$$

$$
= \frac{1}{\pi} \int \frac{d^2 q_\perp}{(\vec{l}_\perp - \vec{q}_\perp)^2} \left[q_\perp^{2(\gamma-1)} e^{in\phi_q} - \frac{l_\perp^2}{2q_\perp^2} l_\perp^{2(\gamma-1)} e^{in\phi_l} \right]. \tag{3.64}
$$

Note that the BFKL equation (3.58) has IR singularities both in the real (first) and virtual (second) terms on its right-hand side. Indeed, the first term is singular at $\vec{q}_\perp = \vec{l}_\perp$, while the second is singular both at $\vec{q}_\perp = \vec{l}_\perp$ and at $\vec{q}_\perp = 0$. As can also be seen from Eq. (3.58) the singularities cancel each other, making the result of the integration IR-finite.

To evaluate (3.64) we first note that[6]

$$
\frac{1}{q_\perp^2 (\vec{l}_\perp - \vec{q}_\perp)^2} = \frac{1}{q_\perp^2 \left[q_\perp^2 + (\vec{l}_\perp - \vec{q}_\perp)^2 \right]} + \frac{1}{(\vec{l}_\perp - \vec{q}_\perp)^2 \left[q_\perp^2 + (\vec{l}_\perp - \vec{q}_\perp)^2 \right]}, \tag{3.65}
$$

so that we obtain, using the substitution $\vec{q}_\perp \to \vec{l}_\perp - \vec{q}_\perp$ in the first term,

$$
\int \frac{d^2 q_\perp}{q_\perp^2 (\vec{l}_\perp - \vec{q}_\perp)^2} = 2 \int \frac{d^2 q_\perp}{(\vec{l}_\perp - \vec{q}_\perp)^2 \left[q_\perp^2 + (\vec{l}_\perp - \vec{q}_\perp)^2 \right]}. \tag{3.66}
$$

After a little more algebra Eq. (3.64) can be written as

$$
\int d^2 q_\perp K_{BFKL}(l, q) \, q_\perp^{2(\gamma-1)} e^{in\phi_q}
$$

$$
= \frac{1}{\pi} \int d^2 q_\perp \left\{ \frac{q_\perp^{2(\gamma-1)} e^{in\phi_q}}{(\vec{l}_\perp - \vec{q}_\perp)^2} - \frac{l_\perp^{2\gamma} e^{in\phi_l}}{q_\perp^2} \left[\frac{1}{(\vec{l}_\perp - \vec{q}_\perp)^2} - \frac{1}{q_\perp^2 + (\vec{l}_\perp - \vec{q}_\perp)^2} \right] \right\}. \tag{3.67}
$$

Taking $l_\perp^{2(\gamma-1)} e^{in\phi_l}$ outside the integral we obtain

$$
\int d^2 q_\perp K_{BFKL}(l, q) q_\perp^{2(\gamma-1)} e^{in\phi_q} = \chi(n, \gamma) l_\perp^{2(\gamma-1)} e^{in\phi_l}, \tag{3.68}
$$

where

$$
\chi(n, \gamma) = \int\limits_0^\infty dt \left[\frac{1}{2\pi} \int\limits_0^{2\pi} \frac{d\phi_q}{1 + t - 2\sqrt{t}\cos(\phi_q - \phi_l)} t^{\gamma-1} e^{in(\phi_q - \phi_l)} \right.
$$

$$
\left. - \frac{1}{t} \left(\frac{1}{|t-1|} - \frac{1}{\sqrt{4t^2 + 1}} \right) \right] \tag{3.69}
$$

with $t = q_\perp^2 / l_\perp^2$. In arriving at Eq. (3.69) we have used Eqs. (A.13) and (A.15) from appendix section A.2 to do the angular integration of the second term in Eq. (3.67).

We see from Eq. (3.68) that $l_\perp^{2(\gamma-1)} e^{in\phi_l}$ is indeed an eigenfunction of the BFKL kernel K_{BFKL}, with $\chi(n, \gamma)$ the corresponding eigenvalue.

---

[6] Our evaluation of the BFKL eigenvalue follows the strategy outlined in the review article by Del Duca (1995).

To perform the remaining angular integral in Eq. (3.69) we define a new complex variable $z = e^{i(\phi_q - \phi_l)}$ and write the integral as

$$\frac{i}{2\pi\sqrt{t}} \oint dz \frac{t^{\gamma-1} z^{|n|}}{\left(z - \sqrt{t}\right)\left(z - \frac{1}{\sqrt{t}}\right)}, \tag{3.70}$$

where the z-integral runs clockwise along a unit circle around the origin in the complex z-plane. To arrive at Eq. (3.70) we also noticed that the angular integral in Eq. (3.69) is an even function of n (and hence is a function of $|n|$), so that, to simplify the z-integration in Eq. (3.70), we can replace n by $|n|$. Performing the z-integral in Eq. (3.70) by the method of residues, we obtain for Eq. (3.69)

$$\chi(n,\gamma) = \int_0^\infty dt \left[\theta(1-t)\frac{t^{\gamma-1+|n|/2}}{1-t} + \theta(t-1)\frac{t^{\gamma-1-|n|/2}}{t-1} - \frac{1}{t}\left(\frac{1}{|t-1|} - \frac{1}{\sqrt{4t^2+1}}\right)\right]. \tag{3.71}$$

Employing the variable substitution $t \to 1/t$ for $t > 1$, we can rewrite Eq. (3.71) as

$$\chi(n,\gamma) = \int_0^1 dt \frac{t^{\gamma-1+|n|/2}}{1-t} + \int_0^1 dt \frac{t^{-\gamma+|n|/2}}{1-t} - 2\int_0^1 \frac{dt}{1-t} - \int_0^1 \frac{dt}{t} + \int_0^\infty \frac{dt}{t\sqrt{4t^2+1}}. \tag{3.72}$$

Regulating the last two integrals in Eq. (3.72) by multiplying their integrands by t^ϵ, performing the integrations, and taking the limit $\epsilon \to 0$ one can see that they cancel each other. For the first three integrals in (3.72) we use the integral representation of the logarithmic derivative of the gamma function (see e.g. Gradshteyn and Ryzhik (1994), formula 8.361.7),

$$\psi(z) = \frac{d}{dz} \ln \Gamma(z) = \int_0^1 dt \frac{t^{z-1}-1}{t-1} + \psi(1), \quad \mathrm{Re}\, z > 0, \tag{3.73}$$

to write (Balitsky and Lipatov 1978)

$$\chi(n,\gamma) = 2\psi(1) - \psi\left(\gamma + \frac{|n|}{2}\right) - \psi\left(1 - \gamma + \frac{|n|}{2}\right). \tag{3.74}$$

Note that the sum of integrals in Eq. (3.72) gives a finite answer only for $0 < \mathrm{Re}\,\gamma < 1$. Therefore, strictly speaking, the functions $l_\perp^{2(\gamma-1)} e^{in\phi_l}$ are the eigenfunctions of the BFKL kernel with eigenvalues $\chi(n,\gamma)$, (3.74), only for $0 < \mathrm{Re}\,\gamma < 1$.

Expanding the general solution of the BFKL equation (3.58) over the eigenfunctions of the BFKL kernel and using the fact that the BFKL Green function is symmetric, $G\left(\vec{l}_\perp, \vec{l}_\perp', Y\right) = G\left(\vec{l}_\perp', \vec{l}_\perp, Y\right)$, which follows from its definition (see Fig. 3.5 and Eq. (3.26)), we write

$$G\left(\vec{l}_\perp, \vec{l}_\perp', Y\right) = \sum_{n=-\infty}^\infty \int_{a-i\infty}^{a+i\infty} \frac{d\gamma}{2\pi i} C_{n,\gamma}(Y) l_\perp^{2(\gamma-1)} l_\perp'^{2(\gamma^*-1)} e^{in(\phi-\phi')}, \tag{3.75}$$

where the $C_{n,\gamma}(Y)$ are some unknown functions, γ^* is the complex conjugate of γ, and ϕ and ϕ' are the angles between $\vec{l}_\perp$ and $\vec{l}'_\perp$ and some arbitrary axis in the transverse plane. The integral over γ runs along a contour which is a straight line parallel to the imaginary axis in the γ-plane, defined by Re $\gamma = a$ such that $0 < a < 1$.

Substituting Eq. (3.75) into Eq. (3.58) and using the eigenvalues from Eq. (3.74) we find

$$C_{n,\gamma}(Y) = C^0_{n,\gamma} \exp\left\{\frac{\alpha_s N_c}{\pi}\chi(n,\gamma)Y\right\}, \tag{3.76}$$

where the coefficient $C^0_{n,\gamma}$ is fixed by the initial condition (3.59):

$$C^0_{n,\gamma} = \frac{1}{\pi}. \tag{3.77}$$

The initial condition (3.59) also fixes a as 1/2. Using this along with Eqs. (3.76) and (3.77), in Eq. (3.75) we obtain the solution of the BFKL equation (3.58):

$$G\left(\vec{l}_\perp, \vec{l}'_\perp, Y\right) = \sum_{n=-\infty}^{\infty} \int_{1/2-i\infty}^{1/2+i\infty} \frac{d\gamma}{2\pi^2 i} \exp\left\{\frac{\alpha_s N_c}{\pi}\chi(n,\gamma)Y\right\} l_\perp^{2(\gamma-1)} l'^{2(\gamma^*-1)}_\perp e^{in(\phi-\phi')}. \tag{3.78}$$

Since the integration contour in Eq. (3.78) runs along the Re $\gamma = 1/2$ line, if we define

$$\gamma = \frac{1}{2} + i\nu \tag{3.79}$$

with real ν we can rewrite Eq. (3.78) as

$$G\left(\vec{l}_\perp, \vec{l}'_\perp, Y\right) = \sum_{n=-\infty}^{\infty} \int_{-\infty}^{\infty} \frac{d\nu}{2\pi^2} \exp\left\{\frac{\alpha_s N_c}{\pi}\chi(n,\nu)Y\right\} l_\perp^{-1+2i\nu} l'^{-1-2i\nu}_\perp e^{in(\phi-\phi')}, \tag{3.80}$$

where

$$\chi(n,\nu) = 2\psi(1) - \psi\left(\frac{1+|n|}{2} + i\nu\right) - \psi\left(\frac{1+|n|}{2} - i\nu\right). \tag{3.81}$$

Note that Eqs. (3.81) and (3.74) are related by the substitution (3.79).

Unfortunately an exact analytic evaluation of Eq. (3.80) does not appear to be feasible. We will therefore construct approximate solutions below.

Diffusion approximation   Consider the case $l_\perp \sim l'_\perp$, i.e., the two transverse momentum scales involved in the problem are not very different from each other. To evaluate the ν-integral in Eq. (3.80) we will now employ the saddle point method, which we have already used in Sec. 2.4.6. A simple analysis of the saddle points of the function $\chi(n,\nu)$ at $\nu = 0$ allows one to conclude that at high energy, i.e., large rapidity Y, the dominant contribution to the amplitude is given by the $n = 0$ term in the sum in Eq. (3.80), as $\chi(n=0,\nu=0) >$

$\chi(n \neq 0, \nu = 0)$. We will therefore keep only the $n = 0$ term in Eq. (3.80) and write

$$G\left(\vec{l}_\perp, \vec{l}'_\perp, Y\right) \approx \int_{-\infty}^{\infty} \frac{d\nu}{2\pi^2 l_\perp l'_\perp} \, \exp\left\{\bar{\alpha}_s \chi(0, \nu)Y + 2i\nu \ln\frac{l_\perp}{l'_\perp}\right\}; \qquad (3.82)$$

here

$$\bar{\alpha}_s \equiv \frac{\alpha_s N_c}{\pi}. \qquad (3.83)$$

Expanding $\chi(n = 0, \nu)$ around the saddle point at $\nu = 0$ we get

$$\chi(0, \nu) \approx 4\ln 2 - 14\zeta(3)\nu^2, \qquad (3.84)$$

where $\zeta(z)$ is the Riemann zeta function. Using Eq. (3.84) in Eq. (3.82) we perform the ν-integration, obtaining (Balitsky and Lipatov 1978)

$$G\left(\vec{l}_\perp, \vec{l}'_\perp, Y\right) \approx \frac{1}{2\pi^2 l_\perp l'_\perp} \sqrt{\frac{\pi}{14\zeta(3)\bar{\alpha}_s Y}} \, \exp\left\{(\alpha_P - 1)Y - \frac{\ln^2(l_\perp/l'_\perp)}{14\zeta(3)\bar{\alpha}_s Y}\right\}, \qquad (3.85)$$

where we have used, for the intercept of the perturbative BFKL pomeron,

$$\alpha_P - 1 = \frac{4\alpha_s N_c}{\pi} \ln 2. \qquad (3.86)$$

The essential feature of Eq. (3.85) is that it shows that cross sections mediated by the BFKL ladder exchange grow as a power of the energy:

$$\sigma \sim e^{(\alpha_P - 1)Y} \sim s^{\alpha_P - 1}. \qquad (3.87)$$

This behavior is reminiscent of pomeron exchange in pre-QCD language (see Eq. (3.20)). The BFKL ladder from Fig. 3.12 is therefore referred to as the "hard" (perturbative) pomeron or as the BFKL pomeron. We see that BFKL evolution modifies the energy-independent Low–Nussinov pomeron, which simply corresponds to a two-gluon exchange and has $\alpha_P - 1 = 0$, which makes the perturbative pomeron intercept $\alpha_P > 1$ as seen from Eq. (3.86). The numerical value of the BFKL intercept (3.86) is rather large: for $\alpha_s = 0.3$ one gets $\alpha_P - 1 \approx 0.79$, which is much larger than the "soft" pomeron intercept of 0.08 observed, say, for the total proton–proton scattering cross section (Donnachie and Landshoff 1992).

Double logarithmic approximation   Let us consider the case $l_\perp \gg l'_\perp$. Now $\ln(l_\perp/l'_\perp)$ is large, and this may affect the location of the saddle point of the ν-integral in Eq. (3.80). The way the saddle point is shifted is shown in Fig. 3.14 for the $n = 0$ term in the series (3.80). As one can show analytically and as can be seen from Fig. 3.14, the effect of $(l_\perp/l'_\perp)^{2i\nu}$ in (3.80) is to shift the saddle point in the imaginary ν direction, moving it closer to the singularity of $\chi(0, \nu)$ at $\nu = i/2$. One can also show that the same is true for any integer n: the saddle point in the nth term in Eq. (3.80) is shifted toward the singularity of $\chi(n, \nu)$

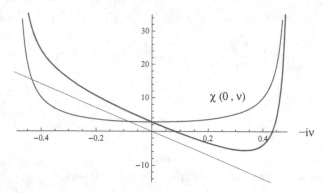

Fig. 3.14. The eigenvalue of the BFKL kernel $\chi(0, \nu)$ plotted as a function of $-i\nu$ (medium-bold line) for Re $\nu = 0$. The thin straight line is due to the linear term $2i\nu \ln(l_\perp / l_\perp')$ in the exponent of Eq. (3.82). The boldest curve is a sum of the medium-bold line and the thin straight line: it represents the complete expression in the exponent of Eq. (3.82). A color version of this figure is available online at www.cambridge.org/9780521112574.

at $\nu = i(|n| + 1)/2$. However, near these saddle points the nth term in the series (3.80) scales as

$$\frac{1}{l_\perp^2} \left(\frac{l_\perp'^2}{l_\perp^2} \right)^{|n|} ; \tag{3.88}$$

we see that terms with $|n| > 0$ are suppressed by powers of $l_\perp'^2 / l_\perp^2 \ll 1$ compared with the $n = 0$ term (i.e., they are higher-twist corrections). Therefore the $n = 0$ term dominates again and, as before, we can work with Eq. (3.82).

Expanding the $n = 0$ eigenvalue of the BFKL kernel near $\nu = i/2$, we find that

$$\chi(0, \nu) \approx -\frac{i}{\nu - i/2}, \tag{3.89}$$

and the saddle point of the integral in Eq. (3.82) is then given by

$$\nu_{DLA} \approx \frac{i}{2} - i \sqrt{\frac{\bar{\alpha}_s Y}{\ln(l_\perp^2 / l_\perp'^2)}}. \tag{3.90}$$

Distorting the ν-integration contour to run through ν_{DLA} and expanding the exponent of Eq. (3.82) up to terms of order $(\nu - \nu_{DLA})^2$, we integrate the result over ν, obtaining

$$G\left(\vec{l}_\perp, \vec{l}_\perp', Y\right) \approx \frac{1}{2\pi^{3/2} l_\perp^2} \frac{(\bar{\alpha}_s Y)^{1/4}}{\ln^{3/4}(l_\perp^2 / l_\perp'^2)} \exp\left\{ 2\sqrt{\bar{\alpha}_s Y \ln(l_\perp^2 / l_\perp'^2)} \right\}. \tag{3.91}$$

Comparing the exponential in Eq. (3.91) with that in Eq. (2.143) or, since here we are assuming a fixed coupling constant, with Eq. (2.159), we see that the DLA limit is indeed the same when obtained from the DGLAP or the BFKL equations! Identifying Y in Eq. (3.91) with $\ln 1/x$ in Eq. (2.159) and the transverse logarithm $\ln(l_\perp^2 / l_\perp'^2)$ in Eq. (3.91) with $\ln(Q^2 / Q_0^2)$ in Eq. (2.159), we see complete agreement between the exponents in the two cases. The prefactor of Eq. (3.91) is different from what one would obtain in Eq. (2.159), since here we are calculating a different quantity (the BFKL Green function) from the gluon distribution calculated in Chapter 2.

The agreement thus found between the DLA limits of BFKL and DGLAP is, of course, simply a self-consistency check: since the former resums powers of $\alpha_s \ln 1/x$ keeping the functions of transverse momentum exact while the latter resums powers of $\alpha_s \ln Q^2$ keeping functions of Bjorken x exact, they should include the same powers of $\alpha_s \ln 1/x \ln Q^2$. This result was illustrated in Table 3.1 in Section 3.1; we have proven it explicitly here.

We can also think of BFKL evolution as a property of the hadronic (or onium for the case at hand) light cone wave function, constructed similarly to the DGLAP evolution in the previous chapter. In essence one can absorb the shaded rectangle in Fig. 3.5, which includes the ladder in Fig. 3.12, into the light cone wave function of one onium. While we will demonstrate this explicitly in Chapter 4, here we absorb the BFKL evolution into the onium wave function by defining the *unintegrated gluon distribution* of an onium (cf. Eqs. (3.23) and (3.26)):

$$\phi(x_{Bj}, k_\perp^2) = \frac{\alpha_s C_F}{\pi} \int d^2 x_\perp \int_0^1 dz |\Psi(\vec{x}_\perp, z)|^2$$

$$\times \int \frac{d^2 l_\perp}{l_\perp^2} \left(2 - e^{-i\vec{l}_\perp \cdot \vec{x}_\perp} - e^{i\vec{l}_\perp \cdot \vec{x}_\perp}\right) G(\vec{k}_\perp, \vec{l}_\perp, y = \ln 1/x_{Bj}). \quad (3.92)$$

One can show that in the small-x LLA the unintegrated gluon distribution $\phi(x_{Bj}, k_\perp^2)$ is related to the gluon distribution function (2.66) by

$$\phi(x, Q^2) = \frac{\partial x G(x, Q^2)}{\partial Q^2}. \quad (3.93)$$

This implies that $\phi(x, k_\perp^2)$ counts the number of partons in a hadron at a given value of $k_\perp$ (and a given value of Bjorken x), unlike xG, which counts the number of partons with $k_\perp \le Q$. This provides a physical interpretation of $\phi(x, k_\perp^2)$ as the unintegrated gluon distribution.

Looking at Eq. (3.58), it is clear that the unintegrated gluon distribution $\phi(x_{Bj}, k_\perp^2)$ from Eq. (3.92) obeys the same BFKL evolution equation:

$$\frac{\partial \phi(x, k_\perp^2)}{\partial \ln(1/x)} = \frac{\alpha_s N_c}{\pi^2} \int \frac{d^2 q_\perp}{(\vec{k}_\perp - \vec{q}_\perp)^2} \left[\phi(x, q_\perp^2) - \frac{k_\perp^2}{2q_\perp^2} \phi(x, k_\perp^2)\right]. \quad (3.94)$$

By analogy with how we arrived at Eq. (3.78), for an unpolarized axially symmetric onium state we can write the solution of Eq. (3.94) as

$$\phi(x, k_\perp^2) = \int_{-\infty}^{\infty} \frac{dv}{2\pi} C_v \exp\left\{\frac{\alpha_s N_c}{\pi} \chi(0, v) \ln \frac{1}{x}\right\} k_\perp^{-1+2iv} \Lambda^{-1-2iv}, \quad (3.95)$$

with Λ some typical transverse momentum scale characterizing the onium (e.g. the inverse size of the onium) and C_v an unknown function determined by the initial conditions at $x = x_0$.

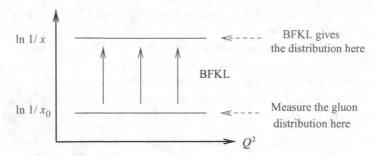

Fig. 3.15. The BFKL evolution in the $(\ln 1/x, Q^2)$-plane. A color version of this figure is available online at www.cambridge.org/9780521112574.

The BFKL equation for the distribution function gives us the evolution in x. This is demonstrated in the $(\ln 1/x, Q^2)$ plane in Fig. 3.15. Given the initial unintegrated gluon distribution $\phi(x_0, k_\perp^2)$ one can find C_ν and then, substituting it into Eq. (3.95), one obtains an expression for the unintegrated gluon distribution $\phi(x, k_\perp^2)$ at other values of x (in the LLA approximation). Comparing Fig. 3.15 with Fig. 2.21 we can see the essential difference between the BFKL evolution in x and the DGLAP evolution in Q^2.

If the transverse momentum of the gluons is not too large, i.e., $k_\perp \sim \Lambda$, while still much larger than Λ_{QCD}, we can evaluate the ν-integral in Eq. (3.95) in the diffusion approximation. Employing (3.84) and integrating yields an expression similar to Eq. (3.85):

$$\phi(x, k_\perp^2) \approx \frac{C_0}{2\pi} \frac{1}{k_\perp \Lambda} \sqrt{\frac{\pi}{14\zeta(3)\,\bar{\alpha}_s \ln(1/x)}} \left(\frac{1}{x}\right)^{\alpha_P - 1} \exp\left\{-\frac{\ln^2(k_\perp/\Lambda)}{14\zeta(3)\,\bar{\alpha}_s \ln(1/x)}\right\}.$$
(3.96)

We see that the unintegrated gluon distribution (and therefore, owing to (3.93), the regular gluon distribution xG as well) generated by the BFKL evolution grows as a power of $1/x$ at small x:

$$\phi(x, k_\perp^2) \sim \left(\frac{1}{x}\right)^{\alpha_P - 1},$$
(3.97)

with $\alpha_P - 1$ given by Eq. (3.86). This should be contrasted with the growth of the gluon distribution at small x resulting from DGLAP evolution, as shown in Eq. (3.3). The small-x growth of a gluon distribution given by BFKL evolution is much faster than that given by the DGLAP equation.

Since BFKL evolution does not impose any transverse momentum ordering, the partons generated by the evolution (3.94) have comparable transverse momenta, as given by Eq. (3.60b), and, according to Eq. (3.61), they are ordered in x,

$$x_1 \ll x_2 \ll \cdots \ll x_n \ll 1,$$
(3.98)

since $y = \ln(P^-/k^-) = \ln 1/x$ for the onium moving in the light cone minus direction.

It is instructive to consider the spatial distribution of the partons generated by BFKL evolution. Since the typical transverse size (wavelength) of a parton is $x_\perp \approx 1/k_\perp$, we see

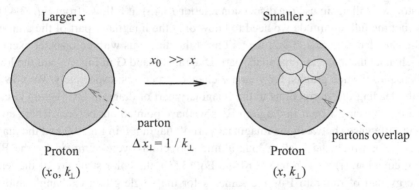

Fig. 3.16. Transverse coordinate space representation of BFKL evolution. The blobs indicate gluons. (Reprinted from Jalilian-Marian and Kovchegov (2006), with permission from Elsevier.) A color version of this figure is available online at www.cambridge.org/9780521112574.

that Eq. (3.60b) implies that

$$x_{1\perp} \sim x_{2\perp} \sim \cdots \sim x_{n\perp}, \tag{3.99}$$

that is, BFKL evolution in, say, a proton wave function generates partons of roughly the same transverse size. At the same time, owing to Eq. (2.56) the ordering in Bjorken x simply indicates that gluons with smaller x have a larger typical longitudinal spread. The BFKL evolution in transverse coordinate space is shown in Fig. 3.16. As Bjorken x decreases, more gluons are generated in the hadron wave function. The partons (gluons) produced are of roughly the same transverse size. (Compare this with the DGLAP evolution in Fig. 2.22.) As a result, after a sufficient number of gluons is generated they may start to overlap with each other in transverse space, as shown in the right-hand panel of Fig. 3.16. This leads to the creation of regions of high parton density in the hadronic wave functions. Therefore one can think of the BFKL equation as a "high parton density machine": it results in high parton density in the hadronic wave functions. This is what makes BFKL (and small-x physics in general) interesting, but it also leads to some problems for BFKL evolution, which we will discuss below.

3.3.5 Bootstrap property of the BFKL equation*

Let us now explain the reggeization of a t-channel gluon in the BFKL ladder. There is a widespread belief in the community that gluon reggeization may be a fundamental property of high energy QCD: thus, while gluon reggeization is not important for the material we present below, no book on high energy QCD would be complete without the presentation of this remarkable feature of strong interactions.

In Sec. 3.3.2 we found the lowest-order correction to the one-gluon exchange amplitude in Fig. 3.3. This correction turned out to be rather simple and is given by Eqs. (3.51) and (3.52). To incorporate the higher-order leading-ln s corrections into the t-channel gluon propagator, we will again use the dispersion relation (3.43): it follows from Eq. (3.43) that to construct the full amplitude we need to know only the imaginary part of the amplitude. As in the calculation in Sec. 3.3.2, we will need the diagrams with color-octet exchanges in the t-channel that have an imaginary part. Diagrams F and G in Fig. 3.9 are the lowest-order examples of such graphs, for s- and u-channel processes respectively. We now need to find the leading-ln s corrections to the imaginary part of diagram F (diagram G can be obtained by replacing s by u in diagram F, as we have pointed out before). There are two important differences between the imaginary part of diagram F in Fig. 3.9 and the diagram in Fig. 3.4, for which the leading logarithmic corrections were summed by the BFKL evolution equation. (i) As can be seen from Eq. (3.50), the color structure of the leading high energy part of diagram F is the same as for the single-gluon exchange amplitude (3.17), which implies that at high energy the two t-channel gluons in diagram F are in the color-octet state. This is to be contrasted with the diagram in Fig. 3.4, which is a diagram for the total cross section; therefore, the two t-channel gluons in it are in the color-singlet state. (ii) Since the diagram in Fig. 3.4 is for the total cross section, there is no net momentum transfer between the onium states in the corresponding forward amplitude. At the same time, in diagram F there is no restriction on the momentum transfer, which is in general nonzero.

One may argue (see Fadin, Kuraev, and Lipatov (1976)) that, despite the two differences stated above, the leading-ln s corrections to diagram F can still be resummed using the BFKL-like ladder from Fig. 3.12. Items (i) and (ii) above simply require that now the ladder should be nonforward, with the t-channel state projected on the color-octet configuration. We start by rewriting the imaginary part of the color-octet t-channel-exchange $qq \to qq$ amplitude $M^{(8)}_{qq \to qq}$ as (cf. Eq. (3.46))

$$\text{Im}\, M^{(8)}_{qq \to qq}(\vec{q}_\perp) = 4\alpha_s^2 (t^b t^a)_{i'i} (t^b t^a)_{j'j} \delta_{\sigma_1 \sigma'_1} \delta_{\sigma_2 \sigma'_2} s$$

$$\times \int \frac{d^2 l_\perp}{l_\perp^2 (\vec{l}_\perp - \vec{q}_\perp)^2} \frac{d^2 l''_\perp}{l''^2_\perp (\vec{l}''_\perp - \vec{q}_\perp)^2} G^{(8)}\left(\vec{l}_\perp, \vec{l}''_\perp; \vec{q}_\perp; Y\right), \quad (3.100)$$

where $G^{(8)}(\vec{l}_\perp, \vec{l}''_\perp; \vec{q}_\perp; Y)$ is the color-octet nonforward analogue of the BFKL Green function defined in Eq. (3.26) above. The octet nonforward Green function $G^{(8)}$ is illustrated on the left-hand side of Fig. 3.17. Equation (3.46) representing diagram F in Fig. 3.9 is recovered by inserting

$$G^{(8)}\left(\vec{l}_\perp, \vec{l}''_\perp; \vec{q}_\perp; Y = 0\right) = l''^2_\perp (\vec{l}''_\perp - \vec{q}_\perp)^2 \delta^2 (\vec{l}_\perp - \vec{l}''_\perp) \qquad (3.101)$$

into Eq. (3.100).

By analogy with the construction of the BFKL equation we can construct the evolution equation for the octet Green function, following the prescription indicated in Fig. 3.17. First we note that the virtual corrections in the color-octet case are the same as they were in the color-singlet case of BFKL evolution, given by Eq. (3.55). The real emission in the case of BFKL evolution was obtained by squaring the Lipatov vertex (3.36); however, since now the process is nonforward, the t-channel gluons' momenta should be different in the two

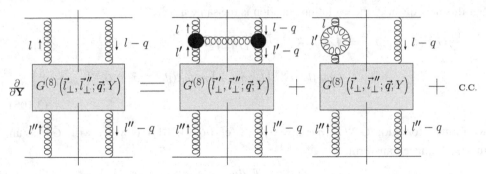

Fig. 3.17. Diagrammatic representation of the leading-ln s evolution of the color-octet nonforward Green function. The notation is the same as in Fig. 3.11.

factors of the Lipatov vertex. In the end we obtain the following evolution equation:

$$\frac{\partial}{\partial Y} G^{(8)}\left(\vec{l}_{\perp}, \vec{l}_{\perp}''; \vec{q}_{\perp}; Y\right) = \int d^2 l_{\perp}' K_{NF}^{(8)}\left(\vec{l}_{\perp}, \vec{l}_{\perp}'; \vec{q}_{\perp}\right) G^{(8)}\left(\vec{l}_{\perp}', \vec{l}_{\perp}''; \vec{q}_{\perp}; Y\right)$$

$$+ \omega_G(l_{\perp}) G^{(8)}\left(\vec{l}_{\perp}, \vec{l}_{\perp}''; \vec{q}_{\perp}; Y\right)$$

$$+ \omega_G(|\vec{l}_{\perp} - \vec{q}_{\perp}|) G^{(8)}\left(\vec{l}_{\perp}, \vec{l}_{\perp}''; \vec{q}_{\perp}; Y\right). \tag{3.102}$$

The terms containing $\omega_G(l_{\perp})$, which is defined in Eq. (3.52), are due to virtual corrections. The part of the nonforward color-octet BFKL kernel coming from real-gluon emission,

$$K_{NF}^{(8)}\left(\vec{l}_{\perp}, \vec{l}_{\perp}'; \vec{q}_{\perp}\right) = \frac{\alpha_s N_c}{4\pi^2}\left[\frac{l_{\perp}^2}{l_{\perp}'^2(\vec{l}_{\perp} - \vec{l}_{\perp}')^2} + \frac{(\vec{l}_{\perp} - \vec{q}_{\perp})^2}{(\vec{l}_{\perp}' - \vec{q}_{\perp})^2(\vec{l}_{\perp} - \vec{l}_{\perp}')^2} - \frac{q_{\perp}^2}{l_{\perp}'^2(\vec{l}_{\perp}' - \vec{q}_{\perp})^2}\right],$$

$$\tag{3.103}$$

results from the (nonforward) square of the Lipatov vertex (3.36) divided by the propagators of the two t-channel gluons below the vertex (the l'- and $(l' - q)$-lines in the first diagram on the right-hand side of Fig. 3.17). In arriving at Eq. (3.102) we have also employed the fact that, for the color-octet two-gluon state in the t-channel, the color factor generated by the real part of the kernel is equal to $N_c/2$ instead of N_c, which one has for the color-singlet (BFKL) evolution. The initial condition for Eq. (3.102) is given by Eq. (3.101).

While the nonforward ($t \neq 0$) BFKL equation (3.102) appears to be more complicated than the forward ($t = 0$) BFKL equation (3.58), in fact its solution is rather straightforward if one is just interested in the resulting scattering amplitude. We begin by defining

$$\bar{G}^{(8)}\left(\vec{l}_{\perp}; \vec{q}_{\perp}; Y\right) \equiv \int \frac{d^2 l_{\perp}''}{l_{\perp}''^2(\vec{l}_{\perp}'' - \vec{q}_{\perp})^2} G^{(8)}\left(\vec{l}_{\perp}, \vec{l}_{\perp}''; \vec{q}_{\perp}; Y\right). \tag{3.104}$$

For this new quantity the evolution equation is the same as (3.102):

$$\frac{\partial}{\partial Y}\bar{G}^{(8)}\left(\vec{l}_\perp;\vec{q}_\perp;Y\right) = \int d^2l'_\perp K_{NF}^{(8)}\left(\vec{l}_\perp,\vec{l}'_\perp;\vec{q}_\perp\right)\bar{G}^{(8)}\left(\vec{l}'_\perp;\vec{q}_\perp;Y\right)$$

$$+\omega_G(l_\perp)\bar{G}^{(8)}\left(\vec{l}_\perp;\vec{q}_\perp;Y\right) + \omega_G(|\vec{l}_\perp - \vec{q}_\perp|)\bar{G}^{(8)}\left(\vec{l}_\perp;\vec{q}_\perp;Y\right)$$

(3.105)

with initial condition $\bar{G}^{(8)}\left(\vec{l}_\perp;\vec{q}_\perp;Y=0\right) = 1$ (cf. Eq. (3.101)). Now we write $\bar{G}^{(8)}$ as an inverse Laplace transform,

$$\bar{G}^{(8)}\left(\vec{l}_\perp;\vec{q}_\perp;Y\right) = \int \frac{d\omega}{2\pi i}e^{\omega Y}\bar{G}_\omega^{(8)}\left(\vec{l}_\perp;\vec{q}_\perp\right),$$

(3.106)

where, as usual, the ω-integration runs parallel to the imaginary axis in the complex ω-plane to the right of all the singularities of the integrand. Substituting (3.106) into Eq. (3.105) and performing a Laplace transform on both sides of the resulting equation yields the *bootstrap equation*

$$\left[\omega - \omega_G(l_\perp) - \omega_G(|\vec{l}_\perp - \vec{q}_\perp|)\right]\bar{G}_\omega^{(8)}\left(\vec{l}_\perp;\vec{q}_\perp\right)$$

$$= 1 + \int d^2l'_\perp K_{NF}^{(8)}\left(\vec{l}_\perp,\vec{l}'_\perp;\vec{q}_\perp\right)\bar{G}_\omega^{(8)}\left(\vec{l}'_\perp;\vec{q}_\perp\right).$$

(3.107)

(We have also made use of the initial condition $\bar{G}^{(8)}\left(\vec{l}_\perp;\vec{q}_\perp;Y=0\right) = 1$.) It is easy to see that

$$\bar{G}_\omega^{(8)}\left(\vec{l}_\perp;\vec{q}_\perp\right) = \frac{1}{\omega - \omega_G(q_\perp)}$$

(3.108)

in fact solves Eq. (3.107) for the kernel given by Eq. (3.103). Substituting Eq. (3.108) into Eq. (3.106) we obtain the solution of Eq. (3.105):

$$\bar{G}^{(8)}\left(\vec{l}_\perp;\vec{q}_\perp;Y\right) = e^{\omega_G(q_\perp)Y}.$$

(3.109)

Using Eqs. (3.109) and (3.104) in Eq. (3.100) we obtain the imaginary part of the color-octet exchange amplitude,

$$\text{Im}M_{qq\to qq}^{(8)}(\vec{q}_\perp) = 4\alpha_s^2(t^bt^a)_{i'i}(t^bt^a)_{j'j}\delta_{\sigma_1\sigma'_1}\delta_{\sigma_2\sigma'_2}s\int\frac{d^2l_\perp}{l_\perp^2(\vec{l}_\perp - \vec{q}_\perp)^2}e^{\omega_G(q_\perp)Y},$$

(3.110)

where, as usual in the leading-ln s approximation, we have $Y = \ln(s/m_\perp^2)$ with $m_\perp$ some transverse momentum scale, the exact value of which is outside the precision of the approximation.

To obtain the complete color-octet exchange amplitude we substitute Eq. (3.110) into Eq. (3.43) (adding the u-channel contribution obtained from Eq. (3.110) on replacing s by u). Integrating in the leading-logarithmic approximation, we obtain the high energy color-octet exchange amplitude

$$M_{qq\to qq}^{(8)}(\vec{q}_\perp) = M_{qq\to qq}^0(\vec{q}_\perp)e^{\omega_G(q_\perp)Y},$$

(3.111)

where M^0 is the single-gluon exchange amplitude given in Eq. (3.17). Note that Eq. (3.111) is in effect the same as Eq. (3.53), which we employed in deriving the BFKL evolution equation.

Equation (3.107) represents the so-called "bootstrap" idea, which states that the evolution of the reggeized gluon is given by the evolution of the color-octet t-channel state of two gluons. It is referred to as the *bootstrap equation*. It has been conjectured that the bootstrap equation (3.107) with its rather simple solution (3.108) holds at any order in α_s. (Indeed both $K_{NF}^{(8)}$ and ω_G receive corrections at higher orders in α_s: the conjecture states that these corrections leave Eqs. (3.107) and (3.108) in exactly the same form as shown above, only modifying $K_{NF}^{(8)}$ and ω_G in them.) So far the conjecture has been verified at the two lowest orders in α_s: the leading order-α_s result is presented here, and the validity of the bootstrap equation at order α_s^2 has been shown by Fadin, Kotsky, and Fiore (1995, 1996).

We see that the bootstrap equation (3.107) with its solution leading to Eq. (3.111) implies that the nonforward BFKL equation for the color-octet state of two gluons should lead to a reggeized gluon with the Regge trajectory $\alpha_G(q_\perp) = 1 + \omega_G(q_\perp)$. This observation completes the proof that in high energy scattering a t-channel gluon should be treated as a reggeized gluon whose spin depends on its transverse momentum.

3.3.6 Problems of BFKL evolution: unitarity and diffusion

The BFKL equation represents an important step towards understanding the high energy asymptotics of QCD. Nonetheless, as for every major scientific advance, the BFKL equation raises some important questions, which we will describe in this section. In particular we will show that as the collision energy increases (i) the leading-logarithmic BFKL equation violates unitarity and (ii) the transverse momenta of the gluons inside the BFKL ladder tend to drift to both the UV and IR, the latter drift eventually leading to a violation of the assumption of the perturbative nature of the interactions.

The Froissart–Martin bound

We begin our presentation of the unitarity bound with a discussion of the black disk limit. Imagine the high energy scattering of a point particle on a "black disk" of radius R. Using the language of nonrelativistic quantum mechanics one can think of the black disk as an infinite potential well occupying a spherical region of space. It can then be shown that the total cross section for the scattering of the point particle from the disk is limited from above by

$$\sigma_{tot} \leq 2\pi R^2 \tag{3.112}$$

(see Landau and Lifshitz (1958), vol. 3, Chapter 131, and the discussion in Appendix B). The total cross section can be as large as twice the geometric cross sectional area of the disk: this doubling is due to Babinet's principle in optics, which states that the diffractive patterns of complementary screens are identical (see Jackson (1998) or Landau and Lifshitz (1958), vol. 2, Chapter 61). In optics Babinet's principle implies that the amount of light diffracted by a screen is equal to the amount of light it absorbs. For very high energy scattering it implies that, when the scattering occurs from a black disk, the elastic (σ_{el}) and inelastic (σ_{inel}) cross sections are equal. Since the inelastic cross section is equal to the cross sectional area of the disk πR^2, we have $\sigma_{el} = \sigma_{inel} = \pi R^2$, so that the total cross section is $\sigma_{tot} = \sigma_{inel} + \sigma_{el} = 2\pi R^2$.

Our derivation of the Froissart–Martin bound will incorporate the argument put forward by Heisenberg (1952, 1939) with the proof devised by Froissart (1961) and Martin (1969). Consider hadron–hadron scattering at impact parameter b. Let us assume that b is large enough that the black-disk limit described above has not been reached. Inspired by the above examples of the BFKL equation and the Low–Nussinov pomeron, we may assume that the interaction between the hadrons is accomplished though an exchange of one or several particles, so that the cross section grows as some positive power Δ of the energy: $\sigma \sim s^\Delta$. At the same time the strength of the interaction should fall off as we increase b: the slowest physically possible fall off is the exponential $e^{-2m_\pi b}$, where m_π is the mass of the lightest QCD bound state, the pion.[7] (As the pion has negative parity, the exchange of a single pion cannot contribute to the total cross section, hence we need to exchange two pions, one in the amplitude and the other in the complex-conjugate amplitude.) We thus have a probability p of interaction that scales with energy and impact parameter as follows:

$$p \sim s^\Delta e^{-2m_\pi b}. \tag{3.113}$$

The interaction gets strong when the probability is of order 1. In fact, for p of order 1 the black-disk-limit behavior should begin to set in. Thus the upper limit on the radius of the black disk can be determined by requiring that $p \approx 1$, which, as we can see from Eq. (3.113), occurs at impact parameter b^* defined by

$$s^\Delta e^{-2m_\pi b^*} = 1, \tag{3.114}$$

which gives

$$R = b^* \sim \frac{\Delta}{2m_\pi} \ln s. \tag{3.115}$$

Since b^* is the upper bound on the black-disk radius, the total cross section, dominated by the black-disk contribution, is then limited by $2\pi R^2 = 2\pi b^{*2}$, yielding eventually

$$\sigma_{tot} \leq \frac{\pi \Delta^2}{2m_\pi^2} \ln^2 s. \tag{3.116}$$

We conclude that the total cross section in QCD cannot grow faster than the logarithm of energy squared. Equation (3.116) is known at *the Froissart–Martin bound* and was first rigorously proven by Froissart (1961) and Martin (1969) (see also Lukaszuk and Martin (1967)).

---

[7] As discussed by Nussinov (2008), it is possible that for realistic estimates the pion mass in this exponential should in fact be replaced by the mass of the lightest glueball (a QCD bound state having no valence quarks) since, as we have seen above, gluons dominate in high energy interactions.

As we saw above (see e.g. Eq. (3.87)), the BFKL equation in the diffusion approximation implies that the total cross section grows as a power of the energy,

$$\sigma_{tot}^{BFKL} \sim s^{\alpha_P - 1}, \tag{3.117}$$

which clearly violates the Froissart–Martin bound (3.116). Things do not get much better in the double logarithmic limit of the BFKL equation, which, according to Eq. (3.91), gives

$$\sigma_{tot}^{DLA\ BFKL} \sim \exp\left\{ 2\sqrt{\bar{\alpha}_s \ln s \ln(l_\perp^2/l_\perp'^2)} \right\}, \tag{3.118}$$

where $l_\perp$ and $l'_\perp$ are the momentum scales at the two ends of the BFKL ladder, with $l_\perp \gg l'_\perp$ in the DLA. The energy growth of the cross section in Eq. (3.118) is exponential in $\sqrt{\ln s}$, and, as such, is much faster than any power of $\ln s$: therefore, DLA BFKL also violates the Froissart–Martin bound. Note that since DLA BFKL is equivalent to DLA DGLAP, the DGLAP evolution also violates unitarity, making this an inherent problem of standard perturbative QCD: no matter how high the larger perturbative scale $l_\perp$ is, at sufficiently high energy s, unitarity will still be violated, as follows from Eq. (3.118). We thus conclude that unitarity violation happens at perturbatively large momentum scales, where perturbative QCD is still applicable. Thus, it is natural to expect that the resolution of this problem should also happen through a QCD perturbative mechanism. We will discuss shortly how a nonlinear evolution equation was proposed by Gribov, Levin, and Ryskin (GLR) to remedy this problem of the BFKL evolution (Gribov, Levin, and Ryskin 1983).

One may indeed argue that in deriving the Froissart–Martin bound above we have used the fact that QCD is a confining theory with bound states such as pions, which one certainly does not see in the perturbative calculations leading to the BFKL equation. Therefore, since the QCD mass gap m_π is not present in perturbation theory (which has a zero mass gap), one should not expect BFKL evolution to satisfy the Froissart–Martin bound. This argument is indeed correct; however, as we will show below, the BFKL equation can also be written in impact parameter space. As we argue in Appendix B, in impact parameter space the high energy cross sections are given by

$$\sigma_{tot} = 2 \int d^2 b \left[1 - \mathrm{Re}\, S(s, \vec{b}_\perp) \right], \tag{3.119a}$$

$$\sigma_{el} = \int d^2 b \left| 1 - S(s, \vec{b}_\perp) \right|^2, \tag{3.119b}$$

$$\sigma_{inel} = \int d^2 b \left[1 - |S(s, \vec{b}_\perp)|^2 \right], \tag{3.119c}$$

with $S(s, \vec{b}_\perp)$ the forward matrix element of the scattering S-matrix. Since at high energy $S \to 0$ we see that $d\sigma_{tot}/d^2 b_\perp \leq 2$, which is equivalent to the black-disk limit (3.112). The BFKL-dominated total cross section at a fixed impact parameter b grows as a power of the energy s, eventually violating the bound $d\sigma_{tot}/d^2 b_\perp \leq 2$ at very high energy. Thus the BFKL equation violates unitarity not only through the fast growth of the black-disk radius but also by the fact that the cross section at each impact parameter becomes larger than the black-disk bound $d\sigma_{tot}/d^2 b_\perp \leq 2$. While the former problem cannot be remedied in QCD

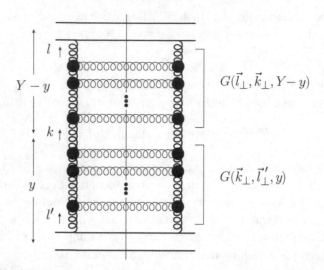

Fig. 3.18. The BFKL ladder, with a t-channel gluon with momentum $\vec{k}_\perp$ singled out for study of its momentum distribution. The vertical soild line denotes a cut.

perturbation theory, one may hope to remedy the latter problem; as we will see below it is cured by the use of nonlinear evolution equations.

We conclude that the BFKL equation violates unitarity (or, more precisely, the black-disk limit). Therefore one would expect it to be modified at very high energies, in order to satisfy those bounds. This is the first problem of BFKL evolution.

Diffusion into the infrared

The second problem of BFKL evolution becomes apparent if we look "inside" the BFKL ladder. To do this, let us pick a particular gluon in the ladder sufficiently far from the ends of the ladder. This is illustrated in Fig. 3.18, where we are considering the transverse momentum distribution of a t-channel gluon with momentum $\vec{k}_\perp$ and rapidity y (of adjacent rungs) inside the BFKL ladder. The ladder can be split into two sub-ladders, each of which is a BFKL ladder in its own right, as indicated by the right-hand square brackets in Fig. 3.18. The BFKL Green function (3.80) can correspondingly be written as a convolution of two BFKL Green functions,

$$G\left(\vec{l}_\perp, \vec{l}_\perp', Y\right) = \int d^2 k_\perp G\left(\vec{l}_\perp, \vec{k}_\perp, Y - y\right) G\left(\vec{k}_\perp, \vec{l}_\perp', y\right), \qquad (3.120)$$

where the net rapidity interval is Y and the transverse momenta at the ends of the original ladder are $\vec{l}_\perp$ and $\vec{l}_\perp'$.

Let us assume that all the transverse momenta involved are comparable, $l_\perp \sim l_\perp' \sim k_\perp$, so that we can use the diffusion approximation (3.85). As follows from Eq. (3.120), the $k_\perp$-distribution $dn/d^2 k_\perp$ for the t-channel gluon in Fig. 3.18 is given by the product of the

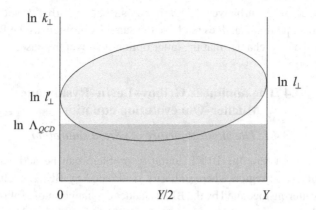

Fig. 3.19. The $\ln k_\perp$-distribution of the momentum of a t-channel gluon inside the BFKL ladder (shaded oval). For a large enough rapidity interval Y the $\ln k_\perp$-distribution overlaps with the nonperturbative region $k_\perp \lesssim \Lambda_{QCD}$ (the shaded rectangle). A color version of this figure is available online at www.cambridge.org/9780521112574.

two BFKL Green functions and, in the diffusion approximation (3.85), it is proportional to

$$\frac{dn}{d \ln k_\perp^2} = k_\perp^2 \, G\left(\vec{l}_\perp, \vec{k}_\perp, Y - y\right) G\left(\vec{k}_\perp, \vec{l}_\perp', y\right)$$

$$\sim \exp\left\{-\frac{\ln^2(l_\perp/k_\perp)}{14\zeta(3)\,\bar{\alpha}_s(Y-y)} - \frac{\ln^2(k_\perp/l_\perp')}{14\zeta(3)\,\bar{\alpha}_s\,y}\right\}. \tag{3.121}$$

We see that the distribution in $\ln k_\perp^2$ is Gaussian. The width of the distribution depends on the rapidity, reaching a maximum at $y = Y/2$. As we increase the total energy (or, equivalently, the net rapidity interval Y), the $\ln k_\perp^2$-distribution at $y \approx Y/2$ gets broader: the values of $k_\perp$ deviate more and more from the original momenta $l_\perp$ and $l_\perp'$ at the ends of the ladder. This feature of BFKL evolution is potentially dangerous in QCD, for the following reason. Suppose that to justify the use of perturbative QCD we choose $l_\perp \sim l_\perp' \gg \Lambda_{QCD}$. As we increase the net rapidity interval Y the transverse momentum in the ladder diffuses away from $l_\perp \sim l_\perp'$, fluctuating both into the UV and IR, so that at $y = Y/2$ we have

$$l_\perp e^{-\text{const} \times \sqrt{Y/2}} \lesssim k_\perp \lesssim l_\perp e^{\text{const} \times \sqrt{Y/2}}. \tag{3.122}$$

(We have assumed for simplicity that $l_\perp = l_\perp'$ and that $\text{const} = \sqrt{(7/2)\zeta(3)\bar{\alpha}_s}$ in Eq. (3.122).) Clearly, for large enough energies or rapidities Y, the lower limit of the $k_\perp$-range in Eq. (3.122) may reach Λ_{QCD}, invalidating the applicability of perturbative QCD to the problem (Bartels 1993b, Bartels, Lotter, and Vogt 1996). The $\ln k_\perp^2$-distribution is plotted in Fig. 3.19; owing to its cigar-like shape, it is sometimes referred to as the *Bartels cigar* (Bartels 1993b).

The problem of BFKL evolution becoming nonperturbative at high Y values implies that we cannot use the BFKL equation at arbitrarily high energies: its applicability is thus limited. Therefore, either the true high energy asymptotics of QCD is nonperturbative and cannot be described by perturbative QCD or it may still be perturbative and described by a different evolution equation that does not have the same IR problem as the BFKL equation. We will show in the next chapter that the latter option is in fact the case.

3.4 The nonlinear Gribov–Levin–Ryskin and Mueller–Qiu evolution equation

3.4.1 The physical picture of parton saturation

In order to understand how the BFKL unitarity problem can be addressed, let us first determine the kinematic region where unitarity is violated by BFKL evolution. Consider onium–onium scattering mediated by the BFKL ladder exchange studied above. The expression for the total onium–onium scattering cross section can be obtained by substituting the LLA Green function (3.85) into Eq. (3.26). At the moment we are not interested in the exact result (see Exercise 3.5 at the end of the chapter) and will only note that in Eq. (3.26) the typical momenta are $l_\perp \approx 1/x_{1\perp}$ and $l'_\perp \approx 1/x_{2\perp}$. The total cross section is then

$$\sigma_{tot}^{onium+onium} \sim \alpha_s^2 x_{1\perp} x_{2\perp} e^{(\alpha_P-1)Y}, \tag{3.123}$$

where we have kept only the most physically important factors, those depending on the total rapidity Y and on the sizes of the onia. We know, however, that the cross section should be bounded from above by the black-disk limit, $\sigma_{tot}^{onium+onium} \leq 2\pi R^2$. While the radius of the black disk R grows logarithmically with energy, as shown in Eq. (3.115), for the moment we will forget about this growth since it is much slower than the power-of-energy growth of the total cross section in Eq. (3.123), and will fix R to be of the order of a typical hadronic radius, $R \approx r_h$. In fact let us assume that our onium–onium scattering models a DIS event, in which the first onium comes from the decay of a virtual photon and the other mimics the proton (we will present this dipole picture of DIS in more detail in the next chapter). Then $x_{2\perp} \approx r_h$ and $x_{1\perp} \approx 1/Q$, with Q the virtuality of the photon (the only scale at the photon end of the ladder). Imposing the black-disk limit on Eq. (3.123) yields

$$\alpha_s^2 \frac{r_h}{Q} e^{(\alpha_P-1)Y} \leq 2\pi r_h^2. \tag{3.124}$$

The equality is reached at the *saturation scale* $Q = Q_s$ given by

$$Q_s \sim \alpha_s^2 \Lambda_{QCD} e^{(\alpha_P-1)Y} = \alpha_s^2 \Lambda_{QCD} \left(\frac{1}{x}\right)^{\alpha_P-1}, \tag{3.125}$$

where, for simplicity, we have replaced the typical hadronic size r_h with $1/\Lambda_{QCD}$: the two scales differ only by a constant coefficient, irrelevant for our rough estimates. The saturation momentum Q_s is a new dimensional scale in the problem (Gribov, Levin, and Ryskin 1983, Mueller and Qiu 1986, McLerran and Venugopalan 1994a).

We conclude that a violation of unitarity occurs for $Q < Q_s$. Note that for very small Bjorken x (large Y) the saturation scale can be much larger than the confinement scale, $Q_s \gg \Lambda_{QCD}$. This implies that the violation of unitarity starts at short distances of the

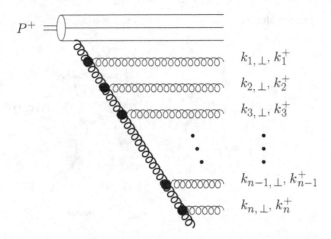

Fig. 3.20. The BFKL evolution of Fig. 3.13 shown here as a time-ordered gluon cascade in the proton wave function, in the IMF/Bjorken or Breit frames.

order of $1/Q_s$, still in the domain of validity of perturbative QCD. Therefore, the unitarity problem of BFKL evolution has to be solved in the framework of perturbative QCD.

To address this unitarity problem we first need to understand what goes wrong with BFKL evolution. This is easier to do if we absorb the BFKL ladder into the wave function of a hadron in DIS in the Breit frame, as for the DGLAP evolution in the previous chapter. (Formally, the BFKL light cone wave function is constructed in the next chapter.) The small-x evolution then appears as a cascade in the proton's wave function, shown in Fig. 3.20 (compare with Figs. 2.25 and 2.10). The fast proton will decay into a system of partons long before the interaction with the virtual photon, which is at rest. The time ordering of emissions is given by Eq. (2.156) in the DLA DGLAP case (the proton is moving in the light cone plus direction)

$$x_1^+ \gg x_2^+ \gg \cdots \gg x_n^+, \tag{3.126}$$

and this would still be valid: in the general BFKL case the DGLAP ordering of transverse momenta (2.149) is replaced by the comparability of all transverse momenta, as given by Eq. (3.60b), with the ordering (2.150) of longitudinal momenta from DLA DGLAP still in place (see Eq. (3.60a)). The latter fact insures that the typical lifetimes of gluon fluctuations, given by $x_i^+ \approx k_i^+/k_{\perp i}^2$, are still ordered as in Eq. (3.126) and as shown in Fig. 3.20. Thus, in terms of time-ordering, the BFKL cascade is quite similar to the DGLAP cascade.

During the long time of parton-cascade evolution a large number of "wee" partons (gluons) are created in the proton's wave function, as in Fig. 3.20, of order $(1/x)^{\alpha_p - 1}$ for BFKL evolution. Each gluon in the cascade is emitted from a pre-existing (larger-x) gluon in the proton's wave function. Each such "wee" parton interacts with the virtual photon over a very short time. This interaction destroys the coherence of the partons (which are mostly gluons at low x). The further fate of the partons is not important to us since any possible interaction in the final state will not change the total cross section of the deep inelastic

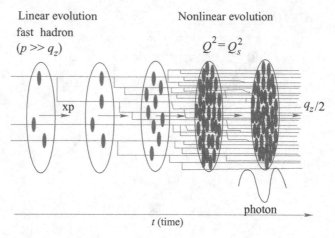

Fig. 3.21. Parton cascade in the Breit frame for DIS (see Sec. 2.3 for the definition of this frame). Both quarks and gluons are denoted by straight lines for simplicity. At the end of the cascade, just before interaction with a virtual photon of wavelength $1/q_z$, the quarks have momentum $q_z/2$ along the collision axis. A color version of this figure is available online at www.cambridge.org/9780521112574.

process. A space–time picture of the QCD evolution that we have just outlined is given in Fig. 3.21; here the straight lines denote both quarks and gluons. The QCD evolution (both DGLAP and BFKL) leads to an increase in the number of "wee" partons with energy. This is a very natural result since both the DGLAP and BFKL equations take into account the emission of partons (along with the virtual corrections). Each emission increases the number of partons at lower energy (i.e., at lower x).

The BFKL evolution creates partons of roughly the same transverse size, as indicated in Fig. 3.16 and discussed in Sec. 3.3.4. As we decrease x the parton density grows. However, at some very small (critical) $x = x_{cr}$, where

$$x_{cr} \sim \left(\frac{Q}{\alpha_s^2 \Lambda_{QCD}} \right)^{1/(\alpha_P - 1)}, \tag{3.127}$$

corresponding to $Q_s(x_{cr}) = Q$, the density of partons in the transverse plane becomes so large that the wave functions of the partons start to overlap, as shown in Fig. 3.21. For such a densely populated system we need to take into account interactions between the partons (Gribov, Levin, and Ryskin 1983).

We conclude that for $Q^2 < Q_s^2$ we should write down a new evolution equation that includes interactions between the partons. This new evolution should slow down and finally stop (*saturate*) the increase in the number of "wee" partons, leading to the saturation of the parton density (Gribov, Levin, and Ryskin 1983). We want to stress again that, since $Q_s^2 \gg \Lambda_{QCD}^2$, the value of the QCD coupling is small and therefore we can apply perturbative QCD methods in the region where unitarity corrections become important.

Our discussion of parton densities is summarized in Fig. 3.22, in our first attempt at constructing a map of high-energy QCD. Figure 3.22 displays the distributions of partons

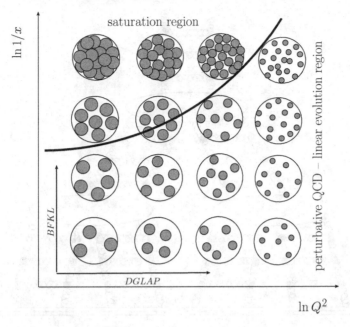

Fig. 3.22. The parton distribution in the transverse plane as a function of $\ln 1/x$ and $\ln Q^2$. The bold curve gives the saturation scale $Q_s(x)$.

in the proton's transverse plane as a function of Q^2 and x, with the saturation scale (3.125) shown by the bold curve. One can see that this saturation momentum curve divides the entire kinematic region in two parts: the region of perturbative QCD with linear evolution for the parton densities (DGLAP, BFKL) and the saturation domain, in which the parton density is large but the running QCD coupling is still small ($\alpha_s(Q_s^2) \ll 1$). Therefore we need to develop a theoretical description for a system of partons that interact with each other weakly but whose number is so large that we cannot apply the regular methods of perturbative QCD with linear evolution for the parton densities to describe them.

3.4.2 The GLR–MQ equation

To tackle the BFKL unitarity problem and to understand how the growth of the gluon distribution can be tamed, Gribov, Levin, and Ryskin (1983) (GLR) considered the gluon distribution functions of a "dense" proton or a nucleus. By a "dense" proton we mean a proton filled with various sources of color charge (sea quarks and gluons) that were pre-created in the proton's wave function by some nonperturbative mechanism. Gribov, Levin, and Ryskin argued that for such systems multiple BFKL ladder exchanges may become important; indeed, as we saw in Eq. (3.124), the BFKL equation violates unitarity when the cross section per unit impact parameter, $d\sigma/d^2b$, becomes of order 1. Hence the contribution of double BFKL ladder exchange at this point should be comparable with that of single BFKL ladder exchange. Moreover, owing to the high density of partons, ladder mergers should also be possible as they evolve away from the proton or nucleus in rapidity.

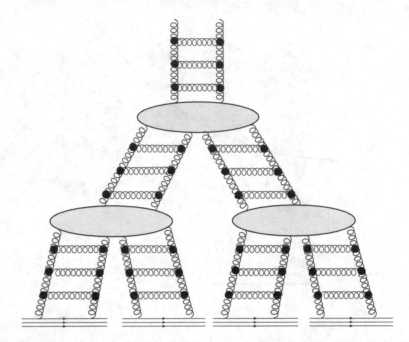

Fig. 3.23. An example of a "fan" diagram resummed by the GLR–MQ equation.

Owing to the large number of partons in the proton or nucleus, the ladders can all connect to different sources of color charge and interact with the proton or nucleus independently. (It may be instructive to mention that in the case of a nuclear target this assumption can be proven to be correct (Schwimmer 1975, Kovchegov 1999, 2000) since the nucleus has a parameter, the atomic number A, allowing one to justify the independent-interaction approximation.) Since we are interested in the gluon distribution, which is a correlation function for two gluon fields, these multiple ladders should all merge in the end into a single ladder, leading to the so-called "fan" diagrams. An example of a fan diagram is shown in Fig. 3.23. There, multiple BFKL ladders start from different quarks and gluons in the proton or nucleus; these are shown by straight lines at the bottom of Fig. 3.23. Owing to the high density of the gluon fields, the ladders cannot stay independent for long. As the energy increases so does the gluon density, eventually leading to *merging* of the ladders, as shown in Fig. 3.23.

Indeed, BFKL ladders can also first split and then recombine, leading to the so-called *pomeron loop* diagrams, one of which is shown in Fig. 3.24. In a purely perturbative picture of the process, such corrections are not enhanced by large parton densities in the proton or nucleus and are therefore subleading. In the original GLR work it was argued that, since the amplitude of a pomeron loop increases with the rapidity interval covered by the loop, the dominant contribution of the pomeron loop diagram as pictured in Fig. 3.24 comes from a loop covering the whole interval in rapidity. It was then argued that in such a contribution the pomeron recombination vertex would be close to the proton in rapidity and, if the proton is nonperturbatively strongly coupled, the vertex can then be effectively absorbed into the

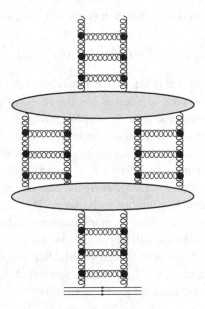

Fig. 3.24. A pomeron loop diagram.

proton. This would make the resulting diagram into a pomeron "fan" diagram, with some complicated nonperturbative interaction with the proton. The latter interaction would then be modeled by independent scatterings of pomeron ladders on different constituents in the proton. We will present the physical parameters justifying the fan diagram approximation in a more systematic way in the next chapter.

Ladder recombination in fan diagrams is described by effective ladder merger vertices, denoted by the ovals in Fig. 3.23. These vertices are called *triple pomeron vertices*, since they connect three different ladders (BFKL pomerons). For their calculation we refer the reader to Bartels and Wusthoff (1995) and Bartels and Kutak (2008) and references therein. If one views the DIS process in the infinite momentum frame described above, the ladder mergers appear to be exactly the partonic interactions that enter when the gluon density becomes high. It turns out that such interactions are not simple individual parton mergers but are, in fact, of a more collective nature, as they consist of ladder mergers, i.e., mergers of different linear parton evolutions.

Note that the diagram in Fig. 3.23 contains only triple pomeron vertices, i.e., $2 \rightarrow 1$ ladder mergers. In the next chapter we will argue that while all higher-order ($3 \rightarrow 1, 4 \rightarrow 1$, etc.) mergers exist, in the LLA they are suppressed by powers of N_c^2 and can therefore be neglected in the large-N_c (fixed-$\alpha_s N_c$) limit (Bartels and Wusthoff 1995, Braun 2000a). Before those results became known, in their original work Gribov, Levin, and Ryskin (1983) had suggested that, before the energy becomes sufficiently high for all nonlinear effects to become important, there could be an intermediate energy region where the physics of gluon distributions is dominated by $2 \rightarrow 1$ ladder recombination. This recombination would bring in a quadratic correction to the linear BFKL equation for the unintegrated gluon distribution

(3.94), leading to the GLR evolution equation (Gribov, Levin, and Ryskin 1983)

$$\frac{\partial \phi(x, k_\perp^2)}{\partial \ln(1/x)} = \frac{\alpha_s N_c}{\pi} \int d^2 l_\perp K_{BFKL}(k, l)\, \phi(x, k_\perp^2) - \frac{\alpha_s^2 N_c \pi}{2 C_F S_\perp} \left[\phi(x, k_\perp^2) \right]^2, \quad (3.128)$$

with the LO BFKL kernel defined in Eq. (3.62). In writing down Eq. (3.128), for simplicity we have assumed that the proton or nucleus has the shape of a cylinder oriented along the beam axis with cross sectional area $S_\perp = \pi R^2$; however, the GLR equation (3.128) can be generalized easily to any shape of proton or nucleus, as the impact parameter ($b_\perp$) integration can be carried out separately for each ladder (Gribov, Levin, and Ryskin 1983). As expected, the linear term in Eq. (3.128) is equivalent to the BFKL equation (3.94), while the quadratic term, responsible for ladder mergers, introduces damping and thus slows down the growth of the gluon distributions with energy. This phenomenon became known as the *saturation* of parton distributions. A more quantitative discussion of the role of the damping term on small-x evolution will be presented in the next chapter.

The GLR equation (3.128) was rederived by Mueller and Qiu (1986) in the double leading-logarithmic approximation (DLA) for the integrated gluon distribution function related to the unintegrated distribution by (cf. Eq. (3.93))

$$x G(x, Q^2) = \int^{Q^2} dk_\perp^2 \phi(x, k_\perp^2). \quad (3.129)$$

Employing the DLA and analyzing diagrams with two merging DGLAP ladders, Mueller and Qiu arrived at the following evolution equation (again written here for a cylindrical proton or nucleus):

$$\frac{\partial^2 x G(x, Q^2)}{\partial \ln(1/x) \partial \ln(Q^2/\Lambda^2)} = \frac{\alpha_s N_c}{\pi} x G(x, Q^2) - \frac{\alpha_s^2 N_c \pi}{2 C_F S_\perp} \frac{1}{Q^2} [x G(x, Q^2)]^2, \quad (3.130)$$

which is known as the GLR–MQ equation.

Equation (3.130) is easily rewritten in terms of the density of gluons (with transverse size $1/Q$) in the transverse plane,

$$\rho_{glue}(x, Q^2) = \frac{x G(x, Q^2)}{S_\perp}; \quad (3.131)$$

we obtain

$$\frac{\partial^2 \rho_{glue}}{\partial \ln(1/x) \partial \ln(Q^2/\Lambda^2)} = \frac{\alpha_s N_c}{\pi} \rho_{glue} - \frac{\alpha_s^2 N_c \pi}{2 C_F Q^2} \rho_{glue}^2. \quad (3.132)$$

This equation has a simple probabilistic interpretation, given in Fig. 3.25. There we show several BFKL gluon cascades in the amplitude, each the same as in Fig. 3.20 but drawn in a time-ordered way (cf. Fig. 2.25) in the IMF/Bjorken or Breit frames. The proton or nucleus is envisioned as being at the top of each diagram though not shown explicitly. The first term on the right of Eq. (3.132) clearly describes the emission of an extra gluon at rapidity $Y = \ln 1/x$ (see the left-hand panel in Fig. 3.25), which leads to an increase in

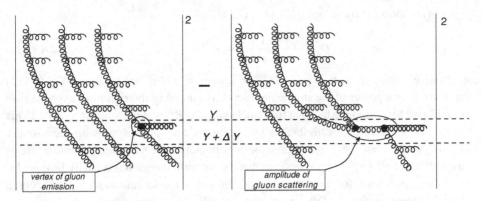

Fig. 3.25. Probabilistic interpretation of the two terms on the right-hand side of the GLR–MQ equation (3.130). The horizontal dashed lines denote one step ΔY of the evolution in rapidity.

the number of s-channel gluons. The gluon can be emitted from any of the three cascades in this panel.

The second term on the right-hand side of Eq. (3.132) corresponds to a process in which two gluon cascades merge together, reducing the number of cascades to two, as shown in the right-hand panel of Fig. 3.25. The s-channel gluon emitted after the merger can only be produced by the two remaining cascades. Thus the cascade merger leads to a decrease in the rate of new gluon production; therefore, it should enter Eq. (3.132) with a minus sign. Since in this process two (t-channel) gluons meet to interact, its contribution is proportional to ρ_{glue}^2 and to the cross section of the $GG \to GG$ process (see the right-hand panel in Fig. 3.25), which, on dimensional and power-counting grounds, should be proportional to $\sigma \sim \alpha_s^2/Q^2$. This is exactly what goes into the second term on the right-hand side of Eq. (3.132). In the next chapter we will put this probabilistic interpretation on a more rigorous basis.

Equation (3.130) allows one to estimate more precisely the saturation scale Q_s^2 at which the nonlinear saturation effects become important. To do that we have to equate the linear and quadratic terms on the right-hand side of Eq. (3.130). This gives

$$Q_s^2 = \frac{\alpha_s \pi^2}{S_\perp 2C_F} xG(x, Q_s^2). \tag{3.133}$$

The gluon distribution near the saturation region grows as a power of $1/x$ at small x (see (3.97)):

$$xG \sim \left(\frac{1}{x}\right)^\lambda, \tag{3.134}$$

with $\lambda > 0$ to be specified in the next chapter in a more detailed estimate. (This behavior of the gluon distribution is supported by experimental data.) Note that for a nucleus consisting of A nucleons we have to multiply this distribution function by the factor A. Using

Eq. (3.134), multiplied by A, in Eq. (3.133) yields

$$Q_s^2 \sim \frac{A}{S_\perp} \left(\frac{1}{x} \right)^\lambda \sim A^{1/3} \left(\frac{1}{x} \right)^\lambda, \tag{3.135}$$

in qualitative agreement with Eq. (3.125). An important additional feature of Eq. (3.135) is that it contains a power of the atomic number A. (In deriving the rightmost expression in Eq. (3.135) we have used the fact that for a nucleus $S_\perp \sim R^2 \sim A^{2/3}$, since $R \sim A^{1/3}$.) This factor $A^{1/3}$ is important: it means that the saturation scale not only grows with decreasing x (increasing energy) but also that it is large for large nuclei. Since nonlinear saturation effects are important for all $Q \lesssim Q_s$, we observe that the saturation region is actually broader for DIS on a nucleus. Also, for the same value of Bjorken x the saturation scale for DIS on a nucleus is larger than that for DIS on a proton, providing a stronger justification for the use of perturbative QCD approach in the former case.

It is instructive to rewrite the saturation condition (3.133) in terms of the transverse gluon density defined in Eq. (3.131). The condition (3.133) implies that

$$\frac{1}{Q_s^2} \rho_{glue}(x, Q_s^2) \sim \frac{1}{\alpha_s}, \tag{3.136}$$

i.e., that the number of gluons in an element of the transverse area comparable with the typical size of the gluon (which is the number of gluons on top of each other in the transverse plane) needs to be of order $1/\alpha_s$ for nonlinear effects to become important. Remembering that $xG \sim \langle A_\mu A_\mu \rangle$, we see from Eq. (3.136) that this implies that the corresponding gluon field should be

$$A_\mu \sim \frac{1}{g}. \tag{3.137}$$

At small coupling this is the strongest that a gluon field can be: the regime of (3.137) sets in when the interaction terms in the QCD Lagrangian (1.1) become parametrically comparable to the kinetic (free) term, as one can see from Eq. (1.4). We have obtained another interpretation of the saturation physics: saturation occurs because the gluon field gets as strong as it can possibly be, leading to the saturation of the gluon field strength and the parton distribution functions. A careful reader might also notice that a strong field of the type (3.137) usually occurs in classical problems, where one is looking for a classical gluon field (e.g. instantons). We will return to this observation and its implications below (McLerran and Venugopalan 1994a, b, c).

When the GLR–MQ equation was originally derived, the quadratic damping term that occurs in both Eqs. (3.128) and (3.130) was believed to be important only near the border of the saturation region, for $Q \sim Q_s$, where nonlinear effects were only starting to become important (Gribov, Levin, and Ryskin 1983, Mueller and Qiu 1986). It was expected that higher-order nonlinear corrections would show up as one goes deeper into the saturation region towards $Q < Q_s$. For instance, a nonlinear AGL evolution equation was proposed by Ayala, Gay Ducati, and Levin (1996, 1997, 1998) that included the suggestion that there should be corrections at all orders in ϕ on the right-hand side of Eq. (3.128). In the next chapter we will present a systematic way of unitarizing the BFKL evolution equation by including it in the light cone wave function, similarly to how we derived the DGLAP evolution above.

Further reading

For a more detailed and extensive discussion of the pre-QCD pomeron and its phenomenology see Collins (1977), Forshaw and Ross (1997), Barone and Predazzi (2002), and Donnachie, Dosch, and Landshoff (2005).

For readers interested in finding out more details on the BFKL equation and its solution we recommend the article by Del Duca (1995). We also recommend the review of Lipatov (1997), which summarizes everything that was then known about the solution to this linear equation. We believe that any difficulty in reading this paper will be compensated by the beauty of the problem. A very nice and detailed presentation of BFKL physics, gluon reggeization and the bootstrap equation, and higher-order corrections to BFKL evolution can be found in the book by Ioffe, Fadin, and Lipatov (2010).

The triple BFKL pomeron vertex has been extensively studied also. We recommend the paper by Bartels and Kutak (2008) and references therein. One can learn a lot from this paper about the classification and summation of Feynman diagrams in high energy scattering.

Exercises

3.1 Find the quark–quark scattering cross section $d\sigma_{qq \to qq}/dt$ at the Born level (order α_s^2), expressing the answer in terms of the Mandelstam variables s and t. (For simplicity assume that quarks are massless.) Take the limit of high energy s keeping t fixed and demonstrate that the expression obtained reduces to the cross section found in Sec. 3.2 above. Explain why the diagram in Fig. 3.3 dominates at high energy.

Repeat the above for the Born-level quark–antiquark scattering cross section $d\sigma_{q\bar{q} \to q\bar{q}}/dt$.

3.2 (a) Derive Eq. (3.24) starting from Eq. (3.18).

(b) Average Eq. (3.24) over the directions of $\vec{x}_{1\perp}$ and $\vec{x}_{2\perp}$ and integrate it over $l_\perp$ to obtain Eq. (3.25).

3.3 Consider onium–onium scattering at fixed impact parameter $\vec{b}_\perp$.

(a) Generalize Eq. (3.24) to the fixed impact parameter case. You should obtain the following cross section per impact parameter:

$$
\frac{d\hat{\sigma}_{tot}^{onium+onium}}{d^2 b} = \frac{\alpha_s^2 C_F}{2\pi^2 N_c} \int \frac{d^2 l_\perp d^2 l'_\perp}{l_\perp^2 l'^2_\perp} \left[e^{i\vec{l}_\perp \cdot (\vec{b}_\perp + \vec{x}_{1\perp}/2)} - e^{i\vec{l}_\perp \cdot (\vec{b}_\perp - \vec{x}_{1\perp}/2)} \right]
$$
$$
\times \left[e^{-i\vec{l}_\perp \cdot \vec{x}_{2\perp}/2} - e^{i\vec{l}_\perp \cdot \vec{x}_{2\perp}/2} \right] \left[e^{-i\vec{l}'_\perp \cdot (\vec{b}_\perp + \vec{x}_{1\perp}/2)} - e^{-i\vec{l}'_\perp \cdot (\vec{b}_\perp - \vec{x}_{1\perp}/2)} \right]
$$
$$
\times \left[e^{i\vec{l}'_\perp \cdot \vec{x}_{2\perp}/2} - e^{-i\vec{l}'_\perp \cdot \vec{x}_{2\perp}/2} \right]. \tag{3.138}
$$

Make sure that integrating Eq. (3.138) over the impact parameter reduces it back to Eq. (3.24).

(b) With the help of Eq. (A.9) integrate Eq. (3.138) over $\vec{l}_\perp$ and $\vec{l}_\perp'$ to obtain

$$\frac{d\hat{\sigma}_{tot}^{onium+onium}}{d^2b} = \frac{2\alpha_s^2 C_F}{N_c} \ln^2 \frac{\left|\vec{b}_\perp + \frac{1}{2}\vec{x}_{1\perp} + \frac{1}{2}\vec{x}_{2\perp}\right|\left|\vec{b}_\perp - \frac{1}{2}\vec{x}_{1\perp} - \frac{1}{2}\vec{x}_{2\perp}\right|}{\left|\vec{b}_\perp + \frac{1}{2}\vec{x}_{1\perp} - \frac{1}{2}\vec{x}_{2\perp}\right|\left|\vec{b}_\perp - \frac{1}{2}\vec{x}_{1\perp} + \frac{1}{2}\vec{x}_{2\perp}\right|}.$$

(3.139)

3.4 Show that the azimuthally symmetric eigenfunction $\chi(0, \nu)$ of the BFKL kernel from Eq. (3.81) reduces to

$$\chi(0, \nu) = \frac{4}{1 + 4\nu^2}$$

(3.140)

if we extract only the leading twist contributions to the BFKL Green function in Eq. (3.82) for both $l_\perp \gg l_\perp'$ and $l_\perp \ll l_\perp'$.

3.5 Use Eq. (3.82) in Eq. (3.26) to find the total onium–onium scattering cross section due to a BFKL pomeron exchange (neglecting the light cone wave functions of the onia).

(a) First the integration over $l_\perp$ and $l_\perp'$. You may find formulas (A.16) and (A.19) handy.

(b) Use the expansion around the saddle point given in (3.84) to perform the remaining ν-integral using the steepest descent method. You should obtain

$$\hat{\sigma}_{tot}^{onium+onium} = \frac{16\alpha_s^2 C_F}{N_c} x_{1\perp} x_{2\perp} \sqrt{\frac{\pi}{14\zeta(3)\bar{\alpha}_s Y}}$$

$$\times \exp\left\{(\alpha_P - 1)Y - \frac{\ln^2(x_{1\perp}/x_{2\perp})}{14\zeta(3)\bar{\alpha}_s Y}\right\}.$$

(3.141)

3.6 Verify explicitly that the BFKL Green function (3.80) satisfies the property (3.120).

4

Dipole approach to high parton density QCD

We are now ready to present more recent developments in high energy QCD. We will consider DIS in the rest frame of a proton or a nucleus. In this frame a virtual photon fluctuates into a quark–antiquark pair, which, in turn, hits the proton or nuclear target. We argue that quark–antiquark dipoles are convenient degrees of freedom for high energy scattering in QCD. We will present a simple model of DIS on a nucleus, due to Glauber, Gribov, and Mueller, in which the $q\bar{q}$ dipole rescatters multiple times on a nuclear target consisting of independent nucleons. We then include quantum corrections to this multiple-rescattering picture: we argue that the initial $q\bar{q}$ dipole may develop a cascade of gluons before hitting the target nucleus. In the large-N_c limit the cascade is described by Mueller's dipole model. When applied to DIS the dipole cascade resummation leads to the Balitsky–Kovchegov (BK) nonlinear evolution equation. We describe approximate analytical and exact numerical solutions of the BK equation and show that it resolves both problems of BFKL evolution: BK evolution is unitary and has no diffusion into the IR. It generates a saturation scale Q_s that grows with energy, justifying the use of perturbative QCD. We conclude the chapter by presenting the Bartels–Kwiecinski–Praszalowicz (BKP) evolution equation for multiple reggeon exchanges, along with the evolution equation for (C-odd) odderon exchange.

4.1 Dipole picture of DIS

Let us begin by considering DIS in the rest frame of the proton or nucleus. While many conclusions in this chapter may also apply to proton DIS, in the strict sense our results would be justified only for DIS on a large nucleus since such a nucleus has a large atomic number parameter A allowing us to make the approximations we will need below. We will therefore only talk about DIS on a nuclear target.

Without any loss of generality we can choose a coordinate axis such that the momentum of the virtual photon is given by

$$q^\mu = \left(q^+, -\frac{Q^2}{q^+}, 0_\perp \right) \tag{4.1}$$

in the $(+, -, \perp)$ light cone notation. The light cone momentum of the virtual photon q^+ is very large (since the (high) photon–nucleus center-of-mass energy is $\hat{s} = mq^+$), so that its

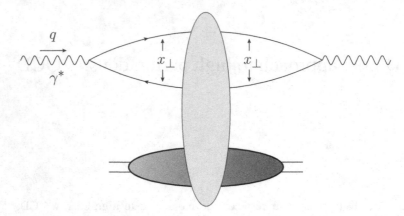

Fig. 4.1. Forward scattering amplitude for DIS on a proton or nuclear target in the rest frame of the target: the virtual photon splits into a $q\bar{q}$ pair which then interacts with the target. The interaction is depicted by the vertical oval. For simplicity the electron that emits the virtual photon is not shown.

coherence length in the longitudinal plus direction (see Sec. 2.3),

$$x^+ \approx \frac{2}{|q^-|} = \frac{2q^+}{Q^2},\tag{4.2}$$

is much larger than the size of the nucleus. If the virtual photon fluctuates into a quark–antiquark pair, the typical lifetime of such a $q\bar{q}$ fluctuation would also be much longer than the nuclear diameter. Therefore, a DIS process in the nuclear rest frame occurs when a virtual photon fluctuates into a $q\bar{q}$ pair (which we will also refer to as a color dipole or simply a dipole); the $q\bar{q}$ pair proceeds to interact with the target (Gribov 1970, Bjorken and Kogut 1973, Frankfurt and Strikman 1988). The forward scattering amplitude for the process is pictured in Fig. 4.1, with the $q\bar{q}$ dipole–nucleus interaction represented by the vertical oval. This is the dipole picture of DIS (Kopeliovich, Lapidus, and Zamolodchikov 1981, Bertsch *et al.* 1981, Mueller 1990, Nikolaev and Zakharov 1991). Note that while the topology of the DIS diagram in Fig. 4.1 is the same as for DIS in the IMF, shown in Fig. 2.2, the time-ordering of the interactions is different in the two figures.

The interaction of a virtual photon with a nucleus can be viewed as a two-stage process: the virtual photon decays into a colorless dipole consisting of a quark and an antiquark and the colorless dipole travels through the nucleus. However, this separation between the time scale for the photon to decay into the $q\bar{q}$ pair and the interaction time is not the only advantage of the dipole picture. Another important simplification comes from the fact that in high energy scattering a colorless dipole, with transverse size $x_\perp$, does not change its size during the interaction and therefore the S-matrix of the interaction is diagonal with respect to the transverse dipole size (Zamolodchikov, Kopeliovich, and Lapidus 1981, Levin and Ryskin 1987, Mueller 1990, Brodsky *et al.* 1994). Indeed, while the colorless dipole is traversing the target, the distance $x_\perp$ between the quark and antiquark can only vary by an amount

$$\Delta x_\perp \approx R \frac{k_\perp}{E} \qquad (4.3)$$

where $E \sim q^0$ denotes the energy of the dipole in the laboratory frame (the target rest frame), R is the longitudinal size of the target, and $k_\perp$ is the relative transverse momentum of the $q\bar{q}$ pair acquired through interaction with the target. In Eq. (4.3) $k_\perp/E$ is the relative transverse velocity of the quark with respect to the antiquark. From Eq. (4.3) we can see already that the change in the dipole size is suppressed by a power of the energy E and is therefore small. To quantify this better let us first remember the definition of Bjorken x, given in (2.2):

$$x = \frac{Q^2}{2P \cdot q} = \frac{Q^2}{mq^+} \approx \frac{Q^2}{2mE}. \qquad (4.4)$$

Using Eq. (4.4) in Eq. (4.3) along with the uncertainty principle $Q \approx k_\perp \approx 1/x_\perp$ yields

$$\frac{\Delta x_\perp}{x_\perp} \approx 2mxR = \frac{4R}{l_{coh}} \ll 1, \qquad (4.5)$$

where $l_{coh} = 2/(mx)$ is the coherence length of the dipole fluctuation (see Eq. (2.56)). We thus see that at small $x \ll 1/(mR)$, when the dipole interacts with the whole nucleus coherently in the longitudinal direction, the transverse recoil of the quark and the antiquark are negligible compared with the size of the dipole. Therefore the transverse size of the dipole is invariant in high energy interactions, as indicated in Fig. 4.1.

We conclude that in calculating the total DIS cross section, along with other high energy QCD observables, it is convenient to work in transverse coordinate space. We will therefore adopt a mixed representation: we will use longitudinal momentum space along with transverse coordinate space. Light cone perturbation theory (LCPT) is a very useful tool here again. Using LCPT to calculate the total DIS $\gamma^* A$ cross section we can factorize the diagram in Fig. 4.1 into the square of the light cone wave function $\Psi^{\gamma^* \to q\bar{q}}(\vec{x}_\perp, z)$ for the splitting of a virtual photon into a $q\bar{q}$ dipole and the total cross section for the scattering of a dipole on a target nucleus $\sigma_{tot}^{q\bar{q}A}(\vec{x}_\perp, Y)$, so that

$$\sigma_{tot}^{\gamma^* A}(x, Q^2) = \int \frac{d^2 x_\perp}{4\pi} \int_0^1 \frac{dz}{z(1-z)} |\Psi^{\gamma^* \to q\bar{q}}(\vec{x}_\perp, z)|^2 \sigma_{tot}^{q\bar{q}A}(\vec{x}_\perp, Y). \qquad (4.6)$$

Here $z = k^+/q^+$, with k^+ the light cone momentum of the quark in the $q\bar{q}$ pair. In general the dipole–nucleus cross section will depend on z too; however, in the eikonal and LLA approximations that we mainly consider below, $\sigma_{tot}^{q\bar{q}A}$ is independent of z. The net rapidity interval for the dipole–nucleus scattering is given by $Y = \ln(\hat{s}x_\perp^2) \approx \ln 1/x$ for $x_\perp \sim 1/Q$.

The reader may have other doubts about the factorization (4.6): after all, the LCPT rules presented in Sec. 1.3 require us to subtract the light cone energy of the incoming state in the energy denominator from each intermediate state's energy. Since the light cone energy of the incoming virtual photon is $q^- = -Q^2/q^+$, it seems that each intermediate state that we have absorbed into $\sigma_{tot}^{q\bar{q}A}(\vec{x}_\perp, Y)$ should "know" about the photon's energy. However, in the rest frame of the nucleus, q^- is equal to $-Q^2/q^+ \sim 1/\hat{s}$ and is therefore negligibly small compared with the typical minus components of momenta involved in dipole–nucleus interactions. The same would be true for dipole–nucleus scattering: the incoming dipole state would have a negligibly small light cone energy compared with the energies involved in the interaction. Therefore, in our eikonal approximation (up to corrections of order $1/\hat{s}$), we can interchange the negligible light cone energy q^- for the light cone energy of the dipole without changing the answer, thus justifying the factorization of Eq. (4.6). (Note that in calculating the light cone wave function $\Psi^{\gamma^* \to q\bar{q}}(\vec{x}_\perp, z)$ we cannot neglect the light cone energies of the virtual photon and the $q\bar{q}$ dipole, since they are the only terms entering the energy denominator.) Another important assumption is that the light cone energy of the target is not modified until the interaction with the dipole: one can show that the time scale of target fluctuations is much shorter than the lifetime of the dipole. Hence the target does not affect the virtual photon's wave function, since in constructing the latter the same light cone energy of the target enters into both the energies of the intermediate states and the initial-state energy, thus canceling in the energy denominators.

The factorization of Eq. (4.6) is very convenient: it allows us to separate the simple $\gamma^* \to q\bar{q}$ QED process from the strong interaction dynamics contained in $\sigma_{tot}^{q\bar{q}A}(\vec{x}_\perp, Y)$.

Note that the virtual photon may have either transverse or longitudinal polarization. Requiring that the photon polarization satisfies $\epsilon \cdot q = 0$ and imposing $\epsilon_T^2 = -1$ for transverse polarization and $\epsilon_L^2 = 1$ for the longitudinal polarization, we obtain for q^μ, Eq. (4.1), the following polarizations:

$$\epsilon_T^\lambda = (0, 0, \vec{\epsilon}_\perp^\lambda), \tag{4.7a}$$

$$\epsilon_L = \left(\frac{q^+}{Q}, \frac{Q}{q^+}, \vec{0}_\perp \right), \tag{4.7b}$$

with $\vec{\epsilon}_\perp^\lambda$ as given in Eq. (1.54). The polarization vectors (4.7) form a complete basis in the space of possible polarizations, so that the numerator of the photon propagator in the Landau gauge can be decomposed in terms of them as

$$g_{\mu\nu} - \frac{q_\mu q_\nu}{q^2} = -\sum_{\lambda=\pm} \epsilon_{T\mu}^\lambda \epsilon_{T\nu}^{\lambda*} + \epsilon_{L\mu} \epsilon_{L\nu}^*. \tag{4.8}$$

Using the polarizations (4.7) along with Eqs. (2.13) and (2.16) one can separate the total DIS cross section into transverse (T) and longitudinal (L) components (see Halzen and Martin 1984):

$$\sigma_T^{\gamma^* A} = \frac{4\pi^2 \alpha_{EM}}{q^0} W^{\mu\nu} \frac{1}{2} \sum_{\lambda=\pm} \epsilon_{T\mu}^\lambda \epsilon_{T\nu}^{\lambda*} = \frac{4\pi^2 \alpha_{EM}}{q^0} W_1 \tag{4.9a}$$

$$\sigma_L^{\gamma^* A} = \frac{4\pi^2 \alpha_{EM}}{q^0} W^{\mu\nu} \epsilon_{L\mu} \epsilon_{L\nu}^* = \frac{4\pi^2 \alpha_{EM}}{q^0} \left[-W_1 + \left(1 + \frac{\nu^2}{Q^2} \right) W_2 \right], \tag{4.9b}$$

with ν as defined in Eq. (2.5) and α_{EM} the fine structure constant. Employing Eqs. (2.18a) and (2.18b), we can rewrite Eqs. (4.9) in the high energy $\nu \gg Q$ limit as expressions for

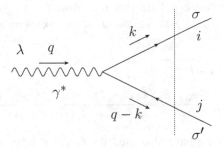

Fig. 4.2. Light cone wave function for a virtual photon fluctuating into a quark–antiquark pair (a dipole). The dotted line denotes the intermediate state.

the dimensionless structure functions:

$$F_2(x, Q^2) = \frac{Q^2}{4\pi^2 \alpha_{EM}} \sigma_{tot}^{\gamma^* A} = \frac{Q^2}{4\pi^2 \alpha_{EM}} \left(\sigma_T^{\gamma^* A} + \sigma_L^{\gamma^* A} \right), \tag{4.10a}$$

$$2x F_1(x, Q^2) = \frac{Q^2}{4\pi^2 \alpha_{EM}} \sigma_T^{\gamma^* A}. \tag{4.10b}$$

It is useful to also define the longitudinal structure function F_L, which measures the violation of the Callan–Gross relation (2.44):

$$F_L(x, Q^2) \equiv F_2(x, Q^2) - 2x F_1(x, Q^2) = \frac{Q^2}{4\pi^2 \alpha_{EM}} \sigma_L^{\gamma^* A}. \tag{4.11}$$

Equations (4.10) and (4.11) allow us to find the DIS structure functions using the transverse and longitudinal cross sections, which, with the help of Eq. (4.6), can be found from the dipole–nucleus scattering via

$$\sigma_{T,L}^{\gamma^* A}(x, Q^2) = \int \frac{d^2 x_\perp}{4\pi} \int_0^1 \frac{dz}{z(1-z)} |\Psi_{T,L}^{\gamma^* \to q\bar{q}}(\vec{x}_\perp, z)|^2 \, \sigma_{tot}^{q\bar{q} A}(\vec{x}_\perp, Y). \tag{4.12}$$

We have defined the transverse, $\Psi_T^{\gamma^* \to q\bar{q}}(\vec{x}_\perp, z)$, and longitudinal, $\Psi_L^{\gamma^* \to q\bar{q}}(\vec{x}_\perp, z)$, light cone wave functions, which differ by the polarization vector of the incoming virtual photon.

Let us now calculate the light cone wave functions $\Psi_{T,L}^{\gamma^* \to q\bar{q}}(\vec{x}_\perp, z)$ for the quark–antiquark fluctuations of a virtual photon. The diagram is shown in Fig. 4.2, in which the vertical dotted line denotes the intermediate state. Using the LCPT rules from Secs. 1.3 and 1.4, we write for the wave functions in momentum space (cf. the calculation in Sec. 2.4.2)

$$\Psi_{T,L}^{\gamma^* \to q\bar{q}}(\vec{k}_\perp, z) = e Z_f \frac{z(1-z)\delta_{ij}}{\vec{k}_\perp^2 + m_f^2 + Q^2 z(1-z)} \bar{u}_\sigma(k)\gamma \cdot \epsilon_{T,L}^\lambda v_{\sigma'}(q-k), \tag{4.13}$$

where σ and σ' are the quark and antiquark helicities, i, j are their colors, m_f is the mass of a quark with flavor f, and Z_f is the quark's electric charge in units of the electron charge e. (Note that q^μ is given in Eq. (4.1).) As mentioned above, we define $z = k^+/q^+$ as the fraction of the photon's light cone momentum carried by the quark.

Starting with the transverse polarization we substitute the polarization vector from Eq. (4.7a) into Eq. (4.13) and evaluate the Dirac matrix element using Appendix A.1, obtaining

$$\Psi_T^{\gamma^* \to q\bar{q}}(\vec{k}_\perp, z) = eZ_f \sqrt{z(1-z)}\, \delta_{ij}$$

$$\times \frac{(1 - \delta_{\sigma\sigma'})\vec{\epsilon}_\perp^\lambda \cdot \vec{k}_\perp (1 - 2z - \sigma\lambda) + \delta_{\sigma\sigma'} m_f (1 + \sigma\lambda)/\sqrt{2}}{\vec{k}_\perp^2 + m_f^2 + Q^2 z(1-z)}. \quad (4.14)$$

In arriving at Eq. (4.14) we have also used the fact that in two transverse dimensions $\vec{\epsilon}_\perp^\lambda \times \vec{k}_\perp = -i\lambda \vec{\epsilon}_\perp^\lambda \cdot \vec{k}_\perp$ for the $\vec{\epsilon}_\perp^\lambda$ from Eq. (1.54).

Since we are interested in using the virtual photon's wave function in transverse coordinate space in Eq. (4.12), we perform a Fourier transform of Eq. (4.14):

$$\Psi_{T,L}^{\gamma^* \to q\bar{q}}(\vec{x}_\perp, z) = \int \frac{d^2 k_\perp}{(2\pi)^2} e^{i\vec{k}_\perp \cdot \vec{x}_\perp} \Psi_{T,L}^{\gamma^* \to q\bar{q}}(\vec{k}_\perp, z) \quad (4.15)$$

and employ Eq. (A.11) along with $K_0'(z) = -K_1(z)$ to obtain

$$\Psi_T^{\gamma^* \to q\bar{q}}(\vec{x}_\perp, z) = \frac{eZ_f}{2\pi} \sqrt{z(1-z)}\, \delta_{ij} \left[(1 - \delta_{\sigma\sigma'})(1 - 2z - \sigma\lambda) i a_f \frac{\vec{\epsilon}_\perp^\lambda \cdot \vec{x}_\perp}{x_\perp} K_1(x_\perp a_f) \right.$$

$$\left. + \delta_{\sigma\sigma'} \frac{m_f}{\sqrt{2}}(1 + \sigma\lambda) K_0(x_\perp a_f) \right], \quad (4.16)$$

where

$$a_f^2 = Q^2 z(1-z) + m_f^2. \quad (4.17)$$

The square of the absolute value of the transverse wave function (4.16), summed over all the outgoing quantum numbers and averaged over the possible polarizations of the incoming transverse photon is (Bjorken, Kogut, and Soper 1971, Nikolaev and Zakharov 1991) given by

$$|\Psi_T^{\gamma^* \to q\bar{q}}(\vec{x}_\perp, z)|^2 = 2N_c \sum_f \frac{\alpha_{EM} Z_f^2}{\pi} z(1-z)$$

$$\times \left\{ a_f^2 \left[K_1(x_\perp a_f) \right]^2 [z^2 + (1-z)^2] + m_f^2 \left[K_0(x_\perp a_f) \right]^2 \right\}. \quad (4.18)$$

To calculate the longitudinal wave function $\Psi_L^{\gamma^* \to q\bar{q}}(\vec{x}_\perp, z)$ we repeat the above steps, now using the longitudinal polarization vector (4.7b) in Eq. (4.13). The transverse momentum space longitudinal wave function is

$$\Psi_L^{\gamma^* \to q\bar{q}}(\vec{k}_\perp, z) = \frac{eZ_f [z(1-z)]^{3/2}\, \delta_{ij}\, 2Q(1 - \delta_{\sigma\sigma'})}{\vec{k}_\perp^2 + m_f^2 + Q^2 z(1-z)}. \quad (4.19)$$

In arriving at Eq. (4.19) we have neglected a term that would have given us a delta function, $\delta^2(\vec{x}_\perp)$, in the transverse coordinate-space wave function; as we will shortly see, zero-transverse-size dipoles do not interact with the nucleus (they have zero scattering cross section) and so such configurations do not contribute to the DIS structure functions.

Fourier-transforming Eq. (4.19) into transverse coordinate space yields

$$\Psi_L^{\gamma^* \to q\bar{q}}(\vec{x}_\perp, z) = \frac{eZ_f}{2\pi}[z(1-z)]^{3/2} \delta_{ij} 2Q(1 - \delta_{\sigma\sigma'})K_0(x_\perp a_f), \qquad (4.20)$$

so that the longitudinal wave function squared, again with all summations performed, is (Bjorken, Kogut, and Soper 1971, Nikolaev and Zakharov 1991)

$$|\Psi_L^{\gamma^* \to q\bar{q}}(\vec{x}_\perp, z)|^2 = 2N_c \sum_f \frac{\alpha_{EM} Z_f^2}{\pi} 4Q^2 z^3 (1-z)^3 \left[K_0(x_\perp a_f) \right]^2. \qquad (4.21)$$

To obtain the phase-space integral in Eqs. (4.6) or (4.12) we remember that the two-particle momentum phase space given in Eq. (1.82) is (remembering that in our case the quarks are not identical)

$$\int \frac{dz}{2z(1-z)} \frac{d^2 k_\perp}{(2\pi)^3}. \qquad (4.22)$$

After Fourier-transforming the wave function into transverse coordinate space the integral becomes

$$\int \frac{dz}{2z(1-z)} \frac{d^2 x_\perp}{2\pi}, \qquad (4.23)$$

in agreement with Eqs. (4.6) and (4.12).

We have now completed the calculation of the QED part of DIS in the dipole picture. Equations (4.18) and (4.21), when used in Eq. (4.12), give us the transverse and longitudinal DIS cross sections, which, in turn, when used in Eqs. (4.10) and (4.11) give us the structure functions. The interesting physics of strong interactions is contained in the dipole–nucleus scattering cross section $\sigma_{tot}^{q\bar{q}A}(\vec{x}_\perp, Y)$: most of this chapter is dedicated to calculating this quantity.

4.2 Glauber–Gribov–Mueller multiple-rescatterings formula

We begin by employing Eq. (3.119a) to rewrite the total dipole–nucleus scattering cross section as

$$\sigma_{tot}^{q\bar{q}A}(\vec{x}_\perp, Y) = 2 \int d^2 b \, N(\vec{x}_\perp, \vec{b}_\perp, Y), \qquad (4.24)$$

where $N(\vec{x}_\perp, \vec{b}_\perp, Y)$ is the imaginary part of the forward scattering amplitude for a dipole of transverse size $\vec{x}_\perp$ interacting with the nucleus at impact parameter $\vec{b}_\perp$ and with net rapidity interval Y. Hence to find the cross section $\sigma_{tot}^{q\bar{q}A}$ we need to calculate $N(\vec{x}_\perp, \vec{b}_\perp, Y)$.

To find $N(\vec{x}_\perp, \vec{b}_\perp, Y)$ let us consider the following (Glauber) model. Assume that the nucleus is very large and dilute and is made out of $A \gg 1$ independent nucleons, where A is

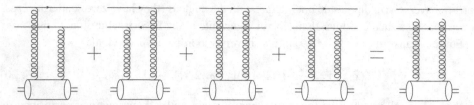

Fig. 4.3. The four diagrams contributing to dipole interaction with a single nucleon at the lowest nontrivial (two-gluon) order in the high energy approximation and an abbreviated notation for their sum.

the atomic number of the nucleus.[1] Any correlations between the nucleons are suppressed by powers of the large parameter A: hence our approximation corresponds to summing the leading powers of A. In evaluating the forward dipole–nucleus scattering amplitude $N(\vec{x}_\perp, \vec{b}_\perp, Y)$ we will follow the strategy originally outlined by Glauber and by Gribov (Glauber 1955, Franco and Glauber 1966, Gribov 1969b, Glauber and Matthiae 1970, Gribov 1970) and implemented in QCD by Mueller (1990).

4.2.1 Scattering on one nucleon

First we consider the case when the dipole interacts with only one nucleon in the nucleus. Assuming that the interaction is entirely perturbative, we see that the lowest-order contribution to the forward high energy scattering amplitude comes from a two-gluon exchange. The relevant diagrams are shown in Fig. 4.3. This lowest-order scattering process was calculated in Sec. 3.2. Employing the results of that section (see Eq. (3.25)) we can write down the total dipole–nucleon cross section as

$$\sigma^{q\bar{q}N} \approx \frac{2\pi \alpha_s^2 C_F}{N_c} x_\perp^2 \ln \frac{1}{x_\perp^2 \Lambda^2}. \tag{4.25}$$

In arriving at Eq. (4.25) we have assumed that the dipole is perturbatively small, $x_\perp \ll 1/\Lambda_{QCD}$, and that the nucleon can be modeled as another dipole of transverse size $1/\Lambda \gg x_\perp$, with Λ some soft QCD scale of order Λ_{QCD}. We have also assumed that the nucleus is sufficiently large that the cross section does not depend on the dipole's orientation in the transverse plane, over which we therefore average.

At the same two-gluon order the unintegrated gluon distribution function of the nucleon can be found using Eq. (3.92) with the lowest-order BFKL Green function (3.59). This gives

$$\phi_{LO}^{onium}(x, k_\perp^2) = \frac{\alpha_s C_F}{\pi} \frac{2}{k_\perp^2}, \tag{4.26}$$

where we have assumed that $k_\perp \gg \Lambda$. The factor 2 on the right-hand side of Eq. (4.26) simply counts the number of quarks in the dipole representing the nucleon. It should be

---

[1] Strictly speaking A is called the mass number of the nucleus; nevertheless, we will follow the standard jargon in the high energy field and refer to it as the atomic number.

replaced by N_c if one wanted to model the nucleon more realistically, as consisting of N_c valence quarks. Using Eq. (3.93) the corresponding lowest-order gluon distribution of an onium (a nucleon) turns out to be

$$xG_{LO}^{onium}(x, Q_\perp^2) = \frac{\alpha_s C_F}{\pi} 2 \ln \frac{Q^2}{\Lambda^2}. \tag{4.27}$$

Comparing Eq. (4.27) and Eq. (4.25), we can rewrite the latter as

$$\sigma^{q\bar{q}N} \approx \frac{\alpha_s \pi^2}{N_c} x_\perp^2 xG_N \left(x, \frac{1}{x_\perp^2} \right), \tag{4.28}$$

where xG_N is the gluon distribution in the nucleon (presently modeled as an onium).[2] Equation (4.28) has an advantage over Eq. (4.25): it is valid for any nonperturbative gluon distribution in the nucleon and is therefore more general. We will use these equations interchangeably, though.

To find the dipole–nucleus scattering cross section at a given impact parameter we need to average the dipole–nucleon scattering amplitude over all possible positions of the nucleon inside the nucleus and to sum over the A nucleons in the nucleus, all of which may participate in the interaction. We have

$$\frac{d\sigma_{LO}^{q\bar{q}A}}{d^2b} = \int db_3' d^2b_\perp' \rho_A(\vec{b}_\perp - \vec{b}_\perp', b_3') \frac{d\sigma^{q\bar{q}N}}{d^2b'}, \tag{4.29}$$

where $db_3' d^2b_\perp' = d^3b$ is the three-dimensional volume element and $\rho_A(\vec{b}_\perp, b_3)$ is the nucleon number density, with $\vec{b}_\perp = (b_1, b_2)$. In a simplified model, the nucleus has a constant nucleon number density $\rho_A = A/V$, where V is the volume of the nucleus in its rest frame. In the general case $\rho_A(\vec{b}_\perp, b_3)$ is given by the Woods–Saxon parametrization of the nuclear density (Woods and Saxon 1954).

Equation (4.29) gives the cross section for a dipole at impact parameter $\vec{b}_\perp$ scattering on a nucleon at impact parameter $\vec{b}_\perp - \vec{b}_\perp'$ (where $\vec{b}_\perp'$ is its transverse distance from the dipole), convoluted with the nucleon density ρ, which, in turn, is proportional to the probability of finding a nucleon at $\vec{b}_\perp - \vec{b}_\perp'$ (see Fig. 4.4). To simplify Eq. (4.29) we note that the perturbative scattering cross section falls off as $d\sigma^{q\bar{q}N}/d^2b' \sim 1/b_\perp'^4$ at large impact parameter, as can be seen for instance from Eq. (3.139) in Exercise 3.3 (after averaging over the azimuthal orientations of one dipole; this mimics an unpolarized nucleon, without any preferred direction). At nonperturbatively large impact parameter $b_\perp' \gtrsim 1/\Lambda_{QCD}$ one expects an even steeper falloff, $d\sigma^{q\bar{q}N}/d^2b' \sim \exp(-2m_\pi b_\perp')$ (cf. Eq. (3.113)). Hence the cross section $d\sigma^{q\bar{q}N}/d^2b'$ is localized at small impact parameters $b_\perp' \lesssim 1/\Lambda_{QCD}$.

In the large-A approximation that we are employing, one assumes that the nuclear wave function and hence the density $\rho_A(\vec{b}_\perp, b_3)$ does not change significantly over distances of order $1/\Lambda_{QCD}$, which is small compared with the size of the nucleus, so that the nucleon has an approximately equal probability of being anywhere within this transverse range.

---

[2] We would like to stress here that in order to conform to the standard notation we write the gluon distribution with Bjorken x in its argument, but throughout this section the gluon distribution is taken at the lowest (two-gluon) order and is therefore x-independent.

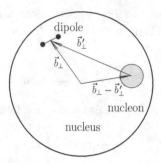

Fig. 4.4. The geometry of dipole–nucleus scattering in the transverse coordinate plane. To illustrate the notation of Eq. (4.29) the dipole is placed far from the nucleon; in reality $b'_\perp \lesssim 1/\Lambda_{QCD}$.

Therefore, for large nuclei we can approximate $\rho_A(\vec{b}_\perp - \vec{b}'_\perp, b'_3)$ as $\rho_A(\vec{b}_\perp, b'_3)$ and recast Eq. (4.29) by integrating over $b'_\perp$ as

$$\frac{d\sigma_{LO}^{q\bar{q}A}}{d^2b} = T(\vec{b}_\perp)\sigma^{q\bar{q}N},\qquad(4.30)$$

where we have defined the *nuclear profile function* $T(\vec{b}_\perp)$ by

$$T(\vec{b}_\perp) \equiv \int\limits_{-\infty}^{\infty} db_3 \rho_A(\vec{b}_\perp, b_3).\qquad(4.31)$$

For a spherical nucleus of radius R with constant nucleon number density $\rho_A = A/V$ one has $T(\vec{b}_\perp) = 2\rho_A\sqrt{R^2 - \vec{b}_\perp^2}$.

Comparing Eq. (4.30) with Eq. (4.24) and employing Eq. (4.28) we obtain

$$N_{LO}(\vec{x}_\perp, \vec{b}_\perp, Y) = \frac{\alpha_s\pi^2}{2N_c}T(\vec{b}_\perp)x_\perp^2 xG_N\left(x, \frac{1}{x_\perp^2}\right)$$

$$= \frac{\pi\alpha_s^2 C_F}{N_c}T(\vec{b}_\perp)x_\perp^2\ln\frac{1}{x_\perp\Lambda},\qquad(4.32)$$

where in the last line we have modeled the nucleon by a single quark with gluon distribution

$$xG(x, Q_\perp^2) = \frac{\alpha_s C_F}{\pi}\ln\frac{Q^2}{\Lambda^2}.$$

We now have the forward dipole–nucleus scattering amplitude for the case when only one nucleon in the nucleus interacts with the dipole. This case has a problem akin to that of linear BFKL evolution: if we increase the dipole size $x_\perp$ in Eq. (4.32), at some point we get $N_{LO} > 1$, violating the *black-disk limit*, which states that

$$N(\vec{x}_\perp, \vec{b}_\perp, Y) \leq 1\qquad(4.33)$$

(see Eq. (B.37) in Appendix B).

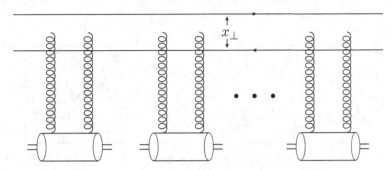

Fig. 4.5. Dipole–nucleus scattering in the Glauber–Gribov–Mueller approximation in the Feynman gauge. The disconnected gluon lines at the top denote the sum over all possible connections of the gluon lines to the dipole, as depicted in Fig. 4.3.

Let us stress again here that the transverse dipole size $x_\perp$ is preserved in high energy interactions. This makes the S-matrix diagonal not only in the impact parameter $\vec{b}_\perp$, as we saw in Eqs. (3.119) and also in Appendix B, but also in the dipole size $\vec{x}_\perp$. Therefore, the relations (3.119) between the cross sections and the S-matrix can also be written down for dipole–nucleus scattering with a fixed dipole size $\vec{x}_\perp$. The unitarity conditions (the optical theorem), which in momentum space are written as complicated convolutions (see e.g. Eq. (B.19)), become simple products of the amplitudes in $(\vec{x}_\perp, \vec{b}_\perp)$-space (see e.g. Eq. (B.30)). For this reason we think of color dipoles (or any other objects in the transverse coordinate representation) as the correct degrees of freedom for high energy scattering.

4.2.2 Scattering on many nucleons

When the probability of interaction with one nucleon becomes large, interactions with multiple nucleons also becomes likely and should be taken into account. Now we will see how multiple rescatterings of the dipole on different nucleons cure the problem of black-disk-limit violation by Eq. (4.32).

Let us consider the case when any number of nucleons can interact, restricting the interaction with each nucleon to the lowest nontrivial order. For this calculation we will be working in the standard Feynman perturbation theory using the Lorenz $\partial_\mu A^\mu = 0$ (Feynman) gauge. (Once we have separated the DIS cross section into the light cone wave function squared and the dipole–nucleus cross section, we can calculate the latter using any technique that is convenient.) We start by stating the diagrammatic answer for the many-nucleon interaction case: in the Feynman gauge, the dipole–nucleus interaction becomes a series of successive independent dipole–nucleon rescatterings, as shown in Fig. 4.5. There each nucleon (denoted by an oval at the bottom, just as in Fig. 4.3) interacts with the dipole via a two-gluon exchange: the disconnected gluon lines at the top of the diagram denote all possible connections to the quark and the antiquark lines in the dipole, as defined in Fig. 4.3.

The diagram in Fig. 4.5 implies that in the covariant gauge there is no direct "cross-talk" between the nucleons and that all the nucleons interact sequentially in the order in which

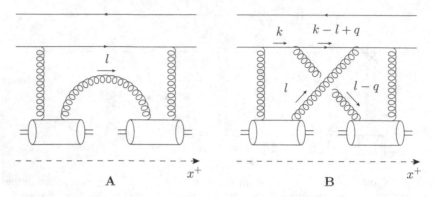

Fig. 4.6. Examples of diagrams that can be neglected for dipole–nucleus scattering in the covariant (Feynman) gauge.

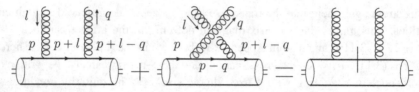

Fig. 4.7. Diagrammatic illustration that for a color-singlet object such as a nucleon, the coupling of two gluons to a single quark line is equivalent to the coupling of each gluon to a quark line that is on mass shell both before and after the quark–gluon interaction. The solid vertical line in the rightmost graph indicates an effective cut.

the dipole encounters them, i.e., according to their ordering along the x^+-axis. The dipole–nucleon interactions in the covariant gauge of Fig. 4.5 are localized inside the nucleons, on distance scales much shorter than the nuclear radius. While for a large dilute nucleus these assertions seem natural, we still need to prove them. To do so, it is convenient to change the frame slightly by giving the nucleus a slight boost, so that it moves along the light cone in the minus direction with a large P^- momentum. At the same time the boost preserves the virtual photon's motion along the plus light cone, with four-momentum as shown in Eq. (4.1). Thus both the dipole and nucleus in this new frame move along their respective light cones, as shown in Fig. 4.8. In the calculations below we will assume that the gluon–nucleon coupling is perturbatively small.

To illustrate why the graphs in Fig. 4.5 dominate the scattering, let us show that the diagrams in Fig. 4.6, demonstrating "cross-talk" (A) and the violation of x^+-ordering (B), are suppressed and can be neglected. Before we do that, let us carry out a simple exercise elucidating the nature of the coupling of two gluons to a nucleon. Consider two gluons coupling to a quark line in a nucleon, as shown in Fig. 4.7. This can be a part of any diagram in Figs. 4.6 and 4.5. Note that one has to include a crossed diagram, as illustrated in Fig. 4.7. Since the nucleon is a color singlet, the color factors of the two graphs on the left in Fig. 4.7 are identical (say, owing to a color trace), so that the difference between the two diagrams is only in the propagators for the internal quark line. Adding the two propagators (using the momentum labels from Fig. 4.7) and remembering that p^- is the largest momentum in the

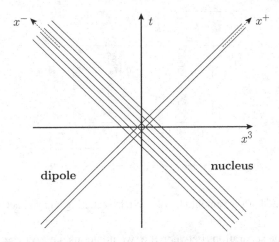

Fig. 4.8. Space–time picture of the dipole–nucleus scattering.

problem, we obtain

$$\frac{i(\not p + \not l + m_q)}{(p+l)^2 - m_q^2 + i\epsilon} + \frac{i(\not p - \not q + m_q)}{(p-q)^2 - m_q^2 + i\epsilon}$$

$$\approx i\not p \left(\frac{1}{p^- l^+ + i\epsilon} + \frac{1}{-p^- q^+ + i\epsilon} \right) = -i\not p 2\pi i \delta(p^- l^+). \tag{4.34}$$

We have used the fact that the outgoing quark is on mass shell, $(p + l - q)^2 = m_q^2$, so that $q^+ = l^+$ with eikonal accuracy (see Sec. 3.2 for similar estimates). We conclude that (with eikonal accuracy) $l^+ = q^+ = 0$. The δ-function in Eq. (4.34) puts the internal quark line in the leftmost diagram of Fig. 4.7 on mass shell. The result (4.34) can be summarized by replacing the internal quark line by the cut line, as shown in the rightmost graph of Fig. 4.7: the cut enforces $l^+ = 0$. What is essential to us is that neither gluon carries any plus momentum.

Now we are ready to evaluate the diagrams in Fig. 4.6. Note that, owing to the large size of the nucleus we are considering, even after the boost the nucleus is still somewhat spread out in the x^+-direction, as demonstrated in Fig. 4.8, where different nucleons correspond to different straight lines parallel to the x^- light cone. Hence each nucleon in the nucleus is located at a different x^+ coordinate. We thus need to Fourier-transform the diagrams in Fig. 4.6 into coordinate x^+-space by integrating over l^-.

Starting with Fig. 4.6A we see that the l^--dependence can be contained only in the propagator of the gluon carrying momentum l that is exchanged between the nucleons there. However, as we have just shown when considering the diagrams in Fig. 4.7, $l^+ = 0$ with eikonal accuracy, so that the diagram in Fig. 4.6A is proportional to

$$\int_{-\infty}^{\infty} dl^- \frac{e^{-il^- \Delta x^+}}{l^2 + i\epsilon} \approx \int_{-\infty}^{\infty} dl^- \frac{e^{-il^- \Delta x^+}}{-\vec{l}_\perp^2} = 0 \tag{4.35}$$

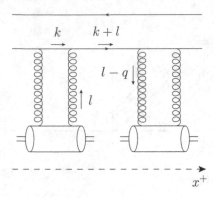

Fig. 4.9. Forward amplitude for a dipole scattering on two nucleons.

for nonzero light cone separations between the two nucleons, i.e., $\Delta x^+ \neq 0$. Hence diagram A is negligible in the covariant gauge in which we are working.[3] Let us stress that in arriving at the result (4.35) we have restricted ourselves to $l^- \ll p^-$: if we relaxed this constraint then the integral in Eq. (4.35) would be nonzero, though it would still be suppressed for large atomic numbers A (Kovchegov 1997).

Similarly, in Fig. 4.6B one has $l^+ = q^+ = 0$, so that the l^--dependence can be contained only in the quark propagator of the $(k - l + q)$-line. Since the light cone momentum of the quark k^+ is large, we see that the diagram in Fig. 4.6B is proportional to

$$\int_{-\infty}^{\infty} dl^- \frac{e^{-il^- \Delta x^+}}{(k - l + q)^2 + i\epsilon} \approx \int_{-\infty}^{\infty} dl^- \frac{e^{-il^- \Delta x^+}}{k^+(k^- - l^- + q^-) - (\vec{k}_\perp - \vec{l}_\perp + \vec{q}_\perp)^2 + i\epsilon} = 0$$

(4.36)

for $\Delta x^+ > 0$, as the pole of the propagator is in the upper half-plane while the contour needs to be closed in the lower half-plane. For $\Delta x^+ < 0$ the integral in Eq. (4.36) is not zero, but then we would obtain zero from the integral over the minus momentum carried by the other pair of t-channel gluon lines. We thus can neglect diagram B as well.

The arguments used in proving that diagrams A and B in Fig. 4.6 are zero can be generalized to more complicated diagrams in the same general categories. We have succeeded in demonstrating that in the covariant gauge and in the approximation of two gluon exchanges per nucleon the dipole–nucleus interaction is given by the graphs in Fig. 4.5. We now need to resum these diagrams. To do this, we first consider dipole scattering on two nucleons ordered in x^+, as shown in Fig. 4.9. Unlike the diagram in Fig. 4.6B, the graph in Fig. 4.9 has the correct x^+-ordering of the nucleons. Instead of giving zero it yields (note that k^- is very small for a quark on a plus light cone)

$$\int_{-\infty}^{\infty} \frac{dl^-}{2\pi} \frac{i e^{-il^- \Delta x^+}}{(k + l)^2 + i\epsilon} \approx \int_{-\infty}^{\infty} \frac{dl^-}{2\pi} \frac{i e^{-il^- \Delta x^+}}{k^+ l^- - (\vec{k}_\perp + \vec{l}_\perp)^2 + i\epsilon} = \frac{1}{k^+}, \qquad (4.37)$$

---

[3] Note that the diagram in Fig. 4.6A is nonzero in the $A^- = 0$ light cone gauge even in the eikonal approximation.

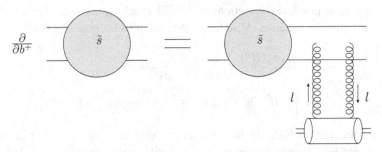

Fig. 4.10. Diagrammatic representation of Eq. (4.40) resumming all the diagrams from Fig. 4.5.

where at the end of the calculation we have neglected the phase of the exponential since it is proportional to $\Delta x^+/k^+ \sim 1/(k^+ p^-)$, which is suppressed by the center-of-mass energy. It is important to note that in picking up the pole in Eq. (4.37) we put the propagator of the quark carrying momentum $k + l$ on mass shell. Therefore, the diagram factorizes into a product of two independent dipole–nucleon scatterings; the quark and the antiquark in the dipole effectively go on shell between the scatterings. (The factor $1/k^+$ left in Eq. (4.37) serves to normalize the dipole–nucleon cross section for the nucleon on the right.) The numerator of the quark propagator can be absorbed into two separate scattering amplitudes using the property that (neglecting the quark mass) $\not{k} + \not{l} = \sum_\sigma u_\sigma(k + l)\bar{u}_\sigma(k + l)$: the factor $u_\sigma(k + l)$ is absorbed into one amplitude, while $\bar{u}_\sigma(k + l)$ is absorbed into the other. Comparing this result with the standard normalization factor for the $2 \to 2$ cross section at high energy (see Eqs. (B.22) and (B.23)), we conclude that to resum the diagrams in Fig. 4.5 we simply need to iterate the dipole–nucleus cross section.

Define the forward matrix element of the S-matrix for the dipole–nucleus scattering by (cf. Eq. (B.34))

$$S(\vec{x}_\perp, \vec{b}_\perp, Y) = 1 - N(\vec{x}_\perp, \vec{b}_\perp, Y). \tag{4.38}$$

Suppressing the arguments $\vec{b}_\perp$ and Y, we can define the S-matrix (the "propagator") $s(\vec{x}_\perp, b^+)$ for a dipole to travel through the nucleus up to a point b^+, so that $S(\vec{x}_\perp) = s(\vec{x}_\perp, L)$ with $b^+ \in (0, L)$, which defines the extent of the nucleus along the x^+ axis. Going to transverse momentum space we have

$$\tilde{s}(\vec{k}_\perp, b^+) = \int d^2x_\perp e^{-i\vec{k}_\perp \cdot \vec{x}_\perp} s(\vec{x}_\perp, b^+), \tag{4.39}$$

with $\vec{k}_\perp$ the relative transverse momentum of the quark and the antiquark in the dipole. As we demonstrated above, all the integrations over the minus momenta in the diagrams in Fig. 4.5 are done straightforwardly. Hence the b^+-evolution of $\tilde{s}(\vec{k}_\perp, b^+)$ is also clear: in one step in b^+ the dipole may interact with one nucleon. Denoting $\tilde{s}(\vec{k}_\perp, b^+)$ by a circle, we illustrate this statement in Fig. 4.10.

Summing over all possible connections of the t-channel gluons to the dipole in Fig. 4.10 we obtain the following equation (Mueller (1990), see also Baier *et al.* (1997)):

$$\frac{\partial \tilde{s}(\vec{k}_\perp, b^+)}{\partial b^+} = -\frac{\rho_A(\vec{b}_\perp, b^+)}{2} \int \frac{d^2 l_\perp}{(2\pi)^2} \frac{d\sigma^0_{qq \to qq}}{d^2 l}$$

$$\times \left[2\, \tilde{s}(\vec{k}_\perp, b^+) - \tilde{s}(\vec{k}_\perp - \vec{l}_\perp, b^+) - \tilde{s}(\vec{k}_\perp + \vec{l}_\perp, b^+) \right], \qquad (4.40)$$

where the minus signs outside the last two terms come from the coupling of one gluon to the quark and the other to the antiquark. The differential cross section $d\sigma^0_{qq \to qq}/d^2 l$ is the momentum-space expression for the two-gluon exchange cross section in quark–quark scattering, as given in Eq. (3.18), and the factor $1/2$ is needed to convert it to the forward amplitude. (Again we are modeling the nucleons as single valence quarks.) The nucleon density factor $\rho_A(\vec{b}_\perp, b^+)$ (now, in the boosted-nucleus frame, equal to the number of nucleons per unit volume element $db^+ d^2 b_\perp$) gives the probability of finding a nucleon at a given location in the nucleus. Again we assume $\rho_A(\vec{b}_\perp, b^+)$ to be constant on the perturbatively short transverse distance scales relevant to Eq. (4.40). The overall minus sign in Eq. (4.40) reflects the fact that we are calculating a variation of the S-matrix that differs from the variation of the forward amplitude by a sign, as follows from Eq. (4.38).

Fourier-transforming Eq. (4.40) into transverse coordinate space we obtain

$$\frac{\partial s(\vec{x}_\perp, b^+)}{\partial b^+} = -\frac{\rho_A(\vec{b}_\perp, b^+)}{2} \sigma^{q\bar{q}N} s(\vec{x}_\perp, b^+), \qquad (4.41)$$

$$\sigma^{q\bar{q}N} = \int \frac{d^2 l_\perp}{(2\pi)^2} \frac{d\sigma^0_{qq \to qq}}{d^2 l} \left(2 - e^{i\vec{l}_\perp \cdot \vec{x}_\perp} - e^{-i\vec{l}_\perp \cdot \vec{x}_\perp} \right), \qquad (4.42)$$

exactly the dipole–nucleus cross section of Eqs. (4.25) and (4.28). (The factor 2 difference in comparison to Eq. (4.25) is due to the fact that in Eq. (4.25) the nucleon is modeled as a dipole whereas in our present calculation it is taken to be a single quark for simplicity.)

One can readily see from Eq. (4.41) that in transverse coordinate space Eq. (4.40) becomes trivial. An important consequence of this triviality is that, for the first time, we see explicitly that the transverse size of the dipole $x_\perp$ does not change in the interactions with the nucleons (and the nucleus). This demonstrates the argument presented in Sec. 4.1.

Equation (4.41) has the following simple physical meaning: as the dipole propagates through the nucleus it may encounter nucleons, with the probability of interaction per unit path length db^+ given by the product of the nucleon density ρ_A and the interaction probability $\sigma^{q\bar{q}N}$ from Eq. (4.28), with another factor one-half inserted owing to the optical theorem (B.23). The initial condition for Eq. (4.41) is given by a freely propagating dipole without interactions, for which $s(\vec{x}_\perp, b^+ = 0) = 1$. Solving Eq. (4.41) with this initial condition yields

$$s(\vec{x}_\perp, b^+) = \exp\left\{ -\int_0^{b^+} db'^+ \frac{\rho_A(\vec{b}_\perp, b'^+)}{2} \sigma^{q\bar{q}N} \right\}. \qquad (4.43)$$

Going back to the nuclear rest frame and remembering that $S(\vec{x}_\perp) = s(\vec{x}_\perp, L)$, we obtain

$$S(\vec{x}_\perp, \vec{b}_\perp, Y = 0) = \exp\left\{-\frac{\sigma^{q\bar{q}N}}{2} T(\vec{b}_\perp)\right\}. \tag{4.44}$$

Note that $\sigma^{q\bar{q}N}$ does not depend on the energy of the collision (and therefore on its net rapidity): to underscore this we have put $Y = 0$ in the argument of the S-matrix in Eq. (4.44). This will delineate this expression from the energy-dependent version that results from incorporating small-x evolution into this picture.

Using Eq. (4.44) along with Eq. (4.28) in Eq. (4.38), the imaginary part of the forward scattering amplitude in the Glauber–Gribov–Mueller (GGM) model (Mueller 1990) is given by

$$N(\vec{x}_\perp, \vec{b}_\perp, Y = 0) = 1 - \exp\left\{-\frac{\alpha_s \pi^2}{2N_c} T(\vec{b}_\perp) x_\perp^2 x G_N\left(x, \frac{1}{x_\perp^2}\right)\right\}. \tag{4.45}$$

This is the GGM multiple–rescattering formula. Note again that the nucleon's gluon distribution function $x G_N$ in Eq. (4.45) is taken at the lowest, two-gluon, level and is thus independent of x, so that the amplitude N in Eq. (4.45) is independent of energy.

Equation (4.45) has a remarkable property: one can see that it implies $N \leq 1$ for all (perturbative) $x_\perp$. This means that the resulting forward scattering amplitude obeys the black-disk limit constraint (4.33), correcting the problem of the single rescattering amplitude from Eq. (4.32). We see that multiple rescatterings unitarize the scattering cross section, preserving the black-disk limit. The lesson we learn from the Glauber–Gribov–Mueller model is that to unitarize a cross section it is important to include multiple rescatterings!

Equation (4.45) allows us to determine the parameter corresponding to resummation of the diagrams like that shown in Fig. 4.5. Using the gluon distribution from Eq. (4.27) in Eq. (4.45), and noting that for large nuclei the profile function $T(\vec{b}_\perp)$ scales as $A^{1/3}$, we conclude that the resummation parameter of multiple rescatterings is (Kovchegov 1997)

$$\alpha_s^2 A^{1/3}. \tag{4.46}$$

The physical meaning of the parameter $\alpha_s^2 A^{1/3}$ is rather straightforward: at a given impact parameter the dipole interacts with $\sim A^{1/3}$ nucleons, exchanging two gluons with each. Since two-gluon exchange is parametrically of order α_s^2 we obtain $\alpha_s^2 A^{1/3}$ as the resummation parameter.

4.2.3 Saturation picture from the GGM formula

Multiple nucleon interactions become important in Eq. (4.45) when the dipole size becomes of order $x_\perp \sim 1/Q_s$, where the *saturation* scale Q_s is defined by the following implicit equation (cf. Eq. (3.133)):

$$Q_s^2(\vec{b}_\perp) = \frac{\alpha_s \pi^2}{2N_c} T(\vec{b}_\perp) x G_N(x, Q_s^2). \tag{4.47}$$

Note that for a cylindrical nucleus, as considered in Sec. 3.4.2, one has $T(\vec{b}_\perp) = A/S_\perp$ so that, taking into account that the nuclear gluon distribution is $x G_A = A x G_N$ (which is

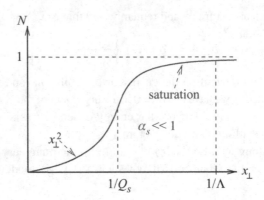

Fig. 4.11. The imaginary part of the forward amplitude of the dipole–nucleus scattering N plotted as a function of the transverse separation between the quark and the anti-quark in a dipole $x_\perp$, using Eq. (4.51). (Reprinted from Jalilian-Marian and Kovchegov (2006), with permission from Elsevier.) A color version of this figure is available online at www.cambridge.org/9780521112574.

true at the two-gluon level considered here), one can recast Eq. (4.47) in almost the exact form of Eq. (3.133). The difference N_c/C_F between the saturation scales in (4.47) and in (3.133) is due to the fact that the saturation scale (4.47) we have just found is that for a quark dipole, whereas the saturation scale in Eq. (3.133) was obtained solely for gluons. If we were to replace the quark dipole in Fig. 4.5 with a gluon dipole, we would need to modify the exponent in Eq. (4.45) by the ratio of the adjoint and fundamental Casimir operators N_c/C_F, putting Eq. (4.47) in exact agreement with Eq. (3.133). With this proviso, we see that, at least at the lowest order considered, Eq. (4.47) gives the same saturation scale as what follows from the GLR–MQ equation.

Inserting the lowest-order single-quark gluon distribution function,

$$xG_{LO}^{quark}(x, Q_\perp^2) = \frac{\alpha_s C_F}{\pi} \ln \frac{Q^2}{\Lambda^2}, \tag{4.48}$$

into Eq. (4.45), we can rewrite it as

$$N(\vec{x}_\perp, \vec{b}_\perp, Y = 0) = 1 - \exp\left\{-\frac{\alpha_s^2 C_F \pi}{N_c} T(\vec{b}_\perp) x_\perp^2 \ln \frac{1}{x_\perp \Lambda}\right\}. \tag{4.49}$$

Defining the quark *saturation* scale (note the factor 4 difference compared with Eq. (4.47) and the absence of a gluon distribution in this definition),

$$Q_s^2(\vec{b}_\perp) \equiv \frac{4\pi \alpha_s^2 C_F}{N_c} T(\vec{b}_\perp), \tag{4.50}$$

we rewrite Eq. (4.49) as

$$N(\vec{x}_\perp, \vec{b}_\perp, Y = 0) = 1 - \exp\left\{-\frac{x_\perp^2 Q_s^2(\vec{b}_\perp)}{4} \ln \frac{1}{x_\perp \Lambda}\right\}. \tag{4.51}$$

The dipole amplitude N from Eq. (4.51) is plotted schematically in Fig. 4.11 as a function of $x_\perp$. One can see that, at small $x_\perp$, i.e., $x_\perp \ll 1/Q_s$, we have $N \sim x_\perp^2$ so that the amplitude is zero for zero dipole size. This result is natural, since in a zero-size dipole the color charges of the quark and the antiquark cancel each other, leading to the disappearance of interactions with the target. This effect is known as *color transparency* (Kopeliovich, Lapidus, and Zamolodchikov 1981, Nikolaev and Zakharov 1991, Heiselberg *et al.* 1991, Blaettel *et al.* 1993, Frankfurt, Miller, and Strikman 1993).

The amplitude (4.51) is a rising function of $x_\perp$ at small dipole size. However, at large dipole size $x_\perp \gtrsim 1/Q_s$, growth stops and the amplitude levels off (*saturates*) at $N = 1$. As mentioned earlier, this regime corresponds to the black-disk limit for dipole–nucleus scattering: for large dipoles the nucleus appears as a black disk. The transition from $N \sim x_\perp^2$ to the black disk-like ($N = 1$) behavior in Fig. 4.11 occurs at around $x_\perp \sim 1/Q_s$. For dipole sizes $x_\perp \gtrsim 1/Q_s$ the amplitude N *saturates* to a constant. This translates into saturation of the quark distribution functions in the nucleus, since $xq + x\bar{q} \sim F_2$ (see Eq. (2.46)) and the saturation of N implies the saturation of F_2, as follows from Eqs. (4.10a), (4.12), and (4.24). Thus the saturation of N can be identified with parton saturation, justifying the name *saturation scale* for Q_s, Eq. (4.50).

Note that since $T(\vec{b}_\perp) \sim A^{1/3}$ the saturation scale grows as

$$Q_s^2 \sim A^{1/3} \tag{4.52}$$

with atomic number A. If A is large enough, Q_s becomes perturbatively large, $Q_s \gg \Lambda_{QCD}$, justifying the use of perturbation theory. The scaling in Eq. (4.52) is consistent with Eq. (3.135), which we obtained from analyzing the GLR equation.

4.3 Mueller's dipole model

The amplitude N given by the Glauber–Gribov–Mueller formula (4.51) is independent of the energy of the collision (see also Eq. (4.49)) and therefore cannot be the final answer for the high energy scattering problem at hand. It turns out that the energy dependence comes into the dipole–nucleus scattering amplitude through quantum evolution corrections, much as the two-gluon exchange amplitude in the onium–onium scattering in Sec. 3.2 acquires energy dependence through the BFKL evolution of Sec. 3.3. To incorporate small-x evolution into dipole–nucleus scattering we begin by rewriting the evolution in the language of LCPT, in which it can be completely absorbed into the light cone wave function, with the help of Mueller's dipole model (Mueller 1994, 1995, Mueller and Patel 1994).

4.3.1 Dipole wave function and generating functional

Let us consider the light cone wave function of an ultrarelativistic meson consisting of a heavy quark and antiquark (an onium), with no sea quarks and gluons present before the small-x evolution, as shown in Fig. 4.12. We can safely apply perturbative QCD to the onium wave function since here typical transverse distance $x_\perp$ is about $1/m_U$, where m_U is the large mass of the heavy quark; the strong coupling constant is clearly small at such distances.

We will denote the "bare" onium light cone wave function by $\Psi_{\sigma\sigma'}^{(0)}(\vec{k}_\perp, z)$, where $\vec{k}_\perp$ is the relative transverse momentum of the $q\bar{q}$ pair, $z = k^+/p^+$ is the fraction of the light cone momentum p^+ of the whole onium carried by the quark, while σ and σ' are the

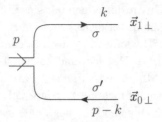

Fig. 4.12. The onium light cone wave function before small-x evolution.

polarizations of the quark and the antiquark (see Fig. 4.12). The onium is moving in the light cone plus direction. As usual we suppress the color and flavor indices, assuming that they will be properly summed over when necessary. As we will shortly see, the transverse size of the dipole remains invariant during the small-x evolution: therefore we will work in a mixed representation where we use the transverse coordinates and longitudinal momenta to describe dipoles. We thus Fourier-transform the onium wave function, using

$$\Psi^{(0)}_{\sigma\sigma'}(\vec{x}_{10}, z) = \int \frac{d^2 k_\perp}{(2\pi)^2} e^{i\vec{k}_\perp \cdot \vec{x}_{10}} \Psi^{(0)}_{\sigma\sigma'}(\vec{k}_\perp, z), \tag{4.53}$$

where $\vec{x}_{10\perp} = \vec{x}_{1\perp} - \vec{x}_{0\perp}$ is the transverse size of dipole, the quark being located at $\vec{x}_{1\perp}$ and the antiquark at $\vec{x}_{0\perp}$ (see Fig. 4.12).

As the initial onium state contains only the $q\bar{q}$ pair its normalization is (cf. Eqs. (1.70) and (1.82))

$$\int_0^1 \frac{dz}{z(1-z)} \int \frac{d^2 k_\perp}{2(2\pi)^3} \sum_{\sigma,\sigma'} \left| \Psi^{(0)}_{\sigma\sigma'}(\vec{k}_\perp, z) \right|^2 = 1, \tag{4.54}$$

which, in transverse coordinate space becomes

$$\int_0^1 \frac{dz}{z(1-z)} \int \frac{d^2 x_{10}}{4\pi} \sum_{\sigma,\sigma'} \left| \Psi^{(0)}_{\sigma\sigma'}(\vec{x}_{10}, z) \right|^2 = 1. \tag{4.55}$$

We are interested in modifications to this wave function under small-x evolution in the LLA approximation; thus we need to resum the terms containing $\alpha_s \ln 1/x$ corrections. Throughout this section we will work in the $A^+ = 0$ light cone gauge. As for the DGLAP evolution of Sec. 2.4.2, one step of LLA small-x evolution consists of the appearance of a single gluon in the wave function: the gluon can be emitted either from the quark line or from the antiquark line, as shown in Fig. 4.13. (Just as in the case of BFKL evolution, quark loops and the emission of $q\bar{q}$ pairs are beyond the LLA, contributing subleading corrections of order $\alpha_s^2 \ln 1/x$.) The corresponding modification of the onium wave function due to the gluon emissions in Fig. 4.13 is easier to calculate than in the DGLAP case. We assume that the light cone momentum k_2^+ of the emitted gluon is small, $k_2^+ \ll k_1^+$, $p^+ - k_1^+$ (see

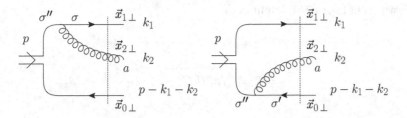

Fig. 4.13. One step of small-x evolution in the onium wave function. The dotted lines denote the intermediate states.

Fig. 4.13 for the explanation of the momentum and coordinate labeling). At the same time we impose no ordering on the transverse momenta of the quarks and the gluon. The kinematics is different from the DGLAP case (cf. Sec. 2.4.2): here the longitudinal momenta are ordered while in the DGLAP case the transverse momenta were ordered. In analogy with Eq. (2.68), we can write down the following expression for the $q\bar{q}G$ (one-gluon) contribution to the onium light cone wave function in the $A^+ = 0$ gauge at order g:

$$\Psi^{(1)}_{\sigma\sigma'}(\vec{k}_{1\perp}, \vec{k}_{2\perp}, z_1, z_2)$$

$$= \frac{g t^a \theta(k_2^+)}{k_2^- + k_1^- + (p - k_1 - k_2)^- - p^-}$$

$$\times \sum_{\sigma''=\pm 1} \left[\frac{\bar{u}_\sigma(k_1)\gamma \cdot \epsilon^*_\lambda(k_2) u_{\sigma''}(k_1 + k_2)}{k_1^+ + k_2^+} \, \Psi^{(0)}_{\sigma''\sigma'}(\vec{k}_{1\perp} + \vec{k}_{2\perp}, z_1 + z_2) \right.$$

$$\left. - \frac{\bar{v}_{\sigma''}(p - k_1)\gamma \cdot \epsilon^*_\lambda(k_2) v_{\sigma'}(p - k_1 - k_2)}{p^+ - k_1^+} \, \Psi^{(0)}_{\sigma''\sigma}(\vec{k}_{1\perp}, z_1) \right]. \qquad (4.56)$$

Here a is the gluon color index, σ, σ', and σ'' are the quark and antiquark polarizations, and λ is the gluon polarization, while $z_2 = k_2^+/p^+$ and $z_1 = k_1^+/p^+$.

To simplify Eq. (4.56) we first remember that we have assumed that $k_2^+ \ll k_1^+, p^+ - k_1^+$ (that is $z_2 \ll z_1, 1 - z_1$) and that all the transverse momenta are comparable. In this kinematics the light cone energy of the gluon, $k_2^- = k_{2\perp}^2/k_2^+$, dominates the energy denominator, just as in the DGLAP case (cf. Eq. (2.69)), only now this is due to longitudinal momentum ordering. We can write

$$\frac{1}{k_2^- + k_1^- + (p - k_1 - k_2)^- - p^-} \approx \frac{1}{k_2^-} = \frac{k_2^+}{k_{2\perp}^2}. \qquad (4.57)$$

To evaluate the Dirac matrix elements in Eq. (4.56) we use Table A.1 along with Eq. (A.2), again keeping in mind that $k_2^+ \ll k_1^+, p^+ - k_1^+$. For instance, the first matrix element in the square brackets of Eq. (4.56) simplifies to

$$\bar{u}_\sigma(k_1)\gamma \cdot \epsilon_\lambda^*(k_2)u_{\sigma''}(k_1 + k_2) \approx \frac{1}{2}\bar{u}_\sigma(k_1)\gamma^+ u_{\sigma''}(k_1 + k_2)\,\epsilon_\lambda^-(k_2)^*$$

$$= 2\delta_{\sigma\sigma''}\sqrt{k_1^+(k_1^+ + k_2^+)}\,\frac{\vec{\epsilon}_\perp^{\lambda*} \cdot \vec{k}_{2\perp}}{k_2^+} \approx 2\delta_{\sigma\sigma''}k_1^+\frac{\vec{\epsilon}_\perp^{\lambda*} \cdot \vec{k}_{2\perp}}{k_2^+}.$$

(4.58)

Performing a similar approximation for the second matrix element in Eq. (4.56) and inserting the result along with Eqs. (4.57) and (4.58) back into Eq. (4.56) yields

$$\Psi_{\sigma\sigma'}^{(1)}(\vec{k}_{1\perp}, \vec{k}_{2\perp}, z_1, z_2)$$

$$\approx 2gt^a\theta(z_2)\frac{\vec{\epsilon}_\perp^{\lambda*} \cdot \vec{k}_{2\perp}}{k_{2\perp}^2}\left[\Psi_{\sigma\sigma'}^{(0)}(\vec{k}_{1\perp} + \vec{k}_{2\perp}, z_1) - \Psi_{\sigma\sigma'}^{(0)}(\vec{k}_{1\perp}, z_1)\right],$$

(4.59)

where we have also neglected z_2 in comparison with z_1 in the argument of one wave function.

In the transverse coordinate space representation, Eq. (4.59) has the form

$$\Psi_{\sigma\sigma'}^{(1)}(\vec{x}_{10}, \vec{x}_{20}, z_1, z_2) = \int \frac{d^2k_{1\perp}d^2k_{2\perp}}{(2\pi)^4}e^{i\vec{k}_{1\perp}\cdot\vec{x}_{10} + i\vec{k}_{2\perp}\cdot\vec{x}_{20}}\Psi_{\sigma\sigma'}^{(1)}(\vec{k}_{1\perp}, \vec{k}_{2\perp}, z_1, z_2)$$

$$= i\frac{gt^a}{\pi}\Psi_{\sigma\sigma'}^{(0)}(\vec{x}_{10}, z_1)\,\vec{\epsilon}_\perp^{\lambda*} \cdot \left(\frac{\vec{x}_{21}}{x_{21}^2} - \frac{\vec{x}_{20}}{x_{20}^2}\right),$$

(4.60)

where $\vec{x}_{20} = \vec{x}_{2\perp} - \vec{x}_{0\perp}$, $\vec{x}_{21} = \vec{x}_{2\perp} - \vec{x}_{1\perp}$, and $x_{ij} = |\vec{x}_{ij}|$ as defined after Eq. (1.87). The gluon has transverse coordinate $\vec{x}_{2\perp}$, as illustrated in Fig. 4.13. We have used Eq. (A.10) to obtain Eq. (4.60) from Eq. (4.59).

Squaring the coordinate-space one-gluon wave function from Eq. (4.60) and summing over the quark and gluon polarizations and colors yields

$$\sum_{\sigma,\sigma',\lambda,a}\left|\Psi_{\sigma\sigma'}^{(1)}\right|^2 = \frac{4\alpha_s C_F}{\pi}\frac{x_{10}^2}{x_{20}^2 x_{21}^2}\sum_{\sigma,\sigma'}\left|\Psi_{\sigma\sigma'}^{(0)}\right|^2.$$

(4.61)

To calculate the probability of finding one extra gluon in the onium wave function we have to integrate Eq. (4.61) over the gluon's phase space, which, in the $z_2 \ll z_1, 1 - z_1 \ll 1$ approximation, is (cf. Eq. (4.23))[4]

$$\int_{z_0}^{\min\{z_1, 1-z_1\}} \frac{dz_2}{z_2}\int\frac{d^2x_2}{4\pi},$$

(4.62)

where z_0 is some lower cutoff on the z_2-integral, imposed to make the integration finite; the exact value of z_0 depends on the physical process incorporating to the wave function we are constructing. The order-α_s contribution to the probability of finding one gluon in

---

[4] One may ask why, if our calculation is valid for $z_2 \ll z_1, 1 - z_1$, we can extend the z_2-integral all the way up to z_1 or $1 - z_1$. While indeed our approximation breaks down for z_2 close to z_1 or $1 - z_1$, putting z_1 or $1 - z_1$ as the upper integration limit gives the correct leading-logarithmic contribution.

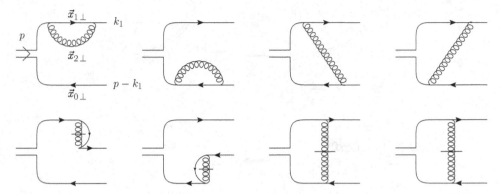

Fig. 4.14. Virtual contribution to small-x evolution in the onium wave function. The quark transverse coordinates in the onium are not changed by the corrections.

the onium wave function is then (Mueller 1994)

$$
\int\limits_{z_0}^{\min\{z_1,1-z_1\}} \frac{dz_2}{z_2} \int \frac{d^2x_2}{4\pi} \sum_{\sigma,\sigma',\lambda,a} \left|\Psi_{\sigma\sigma'}^{(1)}\right|^2 = \int\limits_{z_0}^{\min\{z_1,1-z_1\}} \frac{dz_2}{z_2} \int d^2x_2 \frac{\alpha_s C_F}{\pi^2} \frac{x_{10}^2}{x_{20}^2 x_{21}^2} \sum_{\sigma,\sigma'} \left|\Psi_{\sigma\sigma'}^{(0)}\right|^2 .
$$

$$(4.63)$$

Note that the modified wave function in Eq. (4.63) contains a power of α_s and a logarithmic integral over z_2, which would give us finally $\ln 1/x$. We see that the modification we have calculated brings in a factor $\alpha_s \ln 1/x$. Another feature of Eq. (4.63) is that the $\vec{x}_{21}$-integral in it contains UV divergences at $x_{20} \approx 0$ and $x_{21} \approx 0$. For now we will regulate these divergences by a UV cutoff ρ, such that $x_{20}, x_{21} > \rho$: in the end no physical quantity depends on the value of this cutoff.

Before we proceed let us point out that, as for the Glauber–Gribov–Mueller model (see e.g. Eq. (4.41)), the expression (4.63) completely factorizes transverse coordinate space into the square of the "bare" onium wave function times the probability of emission of the extra gluon. The emission of an extra gluon does not change the coordinates of the initial quark and the antiquark, yet again illustrating our above argument about the convenience of the transverse coordinate representation. This property also gives Eq. (4.61) a very simple physical meaning, resulting from the probabilistic interpretation of the light cone wave functions: the contribution to the onium wave function due to the emission of an extra gluon is equal to the product of the probability of finding a dipole with size x_{10} inside the onium ($\sim |\Psi_{\sigma\sigma'}^{(0)}|^2$) multiplied by the probability that the dipole emits a gluon at $\vec{x}_{21}$.

The one-gluon corrections to the dipole wave function need not be limited to the "real" gluon shown in Fig. 4.13; they should also include virtual corrections, where the gluon is both emitted and absorbed in the onium wave function, again like in the DGLAP case in Sec. 2.4.2. The virtual diagrams giving the LLA contributions are shown in Fig. 4.14, where, in accordance with the LCPT rules introduced in Sec. 1.3, the crossed lines denote instantaneous terms. From the sheer number of graphs in Fig. 4.14 one can see that direct calculation of all the virtual corrections can be a daunting task (see Chen and Mueller

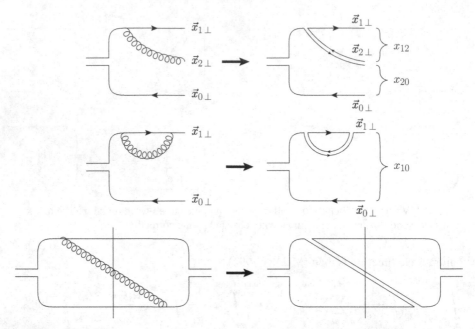

Fig. 4.15. Large-N_c limit in the onium wave function (top two panels) and the wave function squared (bottom panel). The curly brackets in the top panel denote the daughter dipoles generated by the gluon emission. The right-hand brace in the middle panel denotes the parent dipole remaining intact after a virtual correction. The thin vertical line in the bottom panel separates the wave function from its complex conjugate.

(1995) for an outline of the calculation). Instead we will follow Mueller (1994) and use the unitarity argument presented in Sec. 2.4.2 (see Eq. (2.86)) to write down the following expression for the order-α_s virtual correction to the onium wave function:

$$\Psi^{(0)}_{\sigma\sigma'}(\vec{x}_{10}, z_1)\Big|_{O(\alpha_s)} = -\frac{1}{2} \int\limits_{z_0}^{\min\{z_1, 1-z_1\}} \frac{dz_2}{z_2} \int d^2x_2 \frac{\alpha_s C_F}{\pi^2} \frac{x_{10}^2}{x_{20}^2 x_{21}^2} \Psi^{(0)}_{\sigma\sigma'}(\vec{x}_{10}, z_1)\Big|_{O(\alpha_s^0)}$$

$$= -\frac{2\alpha_s C_F}{\pi} \ln \frac{x_{01}}{\rho} \int\limits_{z_0}^{\min\{z_1, 1-z_1\}} \frac{dz_2}{z_2} \Psi^{(0)}_{\sigma\sigma'}(\vec{x}_{10}, z_1)\Big|_{O(\alpha_s^0)}. \quad (4.64)$$

The integral over $\vec{x}_{2\perp}$ is carried out in appendix section A.3 with ρ the UV regulator introduced above.

Having obtained the one-gluon corrections we would now like to derive an equation resumming the higher-order gluon emissions and virtual gluon corrections that bring powers of $\alpha_s \ln 1/x$ into the wave function. (Remember that quark loops do not contribute leading logarithms of x.) This turns out to be a rather difficult problem. A major simplification occurs if we consider the onium wave function in the 't Hooft large-N_c limit ('t Hooft 1974), taking N_c to be very large while keeping $\alpha_s N_c$ constant. In the large-N_c limit the single gluon line is replaced by a double line, corresponding to replacing the gluon by a quark–antiquark pair in the color-octet configuration. This is illustrated in Fig. 4.15. In the

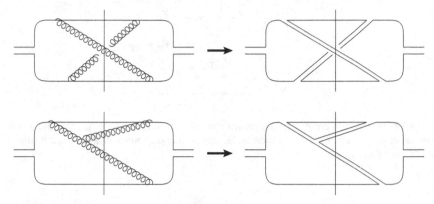

Fig. 4.16. Two steps of small-x evolution in the onium wave function squared (left) and their large-N_c limits (right). The top panel shows a nonplanar diagram, which is N_c^2-suppressed compared with the leading-N_c planar diagram shown in the bottom panel.

large-N_c limit it is convenient to talk about color dipoles instead of gluons. The original onium is a color dipole consisting of a quark at $\vec{x}_{1\perp}$ and an antiquark at $\vec{x}_{0\perp}$. The emission of a gluon in the onium wave function, taken in the large-N_c limit, corresponds to the splitting of the original dipole with size x_{10} into two dipoles with sizes x_{12} and x_{20}: the dipole the size x_{12} consists of the original quark at $\vec{x}_{1\perp}$ and the antiquark part of the gluon line at $\vec{x}_{2\perp}$, while the quark part of the gluon line at $\vec{x}_{2\perp}$ along with the original antiquark at $\vec{x}_{0\perp}$ form the dipole with size x_{20} (see the top and bottom panels of Fig. 4.15). The virtual gluon corrections leave the original dipole intact, as can be seen in the middle panel of Fig. 4.15.

Another important feature of the large-N_c limit is that only *planar diagrams* contribute; the nonplanar diagrams are suppressed by powers of N_c for fixed $\alpha_s N_c$. This means that different color dipoles generated by gluon emissions do not "talk" to each other: subsequent emissions happen independently in each dipole. This is illustrated in Fig. 4.16, where in the top panel we show an example of a diagram where a gluon emitted in one dipole in the amplitude connects to another dipole in the complex conjugate amplitude. As can be seen from the upper panel of Fig. 4.16, such a diagram is indeed nonplanar; hence, it is $1/N_c^2$-suppressed (as can be checked explicitly) and can be neglected in the large-N_c limit. At the same time, the diagram in the lower panel of Fig. 4.16, while of the same order in $\alpha_s \ln 1/x$, is also planar: in it the gluon from one dipole does not interact with the other dipole, remaining instead in its own dipole. This second diagram in Fig. 4.16 is of leading order in N_c and has to be resummed by large-N_c dipole evolution. (Strictly speaking, the diagram in the lower left panel of Fig. 4.16, when written in double-line notation, also contains a nonplanar subleading-N_c correction, in which the quark line in the longer gluon interacts with the quark of the original dipole: this correction is not shown in Fig. 4.16.)

Note that, in order to obtain the leading-ln $1/x$ contribution to the wave function, the softer gluons (those with smaller z) have to be emitted later (to the right in our LCPT diagrams) than the harder gluons, with larger values of z. For instance, let us consider an onium wave function with two gluon emissions, as shown in Fig. 4.17. Assume further that the gluon emitted earlier is softer than the gluon emitted later, i.e., that $z_3 \ll z_2$,

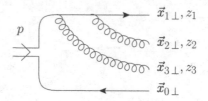

Fig. 4.17. Two gluons emitted in the onium wave function: if one assumes that the gluon emitted earlier is softer, $z_3 \ll z_2$, then the square of this diagram will not give a leading-ln $1/x$ contribution.

where, as usual,

$$z_i = \frac{k_i^+}{p^+}. \tag{4.65}$$

A simple calculation of the wave function in Fig. 4.17, in the $z_3 \ll z_2$ approximation, carried out along the steps outlined above for a single emission would yield a wave function proportional to z_3/z_2 as far as the longitudinal momentum dependence is concerned. Squaring this wave function and integrating the result over z_2 and z_3 with $z_1 \gg z_2 \gg z_3 \gg z_0$ yields an answer proportional to

$$\alpha_s^2 \int\limits_{z_0}^{z_1} \frac{dz_2}{z_2} \int\limits_{z_0}^{z_2} \frac{dz_3}{z_3} \frac{z_3^2}{z_2^2} \approx \frac{1}{2} \alpha_s^2 \ln \frac{z_1}{z_0}. \tag{4.66}$$

We see that we have only one longitudinal logarithm per two powers of the coupling α_s: this is not a leading logarithmic contribution. Hence the square of the diagram in Fig. 4.17 is subleading in ln $1/x$ and does not contribute to the leading-ln $1/x$ evolution we are considering here. It does contribute when one attempts to calculate the NLO corrections to the evolution we are about to construct (see Chapter 6). Using similar arguments, one can show that the diagram in Fig. 4.17 does not contribute to the LLA, even when we take its overlap with the wave function resulting when gluon 3 is emitted after gluon 2. In fact one can also show that no diagram with inverse time-ordering like that in Fig. 4.17 contributes in the LLA approximation. We thus come to another important conclusion: to obtain LLA evolution in the wave function, the gluon emissions with

$$z_2 \gg z_3 \gg \cdots \gg z_n \tag{4.67}$$

must be ordered in time, with the harder (larger-z) gluons emitted *before* the softer (small-z) gluons.

Now the structure of the small-x light cone wave function becomes manifest: in one step of evolution a gluon is emitted. It can be a real gluon, like those in the top and bottom panels of Fig. 4.15, which would split the initial (parent) dipole 10 ("one-zero") into two new (daughter) dipoles 12 and 20. The subsequent $\alpha_s \ln 1/x$ evolution is driven by further gluon emission: this would happen independently (and in parallel) in both daughter dipoles. An example of two-gluon emission is shown in the second panel of Fig. 4.16. Alternatively, the emission in the initial dipole can be virtual, as shown in the middle panel of Fig. 4.15;

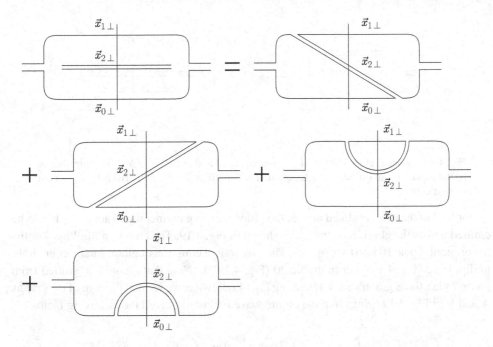

Fig. 4.18. Definition of an abbreviated notation for the sum of all large-N_c diagrams contributing to the real-gluon emission in the square of an onium wave function.

then the initial dipole remains intact, any subsequent evolution occurring within the initial dipole at later times.

As we can see from Eqs. (4.63) and (4.64), in the mixed representation in which we are working, each step of the evolution factorizes from the previous one, simplifying the construction of the gluon wave function. To illustrate this, let us consider two steps of small-x evolution due to two consecutive real-gluon emissions, including all possible LLA diagrams. It is convenient to introduce the shorthand diagram notation presented in Fig. 4.18, where the sum of all four (large-N_c) diagrams corresponding to real-gluon emission in the onium wave function comprises one diagram, that in the upper left of the figure. The diagrams in Fig. 4.18 give us the correction to the dipole wave function in Eq. (4.63). The kernel of this correction can be decomposed as follows:

$$\frac{\alpha_s C_F}{\pi^2} \frac{x_{10}^2}{x_{20}^2 x_{21}^2} = \frac{\alpha_s C_F}{\pi^2} \left(\frac{1}{x_{21}^2} - 2\frac{\vec{x}_{21} \cdot \vec{x}_{20}}{x_{21}^2 x_{20}^2} + \frac{1}{x_{20}^2} \right), \tag{4.68}$$

where the first and the last terms on the right-hand side of Eq. (4.68) correspond to the last two graphs in Fig. 4.18, while the first two (interference) diagrams on the right of Fig. 4.18 give the second term on the right of Eq. (4.68). The very first diagram in Fig. 4.18 corresponds to the full emission kernel on the left of Eq. (4.68).

Using the notation of Fig. 4.18, the square of the large-N_c onium wave function with two real gluons in it in the LLA approximation can be represented simply by the two diagrams depicted in Fig. 4.19, with the gluons ordered in longitudinal momenta such that $z_2 \gg z_3$.

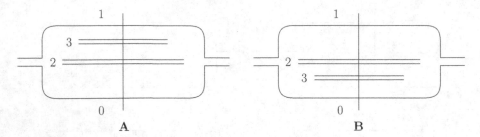

Fig. 4.19. Two real gluons in the LLA approximation and in the large-N_c limit contributing to the square of an onium wave function. The length of the lines is driven by light cone time-ordering.

According to the rules outlined above, the gluon carrying momentum fraction z_2 has to be emitted before the gluon carrying z_3, as shown in Fig. 4.19. The emission of gluon 2 splits the original dipole 10 into two dipoles. The subsequent emission of gluon 3 can occur either in dipole 12 (Fig. 4.19A) or in dipole 20 (Fig. 4.19B). (Note that gluon 3 is emitted from gluon 2 via a three-gluon vertex.) Iterating Eq. (4.63) twice, we see that the sum of the graphs A and B in Fig. 4.19 brings into the onium wave function squared the following factor:

$$\int_{z_0}^{z_1} \frac{dz_2}{z_2} \int_{z_0}^{z_2} \frac{dz_3}{z_3} \int d^2x_2\, d^2x_3 \left(\frac{\alpha_s C_F}{\pi^2} \right)^2 \frac{x_{10}^2}{x_{20}^2 x_{21}^2} \left(\frac{x_{12}^2}{x_{31}^2 x_{32}^2} + \frac{x_{20}^2}{x_{32}^2 x_{30}^2} \right). \tag{4.69}$$

(For simplicity of notation we have put z_1 as the upper cutoff of the z_2-integration, since at LLA accuracy one cannot see any significant difference between z_1 and $1 - z_1$.) Equation (4.69) demonstrates that the small-x evolution in the onium wave function consists of consecutive emissions ordered in rapidity and light cone time, with the transverse dynamics included in a factorized way.

To describe the onium wave function formally including $\alpha_s \ln 1/x$ corrections to all orders it is convenient to define the *dipole generating functional* $Z(\vec{x}_{10}, \vec{b}_{0\perp}, Y; u)$ by

$$Z(\vec{x}_{10}, \vec{b}_{0\perp}, Y; u) \sum_{\sigma\sigma'} |\Psi_{\sigma\sigma'}^{(0)}(\vec{x}_{10}, z_1)|^2 \Big|_{O(\alpha_s^0)}$$

$$= \int d^2r_1 d^2b_1 |\Psi^{[1]}(\vec{r}_{1\perp}, \vec{b}_{1\perp}, Y)|^2 u(\vec{r}_{1\perp}, \vec{b}_{1\perp})$$

$$+ \frac{1}{2!} \int d^2r_1 d^2b_1 d^2r_2 d^2b_2 |\Psi^{[2]}(\vec{r}_{1\perp}, \vec{b}_{1\perp}, \vec{r}_{2\perp}, \vec{b}_{2\perp}, Y)|^2$$

$$\times u(\vec{r}_{1\perp}, \vec{b}_{1\perp}) u(\vec{r}_{2\perp}, \vec{b}_{2\perp}) + \cdots$$

$$= \sum_{n=1}^{\infty} \frac{1}{n!} \int d^2r_1 d^2b_1 \cdots d^2r_n d^2b_n |\Psi^{[n]}(\vec{r}_{1\perp}, \vec{b}_{1\perp}, \ldots, \vec{r}_{n\perp}, \vec{b}_{n\perp}, Y)|^2$$

$$\times u(\vec{r}_{1\perp}, \vec{b}_{1\perp}) \cdots u(\vec{r}_{n\perp}, \vec{b}_{n\perp}). \tag{4.70}$$

We have defined the rapidity variable $Y = \ln(z_1/z_0)$, where now z_0 is the smallest momentum fraction carried by a gluon in the wave function. In Eq. (4.70) the light cone wave functions $\Psi^{[n]}(\vec{r}_{1\perp}, \vec{b}_{1\perp}, \ldots, \vec{r}_{n\perp}, \vec{b}_{n\perp}, Y)$ correspond to the onium state consisting of n dipoles with sizes $\vec{r}_{1\perp}, \ldots, \vec{r}_{n\perp}$ whose centers (in the transverse plane) are located at impact parameters $\vec{b}_{1\perp}, \ldots, \vec{b}_{n\perp}$ (e.g. $\vec{b}_{0\perp} = (1/2)(\vec{x}_{1\perp} + \vec{x}_{0\perp})$). The rapidity interval between these daughter dipoles and the original parent dipole $1\,0$ is less than or equal to Y, i.e., the wave functions squared $|\Psi^{[n]}|^2$ are implicitly integrated over dipole rapidities from 0 to Y. Summation over all appropriate quantum numbers is implied in the square of the wave function $\Psi^{[n]}$. Note that $\Psi^{(0)}_{\sigma\sigma'}(\vec{x}_{10}, z_1)$ taken at order α_s^0 is the bare wave function of the onium before any emissions have taken place. In going from gluons to color dipoles we have changed the notation for the wave functions: while $\Psi^{(n)}$ denotes a wave function with n real gluons in it, $\Psi^{[n]}$ is a wave function with n dipoles (note the use of square brackets rather than parentheses). Since we always have at least one dipole (the original onium), the sum over n in Eq. (4.70) starts at $n = 1$.

The dummy functions $u(\vec{r}_{n\perp}, \vec{b}_{n\perp})$ are introduced so that one can extract the squares of different multi-dipole onium wave functions from the generating functional Z, using

$$|\Psi^{[n]}(\vec{r}_{1\perp}, \vec{b}_{1\perp}, \ldots, \vec{r}_{n\perp}, \vec{b}_{n\perp}, Y)|^2 = \sum_{\sigma\sigma'} |\Psi^{(0)}_{\sigma\sigma'}(\vec{x}_{10}, z_1)|^2 \Big|_{O(\alpha_s^0)}$$

$$\times \frac{\delta^n}{\delta u(\vec{r}_{1\perp}, \vec{b}_{1\perp}) \cdots \delta u(\vec{r}_{n\perp}, \vec{b}_{n\perp})} Z(\vec{x}_{10}, \vec{b}_{0\perp}, Y; u) \Big|_{u=0},$$
(4.71)

where $\delta/\delta u(\vec{r}_\perp, \vec{b}_\perp)$ is a functional derivative. As usual this derivative is defined such that (see e.g. Peskin and Schroeder (1995) for details)

$$\frac{\delta}{\delta u(\vec{r}_\perp, \vec{b}_\perp)} u(\vec{r}'_\perp, \vec{b}'_\perp) = \delta^{(2)}(\vec{r}_\perp - \vec{r}'_\perp) \delta^{(2)}(\vec{b}_\perp - \vec{b}'_\perp),$$
(4.72)

which leads to

$$\frac{\delta}{\delta u(\vec{r}_\perp, \vec{b}_\perp)} \int d^2r' d^2b' \, f(\vec{r}'_\perp, \vec{b}'_\perp) u(\vec{r}'_\perp, \vec{b}'_\perp) = f(\vec{r}_\perp, \vec{b}_\perp)$$
(4.73)

for an arbitrary function $f(\vec{r}_\perp, \vec{b}_\perp)$.

Since $|\Psi^{[n]}|^2$ gives the probability of having n dipoles in the onium wave function in a given transverse space configuration, and since the sum over probabilities of having any number of dipoles in all transverse configurations is 1, we conclude that (see Eq. (1.70))

$$Z(\vec{x}_{10}, \vec{b}_{0\perp}, Y; u = 1) = 1.$$
(4.74)

We want to write down an evolution equation for the generating functional Z summing all powers of $\alpha_s Y$. Before we do so, let us set up the initial condition for such an evolution. When $Y = 0$ we have no evolution and no gluon emissions (neither real nor virtual). Hence

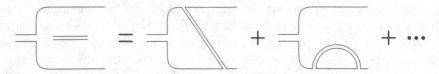

Fig. 4.20. An abbreviated notation for the sum of all large-N_c diagrams contributing to the virtual gluon correction to the onium wave function.

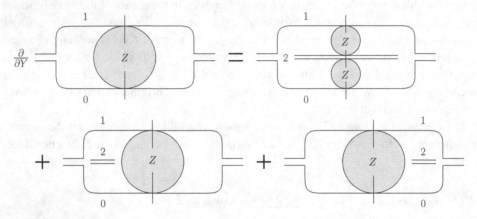

Fig. 4.21. Diagrammatic representation for the evolution equation of the generating functional Z (denoted by a shaded circle).

$|\Psi^{[n>1]}(Y=0)|^2 = 0$ and

$$|\Psi^{[1]}(\vec{r}_{1\perp}, \vec{b}_{1\perp}, Y=0)|^2 = \delta^2\left(\vec{b}_{1\perp} + \frac{\vec{r}_{1\perp}}{2} - \vec{x}_{1\perp}\right)\delta^2\left(\vec{b}_{1\perp} - \frac{\vec{r}_{1\perp}}{2} - \vec{x}_{0\perp}\right), \quad (4.75)$$

such that

$$Z(\vec{x}_{10}, \vec{b}_{0\perp}, Y=0; u) = u(\vec{x}_{10}, \vec{b}_{0\perp}). \quad (4.76)$$

Now that we have the initial conditions for Z-evolution, it is straightforward to write down an evolution equation for Z. The main principle was stated several pages ago: in one step of evolution a gluon is emitted in the dipole wave function: the gluon may be real, splitting the parent dipole into two daughter dipoles, or it may be virtual, leaving the parent dipole intact. In the former case the subsequent evolution continues independently in the two daughter dipoles, while in the latter case evolution continues in the parent dipole. This statement is illustrated diagrammatically in Fig. 4.21, where the generating functional Z is represented by a shaded circle. The first graph on the right of Fig. 4.21 corresponds to real-gluon emission, while the remaining two graphs represent the sum of all virtual corrections, as shown in Fig. 4.20.

Guided by Fig. 4.21, and employing Eqs. (4.63) and (4.64) while replacing C_F by $N_c/2$ in the large-N_c limit, we can write down the following evolution equation for the generating functional Z (Mueller 1994, 1995):

$$\frac{\partial}{\partial Y} Z(\vec{x}_{10}, \vec{b}_{0\perp}, Y; u)$$

$$= \frac{\alpha_s N_c}{2\pi^2} \int d^2 x_2 \frac{x_{10}^2}{x_{20}^2 x_{21}^2}$$

$$\times \left[Z\left(\vec{x}_{12}, \vec{b}_{0\perp} + \frac{\vec{x}_{20}}{2}, Y; u \right) Z\left(\vec{x}_{20}, \vec{b}_{0\perp} + \frac{\vec{x}_{21}}{2}, Y; u \right) - Z(\vec{x}_{10}, \vec{b}_{0\perp}, Y; u) \right].$$

$$(4.77)$$

The first term on the right-hand side of Eq. (4.77) corresponds to the first term on the right of Fig. 4.21, while the last two terms in Fig. 4.21 give rise to the second term on the right of Eq. (4.77). The minus sign in this second term is due to the minus sign in the virtual correction in Eq. (4.64).

Equation (4.77) is a nonlinear evolution equation whose initial condition is given in Eq. (4.76). Solving this evolution equation would allow one to construct the squares of the multi-dipole onium wave functions using Eq. (4.71). Unfortunately the exact analytical solution of Eq. (4.77) is not known. So, let us first connect Eq. (4.77) with results that are already familiar to the reader, such as the BFKL equation.

4.3.2 The BFKL equation in transverse coordinate space

Consider the following functional derivative taken at $u = 1$:

$$\left. \frac{\delta Z(\vec{x}_{10}, \vec{b}_{0\perp}, Y; u)}{\delta u(\vec{r}_\perp, \vec{b}_\perp)} \right|_{u=1} = \sum_{n=1}^{\infty} \frac{n}{n!} \int d^2 r_2 d^2 b_2 \cdots \cdots d^2 r_n d^2 b_n$$

$$\times \left. \frac{|\Psi^{[n]}(\vec{r}_\perp, \vec{b}_\perp, \vec{r}_{2\perp}, \vec{b}_{2\perp}, \ldots, \vec{r}_{n\perp}, \vec{b}_{n\perp}, Y)|^2}{\sum_{\sigma\sigma'} |\Psi_{\sigma\sigma'}^{(0)}(\vec{x}_{10}, z_1)|^2} \right|_{O(\alpha_s^0)}. \qquad (4.78)$$

If instead the value of the derivative had been taken at $u = 0$, the physical meaning of the above object would have been clear from Eq. (4.71): it would have been the single-dipole wave function squared, divided by the original onium's wave function. To understand the physical meaning of the actual object in Eq. (4.78) we note that the probability of having n dipoles in the onium wave function (for an onium of given size $\vec{x}_{10}$ and quark momentum fraction z_1) is given by

$$P_n(Y) = \frac{1}{n!} \int d^2 r_1 d^2 b_1 \cdots d^2 r_n d^2 b_n \left. \frac{|\Psi^{[n]}(\vec{r}_{1\perp}, \vec{b}_{1\perp}, \ldots, \vec{r}_{n\perp}, \vec{b}_{n\perp}, Y)|^2}{\sum_{\sigma\sigma'} |\Psi_{\sigma\sigma'}^{(0)}(\vec{x}_{10}, z_1)|^2} \right|_{O(\alpha_s^0)}, \qquad (4.79)$$

where the factorial is a symmetry factor removing the multiple counting of identical dipole configurations, and where $n > 0$. The condition (4.74) translates into $\sum_{n=1}^{\infty} P_n(Y) = 1$.

The average number of dipoles (at rapidities up to Y) in the onium wave function is

$$\langle n(Y) \rangle = \sum_{n=1}^{\infty} n P_n(Y). \tag{4.80}$$

The series (4.80) is very similar to that in Eq. (4.78), except that in Eq. (4.78) we are keeping the transverse size and impact parameter of one dipole fixed. We have thus arrived at the physical meaning of the object in Eq. (4.78): it is the number of dipoles of size $\vec{r}_{\perp}$ at impact parameter $\vec{b}_{\perp}$ and with rapidities between 0 and Y located in the onium wave function. We denote this object by $n_1(\vec{x}_{10}, \vec{r}_{\perp}, \vec{b}_{\perp} - \vec{b}_{0\perp}, Y)$, so that

$$n_1(\vec{x}_{10}, \vec{r}_{\perp}, \vec{b}_{\perp} - \vec{b}_{0\perp}, Y) = \left. \frac{\delta Z(\vec{x}_{10}, \vec{b}_{0\perp}, Y; u)}{\delta u(\vec{r}_{\perp}, \vec{b}_{\perp})} \right|_{u=1}. \tag{4.81}$$

To construct an equation for $n_1(\vec{x}_{10}, \vec{r}_{\perp}, \vec{b}_{\perp} - \vec{b}_{0\perp}, Y)$ we simply have to differentiate Eq. (4.77) with respect to u, putting $u = 1$ at the end using Eq. (4.74). This yields an equation that we will shortly show to be equivalent to the BFKL equation (Mueller 1994, Mueller and Patel 1994, Mueller 1995, Nikolaev, Zakharov, and Zoller (1994))[5]

$$\frac{\partial}{\partial Y} n_1(\vec{x}_{10}, \vec{r}_{\perp}, \vec{b}_{\perp}, Y) = \frac{\alpha_s N_c}{2\pi^2} \int d^2 x_2 \frac{x_{10}^2}{x_{20}^2 x_{21}^2} \left[n_1\left(\vec{x}_{12}, \vec{r}_{\perp}, \vec{b}_{\perp} - \frac{\vec{x}_{20}}{2}, Y \right) \right.$$

$$\left. + n_1\left(\vec{x}_{20}, \vec{r}_{\perp}, \vec{b}_{\perp} - \frac{\vec{x}_{21}}{2}, Y \right) - n_1(\vec{x}_{10}, \vec{r}_{\perp}, \vec{b}_{\perp}, Y) \right]. \tag{4.82}$$

We have relabeled $\vec{b}_{\perp} - \vec{b}_{0\perp}$ simply as $\vec{b}_{\perp}$, which therefore now has the meaning of the transverse space distance between the centers of the original dipole 10 and the dipole of interest, of size $\vec{r}_{\perp}$. The initial condition for Eq. (4.82) is obtained by differentiating Eq. (4.76) with respect to $u(\vec{r}_{\perp}, \vec{b}_{\perp})$ and afterwards putting $u = 1$:

$$n_1(\vec{x}_{10}, \vec{r}_{\perp}, \vec{b}_{\perp}, Y = 0) = \delta^2(\vec{x}_{10} - \vec{r}_{\perp}) \delta^2\left(\vec{b}_{\perp} \right). \tag{4.83}$$

The distribution of pairs of dipoles in the onium wave function can be defined as a second derivative of the generating functional:

$$n_2(\vec{x}_{10}, \vec{r}_{1\perp}, \vec{b}_{1\perp} - \vec{b}_{0\perp}, \vec{r}_{2\perp}, \vec{b}_{2\perp} - \vec{b}_{0\perp}, Y) = \left. \frac{\delta^2 Z(\vec{x}_{10}, \vec{b}_{0\perp}, Y; u)}{\delta u(\vec{r}_{1\perp}, \vec{b}_{1\perp}) \delta u(\vec{r}_{2\perp}, \vec{b}_{2\perp})} \right|_{u=1}. \tag{4.84}$$

Equation (4.84) gives the number of pairs of dipoles with sizes $\vec{r}_{1\perp}$ and $\vec{r}_{2\perp}$ located at impact parameters $\vec{b}_{1\perp}$ and $\vec{b}_{2\perp}$ and in the rapidity interval $[0, Y]$. The equation for n_2 is constructed in analogy to that for n_1 by the double differentiation of Eq. (4.77) with respect to $u(\vec{r}_{1\perp}, \vec{b}_{1\perp})$ and $u(\vec{r}_{2\perp}, \vec{b}_{2\perp})$, putting $u = 1$ at the end. The main difference

---

[5] Nikolaev and Zakharov (1994) and Nikolaev, Zakharov, and Zoller (1994) were very close to solving these problems. Nikolaev and Zakharov (1994) rewrote the DLA DGLAP evolution in terms of color dipoles. Nikolaev, Zakharov, and Zoller (1994) obtained the BFKL equation in the dipole formulation, though several months later than Mueller (1994). Lipatov (1986) was the first to notice that the BFKL equation has a particularly elegant form in the transverse coordinate representation but his approach lacked the idea of using color dipoles instead of the transverse coordinates of the gluons.

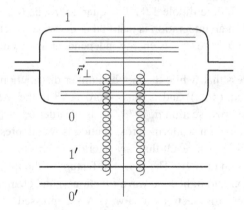

Fig. 4.22. Onium–onium scattering in the BFKL approximation in the dipole model, with the lower onium $1'0'$ at rest.

comes in the initial conditions: the second derivative of Eq. (4.76) with respect to u gives $n_2(\vec{x}_{10}, \vec{r}_{1\perp}, \vec{b}_{1\perp}, \vec{r}_{2\perp}, \vec{b}_{2\perp}, Y = 0) = 0$. The equation for n_2 is also a linear differential equation, though it also contains powers of n_1. We will not write down this equation explicitly and instead refer the reader to the papers by Mueller (1995) and Mueller and Patel (1994). Higher derivatives of the generating functional Z give the number distributions of dipole triplets, quadruplets, etc. The nth-order derivative of Z with respect to u gives the distribution of n dipoles in the onium wave function.

While dipole number distributions are interesting quantities in themselves, they also allow one to calculate scattering cross sections in a physically intuitive way. Consider onium–onium scattering at high energies, where the small-x evolution is important. This is the kinematics in which we studied the BFKL evolution in Sec. 3.3. Let us now try to reproduce the BFKL result in the dipole language. Consider a frame in which one onium is at rest while the other is incident on it at high energy. The total onium–onium scattering cross section per unit impact parameter can then be written as a convolution of the number of dipoles in the incident onium wave function and the scattering cross section of each dipole on the onium when at rest:

$$n(\vec{x}_{10}, \vec{x}_{1'0'}, \vec{b}, Y) = \int d^2r \, d^2b' \, n_1(\vec{x}_{10}, \vec{r}_\perp, \vec{b}'_\perp - \vec{b}_\perp, Y) \frac{d\hat{\sigma}_{tot}^{onium-onium}(\vec{r}_\perp, \vec{x}_{1'0'})}{d^2b'}. \quad (4.85)$$

Here the two colliding onia have transverse sizes $\vec{x}_{10}$ and $\vec{x}_{1'0'}$ (the latter is at rest), $\vec{b}$ is the impact parameter, Y is the net rapidity interval for the onium–onium scattering, and $d\hat{\sigma}_{tot}^{onium-onium}(\vec{r}_\perp, \vec{r}'_\perp)/d^2b$ is the cross section for the scattering of two onia with sizes $\vec{r}_\perp$ and $\vec{r}'_\perp$ mediated by a two-gluon exchange, as calculated in Exercise 3.3 (see Eq. (3.139)). Equation (4.85) is illustrated in Fig. 4.22, where dipole evolution in the onium 10 creates a dipole of size $\vec{r}_\perp$, which then interacts with the onium $1'0'$ via a two-gluon exchange.

Note that the dipole number density n_1 counts all dipoles with rapidities between 0 and Y (with respect to the dipole 10): any of these dipoles (if it has size $\vec{r}_\perp$, over which we will integrate) can interact with the dipole $1'0'$. The quantity n_1 is, therefore, integrated over the dipole rapidities. Such an integration is justified at the wave function level because the Born-level cross section $\hat{\sigma}_{tot}^{onium-onium}$ is energy independent and cannot affect the result of the integration.

It is important to stress that, while the dipole number distribution n_1 in Eq. (4.85) is a function of $\alpha_s N_c$ (see Eq. (4.82)) and is thus of order N_c^0 in the 't Hooft large-N_c limit, the two-gluon exchange cross section $\hat{\sigma}_{tot}^{onium-onium}$ is of order $\alpha_s^2 \sim (\alpha_s N_c)^2/N_c^2 \sim 1/N_c^2$, so that the whole onium–onium scattering cross section is N_c-suppressed. This is indeed in agreement with the well-known result that any interaction cross section is N_c-suppressed at large N_c (see e.g. Witten (1979)). The essential feature of the interactions in the dipole model is the factorization into light cone wave function(s) and elementary scattering cross sections. As the scattering cross sections are always N_c-suppressed, to capture the dominant contribution to the scattering one has to use the leading-N_c wave functions constructed above. The factorization presented in Fig. 4.22 is not unique: in a different reference frame, say the center-of-mass frame, the dipole wave functions of both onia contain small-x evolution; a dipole from one wave function exchanges two gluons with a dipole in another wave function (Mueller and Patel 1994). Such a factorization gives the same answer as the one we will obtain below for Fig. 4.22.

To find the scattering cross section per unit impact parameter,

$$n(\vec{x}_{10}, \vec{x}_{1'0'}, \vec{b}_\perp, Y) = \frac{d\hat{\sigma}_{tot}^{onium-onium}(\vec{x}_{10}, \vec{x}_{1'0'}, Y)}{d^2 b}, \tag{4.86}$$

one can first solve Eq. (4.82) and then use the solution in Eq. (4.85) along with the cross section from Eq. (3.139). Alternatively, one may note that the cross section $n(\vec{x}_{10}, \vec{x}_{1'0'}, \vec{b}, Y)$ itself satisfies Eq. (4.82):

$$\frac{\partial}{\partial Y} n(\vec{x}_{10}, \vec{x}_{1'0'}, \vec{b}_\perp, Y)$$

$$= \frac{\alpha_s N_c}{2\pi^2} \int d^2 x_2 \frac{x_{10}^2}{x_{20}^2 x_{21}^2}$$

$$\times \left[n\left(\vec{x}_{12}, \vec{x}_{1'0'}, \vec{b}_\perp - \frac{\vec{x}_{20}}{2}, Y\right) + n\left(\vec{x}_{20}, \vec{x}_{1'0'}, \vec{b}_\perp - \frac{\vec{x}_{21}}{2}, Y\right) - n(\vec{x}_{10}, \vec{x}_{1'0'}, \vec{b}_\perp, Y) \right] \tag{4.87}$$

with initial condition (cf. Eq. (3.139))

$$n(\vec{x}_{10}, \vec{x}_{1'0'}, \vec{b}_\perp, Y = 0) = \frac{d\hat{\sigma}_{tot}^{onium-onium}(\vec{x}_{10}, \vec{x}_{1'0'})}{d^2 b}$$

$$= \frac{2\alpha_s^2 C_F}{N_c} \ln^2 \frac{x_{11'} x_{00'}}{x_{10'} x_{01'}}. \tag{4.88}$$

Equation (4.87) can be solved exactly: the solution is somewhat involved and will be left for the next section. Instead, we will consider here the simplified case where a cross section is integrated over all impact parameters $\vec{b}_\perp$. In momentum space this corresponds to the $t = 0$ case, of zero momentum transfer. On top of that we will average over the directions of $\vec{x}_{1'0'}$: the resulting cross section does not depend on the directions of $\vec{x}_{10}$ either, since there is no preferred direction left in the transverse space. Defining

$$n(x_{10}, x_{1'0'}, Y) = \int d^2 b \int\limits_0^{2\pi} \frac{d\phi_{1'0'}}{2\pi} n(\vec{x}_{10}, \vec{x}_{1'0'}, \vec{b}_{\perp}, Y),\qquad (4.89)$$

we see that this new quantity satisfies

$$\frac{\partial}{\partial Y} n(x_{10}, x_{1'0'}, Y) = \frac{\alpha_s N_c}{2\pi^2} \int d^2 x_2 \frac{x_{10}^2}{x_{20}^2 x_{21}^2}$$

$$\times \left[n\left(x_{12}, x_{1'0'}, Y\right) + n\left(x_{20}, x_{1'0'}, Y\right) - n(x_{10}, x_{1'0'}, Y) \right] \qquad (4.90)$$

with initial condition (cf. Eq. (3.25))

$$n(x_{10}, x_{1'0'}, Y = 0) = \frac{4\pi\alpha_s^2 C_F}{N_c} x_<^2 \left(\ln \frac{x_>}{x_<} + 1 \right),\qquad (4.91)$$

where $x_{>(<)} = \max(\min)\{|\vec{x}_{10}|, |\vec{x}_{1'0'}|\}$.

The solution of Eq. (4.90) can be found by noticing that in the angular-averaged case the eigenfunctions of the integral kernel are simple powers of the dipole size,

$$\left(x_{01}^2\right)^{1/2+i\nu}\qquad (4.92)$$

with eigenvalues

$$\frac{\alpha_s N_c}{\pi} \chi(0, \nu),\qquad (4.93)$$

where (cf. Eqs. (3.81), (3.74))

$$\chi(0, \nu) = 2\psi(1) - \psi\left(\frac{1}{2} + i\nu \right) - \psi\left(\frac{1}{2} - i\nu \right).\qquad (4.94)$$

To prove this we need to evaluate the following integral:

$$\int d^2 x_2 \frac{x_{10}^2}{x_{20}^2 x_{21}^2} \left[\left(x_{12}^2\right)^{1/2+i\nu} + \left(x_{20}^2\right)^{1/2+i\nu} - \left(x_{10}^2\right)^{1/2+i\nu} \right].\qquad (4.95)$$

This can be done by noticing that the integral (4.95) is equivalent to that in Eq. (3.64) with $n = 0$. Alternatively, one can use the trick presented in appendix section A.3; in order to make each term in Eq. (4.95) finite we insert a UV regulator ρ. After that, with the help of Eqs. (A.18), (A.21), (A.24), and (A.29) one can rewrite Eq. (4.95) as

$$2\pi \left[2^{1+2i\nu} \frac{\Gamma\left(\frac{1}{2} + i\nu\right)}{\Gamma\left(\frac{1}{2} - i\nu\right)} x_{10}^2 \int\limits_0^\infty dk\, k^{-2i\nu} \left(\ln \frac{2}{k\rho} + \psi(1) \right) J_0(kx_{10}) - x_{10}^{1+2i\nu} \ln \frac{x_{10}^2}{\rho^2} \right].\qquad (4.96)$$

Integrating over k in Eq. (4.96) using Eq. (A.18) yields

$$2\pi x_{10}^{1+2i\nu} \chi(0, \nu),\qquad (4.97)$$

as desired.

We see that, as for to the BFKL equation (3.58), the eigenfunctions of Eq. (4.90) are powers (though of the transverse dipole size instead of the transverse momentum), with exactly the same eigenvalues, (4.93) as in that case.[6] We conclude that Eq. (4.90) is equivalent to the BFKL equation!

In fact, the substitution (Levin and Ryskin 1987)

$$n(x_{10}, x_{1'0'}, Y) = \int d^2k \left(1 - e^{i\vec{k}_\perp \cdot \vec{x}_{10}}\right) \frac{1}{k_\perp^2} f(\vec{k}_\perp, x_{1'0'}, Y) \tag{4.98}$$

turns Eq. (4.90) into the BFKL equation (3.58) for the function f (Kovchegov and Weigert 2007b). Verification of this statement is left as an exercise for the reader.

Using the eigenfunctions and the eigenvalues of the integral kernel in Eq. (4.90), we can write down the solution of Eq. (4.90) as

$$n(x_{10}, x_{1'0'}, Y) = \int_{-\infty}^{\infty} dv \, C_v(x_{1'0'}) x_{10}^{1+2iv} e^{\bar{\alpha}_s \chi(0,v)Y}, \tag{4.99}$$

where the coefficient $C_v(x_{1'0'})$ is fixed by the initial conditions (4.91) as follows:

$$C_v(x_{1'0'}) = \frac{16\,\alpha_s^2 C_F}{N_c} \frac{1}{(1+4v^2)^2} x_{1'0'}^{1-2iv}. \tag{4.100}$$

The general solution of Eq. (4.90) is then

$$n(x_{10}, x_{1'0'}, Y) = \frac{16\alpha_s^2 C_F}{N_c} x_{10} \, x_{1'0'} \int_{-\infty}^{\infty} dv \left(\frac{x_{10}}{x_{1'0'}}\right)^{2iv} \frac{e^{\bar{\alpha}_s \chi(0,v)Y}}{(1+4v^2)^2}. \tag{4.101}$$

For $x_{10} \approx x_{1'0'}$ we can use the diffusion approximation from Sec. 3.3.4: expanding $\chi(0, v)$ around $v = 0$ using Eq. (3.84) and integrating over v we obtain

$$n(x_{10}, x_{1'0'}, Y) = \frac{16\alpha_s^2 C_F}{N_c} x_{10} x_{1'0'} \sqrt{\frac{\pi}{14\zeta(3)\bar{\alpha}_s Y}} \tag{4.102}$$

$$\times \exp\left[(\alpha_P - 1)Y - \frac{\ln^2(x_{10}/x_{1'0'})}{14\zeta(3)\bar{\alpha}_s Y}\right].$$

Readers who performed Exercise 3.5 will recognize Eq. (4.102) as the answer for the onium–onium scattering cross section obtained there using the standard Feynman diagram approach. Now we see that a calculation based on LCPT wave functions gives the same result. Note that the single-dipole distribution n_1 is only one component of the onium wave function. This wave function also contains multi-dipole distributions n_2, n_3, etc. Hence, as we will shortly see, the dipole approach, while in a certain limit equivalent to BFKL, in fact contains more information.

---

[6] We have verified this statement so far only in the case where the angular dependence has been integrated out: we will consider the general angular-dependent case in the next section.

4.3.3 The general solution of the coordinate-space BFKL equation*

Let us now construct the solution of the BFKL equation (4.87) without making any simplifying assumptions. The goal now is to construct the most general eigenfunctions of the
kernel of Eq. (4.87). This kernel operates in the transverse plane: it is convenient to think
of this plane as a complex plane, replacing the two-component vectors $\vec{x}_{i\perp}$ by complex
numbers ρ_i, namely

$$\rho_i = x_{i,1} + ix_{i,2}; \quad \rho_i^* = x_{i,1} - ix_{i,2}, \tag{4.103}$$

where the indices 1, 2 denote two transverse axes. In the same way as in the vector notation
we define $\rho_{ij} = \rho_i - \rho_j$ and $\rho_{ij}^* = \rho_i^* - \rho_j^*$, along with the absolute value squared $|\rho_{ij}|^2 = \rho_{ij}\rho_{ij}^*$ and the integration measure $d^2\rho = d\rho d\rho^*$. Using the above complex notation it is
straightforward to check that the kernel of Eq. (4.87), written as

$$\int d^2\rho_2 \frac{|\rho_{10}^2|}{|\rho_{20}^2||\rho_{21}^2|} \tag{4.104}$$

is conformally invariant: it is clearly invariant under rotations, translations, scale transformations, and reflections in the complex plane. It is also invariant under the inversion

$$\rho_i \rightarrow \frac{1}{\rho_i^*}, \quad \rho_i^* \rightarrow \frac{1}{\rho_i}. \tag{4.105}$$

Thus the kernel is invariant under all Möbius transformations

$$z \rightarrow \frac{az+b}{cz+d} \tag{4.106}$$

for arbitrary complex a, b, c, and d with $ad - bc \neq 0$. When $ad - bc = 1$ the group
reduces to SL(2, C).

Consider the functions (Lipatov 1986)

$$E^{n,\nu}(\rho_{1a}, \rho_{2a}) = \left(\frac{\rho_{12}}{\rho_{1a}\rho_{2a}}\right)^{(1+n)/2+i\nu} \left(\frac{\rho_{12}^*}{\rho_{1a}^*\rho_{2a}^*}\right)^{(1-n)/2+i\nu}, \tag{4.107}$$

where ρ_a is an arbitrary point in the complex (transverse) plane, with $\rho_{ia} = \rho_i - \rho_a$ as
before; n is integer and ν is real. It is easy to check by direct differentiation that the functions
$E^{n,\nu}$ are the eigenfunctions of the Casimir operators M^2 and M^{*2} of the conformal Möbius
group (Lipatov 1986, Lipatov 1989, Bartels, Lipatov, and Vacca 2005):

$$M^2 E^{n,\nu}(\rho_{1a}, \rho_{2a}) \equiv \rho_{12}^2 \partial_1 \partial_2 E^{n,\nu}(\rho_{1a}, \rho_{2a}) = -h(h-1)E^{n,\nu}(\rho_{1a}, \rho_{2a}), \tag{4.108a}$$

$$M^{*2} E^{n,\nu}(\rho_{1a}, \rho_{2a}) \equiv \rho_{12}^{*2} \partial_1^* \partial_2^* E^{n,\nu}(\rho_{1a}, \rho_{2a}) = -\bar{h}(\bar{h}-1)E^{n,\nu}(\rho_{1a}, \rho_{2a}), \tag{4.108b}$$

where $\partial_i = \partial/\partial\rho_i$, $\partial_i^* = \partial/\partial\rho_i^*$, and

$$h = \frac{1+n}{2} + i\nu, \quad \bar{h} = \frac{1-n}{2} + i\nu. \tag{4.109}$$

The functions $E^{n,\nu}$ are orthonormal (Lipatov 1986)

$$\int \frac{d^2\rho_1 d^2\rho_2}{|\rho_{12}|^4} E^{n,\nu}(\rho_{1a}, \rho_{2a})E^{m,\mu}(\rho_{1b}, \rho_{2b})$$

$$= a_{n,\nu}\delta_{nm}\delta(\nu - \mu)\delta^2(\rho_{ab}) + b_{n,\nu}|\rho_{ab}|^{-2-4i\nu}\left(\frac{\rho_{ab}}{\rho_{ab}^*}\right)^n \delta_{n,-m}\delta(\nu + \mu), \quad (4.110)$$

where

$$a_{n,\nu} = \frac{\pi^4/2}{\nu^2 + n^2/4} = \frac{|b_{n,\nu}|^2}{2\pi^2}, \quad (4.111a)$$

$$b_{n,\nu} = \pi^3 2^{4i\nu} \frac{\Gamma\left(-i\nu + \frac{1}{2}(1+|n|)\right)}{\Gamma\left(i\nu + \frac{1}{2}(1+|n|)\right)} \frac{\Gamma\left(i\nu + \frac{1}{2}|n|\right)}{\Gamma\left(1 - i\nu + \frac{1}{2}|n|\right)}. \quad (4.111b)$$

(Note that $E^{n,\nu}$ and $E^{-n,-\nu}$ are not orthogonal.) The functions $E^{n,\nu}$ also form a complete basis (Lipatov 1986), so that

$$(2\pi)^4\delta^2(\rho_{11'})\delta^2(\rho_{22'}) = \sum_{n=-\infty}^{\infty} \int_{-\infty}^{\infty} d\nu \int d^2\rho_a \frac{16(\nu^2 + \frac{1}{4}n^2)}{|\rho_{12}|^2|\rho_{1'2'}|^2}$$

$$\times E^{n,\nu}(\rho_{1a}, \rho_{2a})E^{n,\nu*}(\rho_{1'a}, \rho_{2'a}). \quad (4.112)$$

The delta functions on the left of Eq. (4.112) should be understood as acting on the space of well-behaved functions of ρ_1, ρ_2, $\rho_{1'}$, and $\rho_{2'}$ that go to zero in the limits $\rho_1 = \rho_2$ and $\rho_{1'} = \rho_{2'}$.

Since the kernel of Eq. (4.87) is invariant under Möbius transformations, the functions $E^{n,\nu}$ are its eigenfunctions. To see this explicitly we need to find

$$I(\rho_0, \rho_1, \rho_a) \equiv \int d^2\rho_2 \frac{|\rho_{10}^2|}{|\rho_{20}^2||\rho_{21}^2|}[E^{n,\nu}(\rho_{1a}, \rho_{2a}) + E^{n,\nu}(\rho_{2a}, \rho_{0a}) - E^{n,\nu}(\rho_{1a}, \rho_{0a})]. \quad (4.113)$$

Performing the inversion transformation and also reflection with respect to ρ_a, i.e., $\rho_i \to 1/\rho_{ia}$, yields

$$I(1/\rho_0, 1/\rho_1, \infty) = \int d^2\rho_2 \frac{|\rho_{01}|^2}{|\rho_{02}|^2|\rho_{12}|^2}\left(\rho_{20}^h\rho_{20}^{*\bar{h}} + \rho_{12}^h\rho_{12}^{*\bar{h}} - \rho_{10}^h\rho_{10}^{*\bar{h}}\right). \quad (4.114)$$

The integral now becomes equivalent to that in Eq. (3.64), the answer to which is given by Eqs. (3.68) and (3.74). Using those results and reversing the $\rho_i \to 1/\rho_{ia}$ transformation, we write

$$I(\rho_0, \rho_1, \rho_a) = 2\pi \chi(n, \nu)E^{n,\nu}(\rho_{1a}, \rho_{0a}), \quad (4.115)$$

with $\chi(n, \nu)$ given by Eq. (3.81).

Since the $E^{n,\nu}$ are the eigenfunctions of the dipole kernel and form a complete orthonormal basis, we can write the general solution of Eq. (4.87) as

$$n(\vec{x}_{10}, \vec{x}_{1'0'}, \vec{b}_\perp, Y) = n(\rho_1, \rho_0; \rho_{1'}, \rho_{0'}; Y)$$

$$= \sum_{n=-\infty}^{\infty} \int_{-\infty}^{\infty} d\nu \int d^2\rho_a e^{\bar{\alpha}_s \chi(n,\nu) Y} C_{n,\nu} E^{n,\nu}(\rho_{1a}, \rho_{0a}) E^{n,\nu*}(\rho_{1'a}, \rho_{0'a}),$$

$$(4.116)$$

with the coefficients $C_{n,\nu}$ fixed by the initial condition (4.88), which in the complex plane can be written as

$$n(\rho_1, \rho_0; \rho_{1'}, \rho_{0'}; Y = 0) = \frac{2\alpha_s^2 C_F}{N_c} \ln^2 \left| \frac{\rho_{11'} \rho_{00'}}{\rho_{10'} \rho_{01'}} \right|. \quad (4.117)$$

To find the $C_{n,\nu}$ we need to decompose the (Möbius-invariant) logarithm squared into a series over the $E^{n,\nu}$:

$$\ln^2 \left| \frac{\rho_{11'} \rho_{00'}}{\rho_{10'} \rho_{01'}} \right| = \sum_{n=-\infty}^{\infty} \int_{-\infty}^{\infty} d\nu \int d^2\rho_a D_{n,\nu} E^{n,\nu}(\rho_{1a}, \rho_{0a}) E^{n,\nu*}(\rho_{1'a}, \rho_{0'a}). \quad (4.118)$$

The coefficients $D_{n,\nu}$ can be found if we first note that

$$|\rho_{10}|^4 \partial_1 \partial_1^* \partial_0 \partial_0^* \ln^2 \left| \frac{\rho_{11'} \rho_{00'}}{\rho_{10'} \rho_{01'}} \right| = \frac{\pi^2}{2} |\rho_{10}|^2 |\rho_{1'0'}|^2 \left[\delta^2(\rho_{11'}) \delta^2(\rho_{00'}) + \delta^2(\rho_{10'}) \delta^2(\rho_{01'}) \right].$$

$$(4.119)$$

Using Eq. (4.112) along with the following property of the $E^{n,\nu}$ functions,

$$E^{n,\nu}(\rho_{1a}, \rho_{2a}) = (-1)^n E^{n,\nu}(\rho_{2a}, \rho_{1a}), \quad (4.120)$$

yields

$$|\rho_{10}|^4 \partial_1 \partial_1^* \partial_0 \partial_0^* \ln^2 \left| \frac{\rho_{11'} \rho_{00'}}{\rho_{10'} \rho_{01'}} \right| = \frac{1}{\pi^2} \sum_{\text{even } n} \int_{-\infty}^{\infty} d\nu \int d^2\rho_a \left(\nu^2 + \tfrac{1}{4} n^2 \right)$$

$$\times E^{n,\nu}(\rho_{0a}, \rho_{1a}) E^{n,\nu*}(\rho_{0'a}, \rho_{1'a}), \quad (4.121)$$

where the sum runs over all integer even n. However, using Eqs. (4.108) and (4.118) we get

$$|\rho_{10}|^4 \partial_1 \partial_1^* \partial_0 \partial_0^* \ln^2 \left| \frac{\rho_{11'} \rho_{00'}}{\rho_{10'} \rho_{01'}} \right| = \sum_{n=-\infty}^{\infty} \int_{-\infty}^{\infty} d\nu \int d^2\rho_a D_{n,\nu} h(h-1) \bar{h}(\bar{h}-1)$$

$$\times E^{n,\nu}(\rho_{1a}, \rho_{0a}) E^{n,\nu*}(\rho_{1'a}, \rho_{0'a}). \quad (4.122)$$

Comparing Eqs. (4.122) and (4.121) we can read off $D_{n,\nu}$, and substituting it back into Eq. (4.118) we obtain

$$\ln^2 \left| \frac{\rho_{11'}\rho_{00'}}{\rho_{10'}\rho_{01'}} \right| = \frac{1}{\pi^2} \sum_{\text{even } n} \int\limits_{-\infty}^{\infty} d\nu \int d^2\rho_a \frac{\nu^2 + \frac{1}{4}n^2}{\left[\nu^2 + \frac{1}{4}(n+1)^2\right]\left[\nu^2 + \frac{1}{4}(n-1)^2\right]}$$

$$\times E^{n,\nu}(\rho_{0a}, \rho_{1a}) E^{n,\nu*}(\rho_{0'a}, \rho_{1'a}). \tag{4.123}$$

Equations (4.123) and (4.117), when compared with Eq. (4.116), allow us to write for even n

$$C_{n,\nu} = \frac{2\alpha_s^2 C_F}{\pi^2 N_c} \frac{\nu^2 + \frac{1}{4}n^2}{\left[\nu^2 + \frac{1}{4}(n+1)^2\right]\left[\nu^2 + \frac{1}{4}(n-1)^2\right]} \tag{4.124}$$

with $C_{n,\nu} = 0$ for odd n.

Equations (4.116) and (4.124) give us the most general solution of Eq. (4.87) with initial condition (4.88) (cf. Lipatov 1986):

$$n(\rho_1, \rho_0; \rho_{1'}, \rho_{0'}; Y)$$

$$= \frac{2\alpha_s^2 C_F}{\pi^2 N_c} \sum_{\text{even } n} \int\limits_{-\infty}^{\infty} d\nu \, e^{\bar{\alpha}_s \chi(n,\nu) Y}$$

$$\times \frac{\nu^2 + \frac{1}{4}n^2}{\left[\nu^2 + \frac{1}{4}(n+1)^2\right]\left[\nu^2 + \frac{1}{4}(n-1)^2\right]} \int d^2\rho_a E^{n,\nu}(\rho_{0a}, \rho_{1a}) E^{n,\nu*}(\rho_{0'a}, \rho_{1'a}). \tag{4.125}$$

The integral over ρ_a in Eq. (4.125) can be carried out analytically (Lipatov 1997, Navelet and Peschanski 1997), yielding a somewhat simplified expression in terms of hypergeometric functions:

$$n(\rho_1, \rho_0; \rho_{1'}, \rho_{0'}; Y)$$

$$= \frac{\alpha_s^2 C_F}{\pi^4 N_c} \sum_{\text{even } n} \int\limits_{-\infty}^{\infty} d\nu \, e^{\bar{\alpha}_s \chi(n,\nu) Y} \frac{\nu^2 + \frac{1}{4}n^2}{\left[\nu^2 + \frac{1}{4}(n+1)^2\right]\left[\nu^2 + \frac{1}{4}(n-1)^2\right]}$$

$$\times \left[b_{n,-\nu} w^h w^{*\bar{h}} F(h, h; 2h; w) F(\bar{h}, \bar{h}; 2\bar{h}; w^*) \right.$$

$$\left. + b_{n,\nu} w^{1-h} w^{*1-\bar{h}} F(1-h, 1-h; 2(1-h); w) F(1-\bar{h}, 1-\bar{h}; 2(1-\bar{h}); w^*) \right], \tag{4.126}$$

where

$$w = \frac{\rho_{01}\rho_{0'1'}}{\rho_{00'}\rho_{11'}}, \tag{4.127}$$

so that

$$\ln^2 \left| \frac{\rho_{11'}\rho_{00'}}{\rho_{10'}\rho_{01'}} \right| = \ln^2 |1 - w|. \tag{4.128}$$

4.4 The Balitsky–Kovchegov equation

We now return to the DIS process in the dipole picture of Sec. 4.1. As follows from Eqs. (4.12) and (4.24), in order to find the DIS structure function all one needs is to find the imaginary part of the dipole–nucleus forward scattering amplitude $N(\vec{x}_\perp, \vec{b}_\perp, Y)$. In Sec. 4.2 we constructed such an amplitude in the Glauber–Gribov–Mueller multiple rescattering approximation. The resulting forward amplitude has no energy dependence, as one can see from Eq. (4.49), and therefore cannot be a realistic description of the high energy asymptotics of dipole–nucleus scattering. At the same time, the approach of Sec. 4.2 is valid only when the small-x evolution emissions are not important, that is, only for $\alpha_s Y \ll 1$. At higher energies, corresponding to rapidities Y satisfying $\alpha_s Y \gtrsim 1$, small-x evolution becomes important and can no longer be neglected.

We see that we need to resum the LLA corrections to the dipole–nucleus scattering amplitude (4.49). As usual we are interested in quantum evolution corrections that resum the powers of $\alpha_s \ln 1/x \sim \alpha_s Y$.[7] Just as in Sec. 4.2 we will be working in the rest frame of the nucleus, but this time we choose to work in the light cone gauge of the projectile dipole, $A^+ = 0$, if it is moving in the light cone plus direction. One can show by explicit calculation that for the multiple rescatterings in Fig. 4.5 this gauge is equivalent to the covariant gauge ($\partial_\mu A^\mu = 0$, see Sec. 3.3.1); therefore, our discussion in Sec. 4.2 remains valid in this new gauge. As in Sec. 4.2 we will be working either in the nucleus rest frame or in the frame in which the dipole is moving in the light cone plus direction while the target nucleus is moving in the minus direction.

We need to identify radiative corrections that bring in powers of $\alpha_s Y$. As we saw in Sec. 4.2, multiple rescatterings bring in only powers of α_s not enhanced by factors of Y (but enhanced by powers of A; the resummation parameter was $\alpha_s^2 A^{1/3}$). Therefore, additional t-channel gluon exchanges with new nucleons would not generate any powers of Y but would bring in only extra factors of α_s. These are not the corrections we are trying to resum now. Other possible corrections in the light cone gauge of the projectile dipole are modifications of the dipole wave function. The incoming dipole may have some gluons (and "sea" quarks) present in its wave function. For instance, the dipole may emit a gluon before interacting with the target; then the whole system of quark, antiquark, and gluon would rescatter in the nucleus, as shown in the upper panel of Fig. 4.23. The dipole may emit two gluons, which would then interact with the nucleus, along with the original $q\bar{q}$ pair, as shown in the lower panel of Fig. 4.23. In principle there could be many extra gluon emissions, as well as the generation of extra qq pairs in the incoming dipole's wave function. As we will shortly see, these gluonic fluctuations from Fig. 4.23 actually do bring

---

[7] Quantum evolution is defined as the variation of a physical quantity, with Q^2 and/or x, resulting from quantum emissions and absorptions.

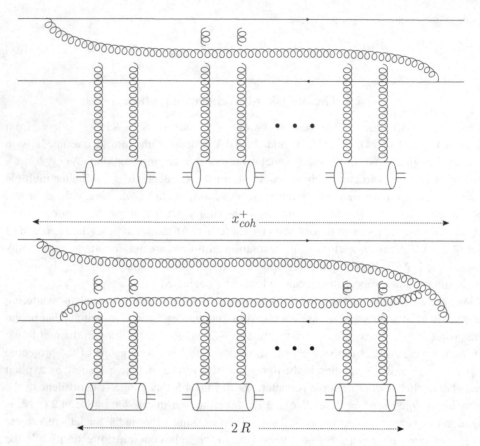

Fig. 4.23. Quantum evolution corrections to dipole–nucleus scattering due to one-gluon (upper panel) and two-gluon (lower panel) emissions. The lower panel also shows the coherence time scale for gluon emission x_{coh}^+ and the nuclear size $2R$. At high energy $x_{coh}^+ \gg 2R$: the figure does not fully reflect this scale difference.

the factors of α_s enhanced by powers of rapidity Y, i.e., they do generate leading logarithmic corrections. Just as with the BFKL evolution, fluctuations leading to the formation of $q\bar{q}$ pairs actually enter at the subleading logarithmic level, bringing in powers of $\alpha_s^2 Y$, and are not important for the leading logarithmic approximation used in this chapter.

Several times above (see the discussion around Eqs. (2.156), (3.126), and (4.2)), we have used the fact that owing to the uncertainty principle, for an incoming dipole moving in the light cone plus direction a gluon with momentum k^μ in its wave function would have coherence length

$$x_{coh}^+ \approx \frac{k^+}{k_\perp^2} \qquad (4.129)$$

along the x^+-axis. Note straight away that t-channel gluon exchanges between the dipole and the nucleons in the nucleus, in the Glauber–Gribov–Mueller approximation of Sec. 4.2, have $k^+ = 0$ with eikonal accuracy (i.e., up to corrections suppressed by powers of the

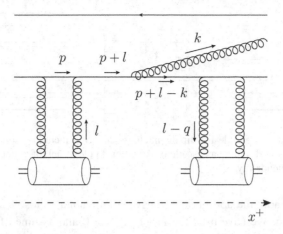

Fig. 4.24. A diagram with a gluon emission between the dipole interactions with two nucleons.

energy). Thus these t-channel gluons have $x^+_{coh} = 0$ and are instantaneous in the x^+ "time" direction in our eikonal picture. These are the instantaneous or Coulomb gluons. The instantaneous nature of these gluons explains why the dipole rescatters on the nucleons sequentially: as the nucleons are assumed to be separated in x^+, the dipole interacts with a nucleon as it crosses the latter's x^+-range, with interactions that are out of order, like that in Fig. 4.6B, not allowed by causality. The nucleons span the whole nucleus; thus the x^+-time interval filled with the instantaneous interactions of Fig. 4.5 is of the order of the nuclear radius R in the nuclear rest frame.

Consider now the gluon modifications to the incoming dipole's wave function shown in Fig. 4.23. If a gluon's k^+ is large enough, as is the case at high energy, the coherence lengths of these gluons would be much larger than the nuclear radius, $x^+_{coh} \gg R$, so that each gluon would coherently rescatter on the nucleons in the nucleus, just like the original dipole in Fig. 4.5. This is indeed what is shown in Fig. 4.23.

Note that gluons are emitted by the incoming dipole only before the multiple rescattering interaction (and absorbed back, after the interaction, into the forward amplitude). Emissions during the interaction are suppressed by the inverse powers of the center-of-mass energy of the scattering system. This can be checked via an explicit calculation in the covariant Feynman perturbation theory. Imagine a diagram with the gluon emitted or absorbed between the rescatterings, as shown in Fig. 4.24. As in our analysis of the graph in Fig. 4.9 above, we concentrate on the contribution of quark propagators to the l^--integral. We see that the diagram is proportional to

$$\int_{-\infty}^{\infty} \frac{dl^-}{2\pi} \frac{e^{-il^- \Delta x^+}}{[(p+l)^2 + i\epsilon][(p+l-k)^2 + i\epsilon]}$$

$$\approx \int_{-\infty}^{\infty} \frac{dl^-}{2\pi} \frac{e^{-il^- \Delta x^+}}{[p^+ l^- - \perp^2 + i\epsilon][(p^+ - k^+)(k^- + l^-) - \perp'^2 + i\epsilon]} \tag{4.130}$$

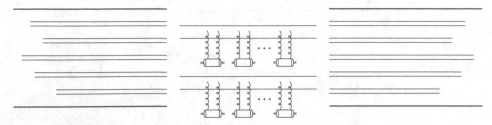

Fig. 4.25. Forward amplitude for dipole–nucleus scattering including small-x evolution: the incoming dipole develops a cascade of daughter dipoles, each of which interacts with the nucleus independently.

where $\perp$ and $\perp'$ denote the appropriate transverse momenta, whose exact values are not important to us here. We have used the fact that $l^+ = 0$ and assumed for simplicity that $p^- = 0$. We also changed the frame to that where the nucleus is moving along the negative light cone. Closing the integration contour in the lower half-plane we obtain

$$\frac{-i}{p^+(p^+ - k^+)} \frac{1}{k^- + \perp^2/p^+ - \perp'^2/(p^+ - k^+)} \left[e^{-i\frac{\perp^2}{p^+}\Delta x^+} - e^{-i\left(-k^- + \frac{\perp'^2}{p^+ - k^+}\right)\Delta x^+} \right]$$

$$\approx \frac{-i}{p^{+2}k^-} \left(1 - e^{ik^-\Delta x^+} \right) \sim \frac{-1}{p^{+2}p'^-} = \frac{-1}{p^+ s}, \tag{4.131}$$

where we have used the fact that $p^+ \gg k^+$ and, more importantly, $\Delta x^+ \sim 1/p'^-$ with p'^- the large light cone momentum of a nucleon in the nucleus (such that $s = p^+ p'^-$ is the dipole–nucleon center-of-mass energy squared). This allowed us to expand the exponential in the second line of Eq. (4.131). Comparing with the rescatterings without gluon emission given in Eq. (4.37) (identifying k^+ in (4.37) with p^+ here), we see that gluon emission between rescatterings brings in suppression by a power of the energy squared s and can thus be neglected.

Alternatively we can consider this calculation in light cone perturbation theory. In this case, the emission of a gluon is allowed and is equally probable at any point throughout the coherence length of the parent dipole $x_{coh}^{q\bar{q}+} = p^+/p_\perp^2$, with p the momentum of the dipole and p^+ very large. The probability of emission of a gluon inside the nucleus (in the nuclear rest frame) is then proportional to $R/x_{coh}^{q\bar{q}+} \sim 1/p^+ \sim 1/s$; i.e., again, just as in Eq. (4.131) it is suppressed by a power of the center-of-mass energy squared s compared with emission outside the nucleus and can be neglected in the eikonal approximation considered here.

Our goal, therefore, is to resum the cascade of long-lived gluons that the dipole in Fig. 4.23 develops before interacting with the nucleus and then to convolute this cascade with the interaction amplitudes of the gluons with the nucleus. To resum the cascade we will assume the large-N_c limit and use Mueller's dipole model, presented in Sec. 4.3. In the large-N_c limit the gluon cascade translates into a dipole cascade, examples of which are shown in Figs. 4.19 and 4.22. As we have seen above, in the LLA gluon emissions do not change the transverse coordinates of the quark and antiquark lines in the parent dipole. Therefore, the color dipoles have the same transverse coordinates throughout the whole process: once they are created their transverse coordinates do not change. Resummation of the dipole cascade reduces to the set of diagrams represented in Fig. 4.25, which is a generalization of Fig. 4.5 to the case of quantum evolution corrections. The incoming dipole develops a cascade of daughter dipoles through evolution according to Mueller dipole model.

The evolved system of dipoles interacts with the nucleus. The interaction is brief and does not change the transverse coordinates of the dipoles. In the large-N_c limit no dipole interacts with any other dipole during the evolution that generates all the dipoles. For a large nucleus the daughter dipole–nucleus interaction was calculated above in the GGM approximation and is given by Eq. (4.51). That result resums powers of $\alpha_s^2 A^{1/3}$. Analyzing the diagrams for the interaction of several dipoles with the nucleus we see that the GGM interaction of, say, two dipoles with a single nucleon is suppressed by extra powers of α_s not enhanced by $A^{1/3}$ and is therefore subleading and can be neglected. The interaction of two dipoles with two nucleons in the large-N_c limit is dominated by diagrams where each dipole interacts with only one nucleon (assuming both dipoles interact). In general one can argue that, in the large-N_c limit and at the leading order in A (or, equivalently, resumming powers of $\alpha_s^2 A^{1/3}$), the interaction of any number of dipoles with the nucleus is dominated by the *independent* interactions of each dipole with a different set of nucleons in the nucleus through multiple rescatterings of the type in Fig. 4.5. This is depicted in Fig. 4.25: when the dipole wave function hits the nucleus, each dipole present in the wave function may interact with different nucleons in the nucleus by the exchange of pairs of gluons. (It can be shown that only some dipoles thus interact.) Therefore, the dipoles are completely mutually noninteracting: they do not exchange gluons in the process of evolution, since those corrections would be suppressed by powers of N_c, and they interact with *different* nucleons in the nucleus; the last statement is correct at leading order in A (Kovchegov 1999).

Summation of the dipole cascade of Fig. 4.25 now becomes straightforward. Instead of calculating the forward dipole–nucleus scattering amplitude $N(\vec{x}_\perp, \vec{b}_\perp, Y)$ we start with the S-matrix $S(\vec{x}_\perp, \vec{b}_\perp, Y)$, which is related to N via Eq. (4.38). We write it here again for completeness:

$$S(\vec{x}_\perp, \vec{b}_\perp, Y) = 1 - N(\vec{x}_\perp, \vec{b}_\perp, Y). \qquad (4.132)$$

As follows from the above discussion, $S(\vec{x}_{10}, \vec{b}_{0\perp}, Y)$ can be written as a convolution of the dipole cascade and the dipole interactions with the target, as shown in Fig. 4.25. Namely, it is a sum of the probability of finding one daughter dipole in the parent dipole, convoluted with the S-matrix for dipole–nucleus scattering in the GGM approximation, and the probability of finding two dipoles, convoluted with their multiple rescattering interactions with the nucleus, etc. We write (Kovchegov 1999)

$$S(\vec{x}_{10}, \vec{b}_\perp, Y) = \sum_{k=1}^\infty \frac{1}{k!} \int d^2 r_1 d^2 b_1 \cdots d^2 r_k d^2 b_k$$

$$\times \left. \frac{\delta^k Z(\vec{x}_{10}, \vec{b}_\perp, Y; u)}{\delta u(\vec{r}_{1\perp}, \vec{b}_{1\perp}) \cdots \delta u(\vec{r}_{k\perp}, \vec{b}_{k\perp})} \right|_{u=0} s_0(\vec{r}_{1\perp}, \vec{b}_{1\perp}) \cdots s_0(\vec{r}_{k\perp}, \vec{b}_{k\perp}).$$

$$(4.133)$$

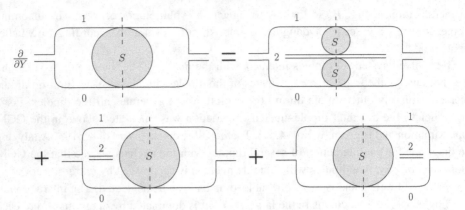

Fig. 4.26. Diagrammatic representation for the evolution equation of the S-matrix for dipole–nucleus scattering, denoted by a shaded circle. The vertical dashed lines denote the interaction with the target.

Here

$$\left. \frac{\delta^k Z(\vec{x}_{10}, \vec{b}_\perp, Y; u)}{\delta u(\vec{r}_{1\perp}, \vec{b}_{1\perp}) \cdots \delta u(\vec{r}_{k\perp}, \vec{b}_{k\perp})} \right|_{u=0} \tag{4.134}$$

gives the probability of finding *exactly* k daughter dipoles in the parent dipole wave function (cf. Eq. (4.71)), and

$$s_0(\vec{r}_\perp, \vec{b}_\perp) \equiv S(\vec{r}_\perp, \vec{b}_\perp, Y = 0) = \exp\left\{ -\frac{x_\perp^2 Q_{s0}^2(\vec{b}_\perp)}{4} \ln \frac{1}{x_\perp \Lambda} \right\}, \tag{4.135}$$

as follows from Eqs. (4.51) and (4.132).

Summing the series in Eq. (4.133) yields (see Eq. (4.70))

$$S(\vec{x}_{10}, \vec{b}_\perp, Y) = Z\left(\vec{x}_{10}, \vec{b}_\perp, Y; u = s_0 \right) \tag{4.136}$$

(Kovchegov 1999). This relation shows that both the dipole–nucleus S-matrix and the generating functional Z obey the same nonlinear evolution equation. The initial condition for Z in (4.76) is replaced by Eq. (4.135).

We see that the evolution of $S(\vec{x}_\perp, \vec{b}_\perp, Y)$ is the same as the evolution of the generation functional Z in Sec. 4.3: it is illustrated in Fig. 4.26 (cf. Fig. 4.21). The dipole cascade and its interaction with the target are denoted by a shaded circle. In one step of the evolution in energy (or rapidity) a soft gluon is emitted in the dipole. If the gluon is real then the original dipole is split into two dipoles, as shown at top right of Fig. 4.26; these dipoles proceed to evolve and interact (or not) independently with the target (the S-matrix includes the noninteraction term, the "1" in Eq. (4.132)). Virtual corrections, given by the two lower diagrams in Fig. 4.26, lead only to the parent dipole's subsequent evolution and interaction with the target. We obtain an evolution equation for the S-matrix (Balitsky 1996, Kovchegov

1999):

$$\frac{\partial}{\partial Y} S(\vec{x}_{10}, \vec{b}_\perp, Y)$$

$$= \frac{\alpha_s N_c}{2\pi^2} \int d^2 x_2 \frac{x_{10}^2}{x_{20}^2 x_{21}^2}$$

$$\times \left[S\left(\vec{x}_{12}, \vec{b}_\perp + \frac{\vec{x}_{20}}{2}, Y\right) S\left(\vec{x}_{20}, \vec{b}_\perp + \frac{\vec{x}_{21}}{2}, Y\right) - S(\vec{x}_{10}, \vec{b}_\perp, Y)\right]. \qquad (4.137)$$

The initial condition for this evolution equation is given by $S(\vec{x}_{10}, \vec{b}_\perp, Y = 0)$ in Eq. (4.135). As usual $\vec{b}_\perp = (\vec{x}_{1\perp} + \vec{x}_{0\perp})/2$.

Using Eq. (4.132) in Eq. (4.137) we derive an evolution equation for the imaginary part of the forward dipole–nucleus scattering amplitude N (Balitsky 1996, Kovchegov 1999):

$$\frac{\partial}{\partial Y} N(\vec{x}_{10}, \vec{b}_\perp, Y) = \frac{\alpha_s N_c}{2\pi^2} \int d^2 x_2 \frac{x_{10}^2}{x_{20}^2 x_{21}^2}$$

$$\times \left[N\left(\vec{x}_{12}, \vec{b}_\perp + \frac{\vec{x}_{20}}{2}, Y\right) + N\left(\vec{x}_{20}, \vec{b}_\perp + \frac{\vec{x}_{21}}{2}, Y\right) - N(\vec{x}_{10}, \vec{b}_\perp, Y)\right.$$

$$\left. - N\left(\vec{x}_{12}, \vec{b}_\perp + \frac{\vec{x}_{20}}{2}, Y\right) N\left(\vec{x}_{20}, \vec{b}_\perp + \frac{\vec{x}_{21}}{2}, Y\right)\right]. \qquad (4.138)$$

This is the Balitsky–Kovchegov (BK) evolution equation. The initial condition for the BK evolution is given by Eq. (4.51):

$$N(\vec{x}_\perp, \vec{b}_\perp, Y = 0) = 1 - \exp\left\{ -\frac{x_\perp^2 Q_{s0}^2(\vec{b}_\perp)}{4} \ln \frac{1}{x_\perp \Lambda}\right\}, \qquad (4.139)$$

where we have replaced $Q_s^2(\vec{b}_\perp)$ from Eq. (4.51) by $Q_{s0}^2(\vec{b}_\perp)$ to underscore that this is the saturation scale in the initial condition for the evolution. (As we will see shortly, the saturation scale is modified by the nonlinear BK evolution equation: in particular it becomes dependent on the rapidity Y.) Equation (4.138) resums all powers of the multiple rescattering parameter $\alpha_s^2 A^{1/3}$, along with the leading logarithms of energy in the large-N_c limit given by powers of $\alpha_s N_c Y$.

Below we will sometimes use a more compact notation for the dipole–nucleus amplitude,

$$N(\vec{x}_{1\perp}, \vec{x}_{0\perp}, Y) \equiv N(\vec{x}_{10}, \vec{b}_\perp, Y). \qquad (4.140)$$

Using this notation, we can rewrite Eq. (4.138) as

$$\frac{\partial}{\partial Y} N(\vec{x}_{1\perp}, \vec{x}_{0\perp}, Y) = \frac{\alpha_s N_c}{2\pi^2} \int d^2 x_2 \frac{x_{10}^2}{x_{20}^2 x_{21}^2}$$

$$\times \left[N(\vec{x}_{1\perp}, \vec{x}_{2\perp}, Y) + N(\vec{x}_{2\perp}, \vec{x}_{0\perp}, Y) - N(\vec{x}_{1\perp}, \vec{x}_{0\perp}, Y)\right.$$

$$\left. - N(\vec{x}_{1\perp}, \vec{x}_{2\perp}, Y) N(\vec{x}_{2\perp}, \vec{x}_{0\perp}, Y)\right]. \qquad (4.141)$$

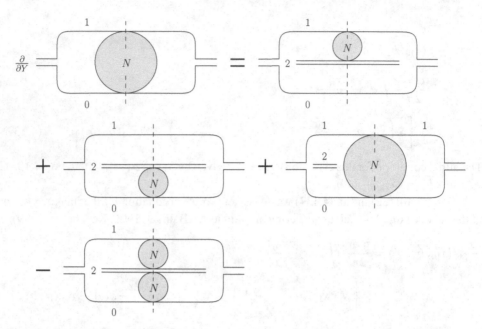

Fig. 4.27. Diagrammatic representation of the BK evolution equation for the forward dipole–nucleus scattering amplitude N, denoted by a shaded circle. Only one virtual term is shown, for brevity.

The BK equation is represented diagrammatically in Fig. 4.27. Balitsky–Kovchegov evolution has a simple physical meaning. At fixed rapidity a colorless dipole with size x_{10} decays into two dipoles with sizes x_{12} and x_{20}. Either one dipole proceeds to evolve and interact with the target while the other dipole remains a spectator (the first two, linear, terms after the equals sign in Fig. 4.27) or both dipoles evolve and interact with the target (the nonlinear term in Fig. 4.27). The minus sign in front of the nonlinear term reflects the fact that taking into account two independent interactions overestimates the result. The nonlinear term corresponds to the shadowing corrections in the GGM approach: for instance, expanding Eq. (4.45) in powers of interactions with the nucleons we see that the quadratic term enters with a minus sign. The reason for that minus sign is the same as the reason for the minus sign in the last term of Eq. (4.138).

Equation (4.138) was originally derived by Balitsky (1996) in the framework of the effective theory of high energy interactions and, independently, by one of the present authors (Kovchegov 1999) using the formalism of Mueller's dipole model (Mueller 1994, 1995). It was rederived by Braun (2000a) using the large-N_c limit of the expression for the triple pomeron vertex from Bartels and Wusthoff (1995) in a resummation of the fan diagrams in Fig. 3.23.

Comparing the linear part of the BK equation (the first three terms on the right of Eq. (4.138)) with Eq. (4.87), we see that the linear terms in the BK equation give the coordinate-space BFKL equation. As already mentioned, the nonlinear term can be obtained from the triple-pomeron vertex in the large-N_c limit (Braun 2000a). Hence Eq. (4.138) has the overall structure of the GLR equation and corresponds to fan diagram resummation in the conventional Feynman perturbation theory. It is natural to expect that the BK evolution leads to the same physical effects as the GLR equation: for a given fixed dipole size, the dipole amplitude N would start out growing with rapidity owing to BFKL evolution (see Eq. (4.102)); the nonlinear term would become important at higher rapidity and lead to saturation and slowdown of the energy growth. In the next section we will see that this is indeed the case.

In solving the BK equation (4.138) one often (but not always) assumes that the variation in the amplitude $N(\vec{x}_{10}, \vec{b}_\perp, Y)$ with the impact parameter $\vec{b}_\perp$ is small when $\vec{b}_\perp$ varies over distance scales comparable with the dipole size $|\vec{x}_{10}|$. This is indeed true for scattering on a very large nucleus far away from its edges. This assumption allows one to neglect the shifts in the impact parameter on the right-hand side of Eq. (4.138). Moreover, assuming that the nucleus is isotropic we may neglect the angular dependence of $\vec{x}_{10}$. We thus may replace $N(\vec{x}_{10}, \vec{b}_\perp, Y)$ approximately by $N(x_{10}, Y)$ in Eq. (4.138), obtaining

$$\frac{\partial}{\partial Y} N(x_{10}, Y) = \frac{\alpha_s N_c}{2\pi^2} \int d^2 x_2 \frac{x_{10}^2}{x_{20}^2 x_{21}^2}$$
$$\times \left[N(x_{12}, Y) + N(x_{20}, Y) - N(x_{10}, Y) - N(x_{12}, Y) N(x_{20}, Y) \right]. \tag{4.142}$$

Performing the Fourier transformation

$$N(x_\perp, Y) = x_\perp^2 \int \frac{d^2 k}{2\pi} e^{i \vec{k}_\perp \cdot \vec{x}_\perp} \tilde{N}(k_\perp, Y), \tag{4.143}$$

we write (Kovchegov 2000)

$$\frac{\partial \tilde{N}(k_\perp, Y)}{\partial Y} = \bar{\alpha}_s \chi \left[0, \frac{i}{2} \left(1 + \frac{\partial}{\partial \ln k_\perp} \right) \right] \tilde{N}(k_\perp, Y) - \bar{\alpha}_s \tilde{N}^2(k_\perp, Y). \tag{4.144}$$

This equation is useful for obtaining approximate solutions for the BK evolution that we will present below. Also, note that making the identification

$$\phi(x, k_\perp^2) = \frac{N_c S_\perp}{\alpha_s \pi^2} \tilde{N}(k_\perp, Y = \ln 1/x) \tag{4.145}$$

in Eq. (4.144) reduces it to the GLR equation (3.128). This is indeed remarkable: however, there exists no physical justification for the Fourier transformation (4.143). At the lowest, two-gluon-exchange, order the relation between the dipole amplitude N and the unintegrated gluon distribution ϕ should be of the form of Eq. (4.98) (with f there proportional to ϕ). In the region where multiple rescatterings and quantum evolution are important, the exact relation between N and ϕ is not clear.

4.5 Solution of the Balitsky–Kovchegov equation

To date there is no exact analytical solution of the BK equation. Below we will present several approximate analytical solutions, along with some numerical results. One of the main conclusions can be derived right away, without doing any work, if we notice that $N = 1$ is the fixed point of BK evolution: for $N = 1$, corresponding to the black-disk limit, the right-hand side of Eq. (4.138) vanishes and the growth of N with rapidity stops. Hence the BK equation does not violate the black-disk limit.

4.5.1 Solution outside the saturation region; extended geometric scaling

Let us begin analyzing BK evolution when the forward amplitude is small, $N \ll 1$. In the multiple rescatterings (GGM) approximation (see Eq. (4.139)) we know that for small dipoles with $x_\perp \ll 1/Q_{s0}$ the amplitude N is also small, and saturation and unitarization effects are not very important yet. The fact that the forward amplitude N goes to zero as $x_\perp \to 0$ is based on a fundamental physical principle of color transparency, which is valid beyond the multiple-rescattering approximation. This allows us to conclude that $N \ll 1$ for small dipole sizes $x_\perp$ even when small-x evolution is included. For $N \ll 1$ we can linearize the BK equation; as we observed earlier, this gives us the coordinate-space BFKL equation. With the approximation used in Eq. (4.142) we can write

$$\frac{\partial N(x_{10}, Y)}{\partial Y} = \frac{\bar{\alpha}_s}{2\pi} \int d^2 x_2 \frac{x_{10}^2}{x_{20}^2 x_{21}^2} \Big[N(x_{12}, Y) + N(x_{20}, Y) - N(x_{10}, Y) \Big]. \qquad (4.146)$$

This is exactly Eq. (4.90), whose solution we can write as (cf. Eq. (4.101))

$$N(x_\perp, Y) = \int_{-\infty}^{\infty} dv \, C_v \, \exp\{\bar{\alpha}_s \chi(0, v) Y + (1 + 2iv) \ln(x_\perp Q_{s0})\}, \qquad (4.147)$$

where we use Q_{s0} as the typical transverse scale characterizing the target nucleus. In the $\vec{b}_\perp$-independent approximation that we are employing, Q_{s0} is not a function of $\vec{b}_\perp$. As usual, C_v is a constant fixed by the initial conditions. Just as in the case of BFKL evolution, the integral (4.147) can be evaluated either in the DLA or in the diffusion approximation depending on the kinematics of the problem.

Double logarithmic approximation

Consider the case of very small dipole size, $x_\perp Q_{s0} \ll 1$, such that transverse logarithms like $\ln(x_\perp Q_{s0})$ become important, leading to a new resummation parameter $\alpha_s Y \ln(x_\perp Q_{s0})$. This is the DLA we considered before. Approximating $\chi(0, v)$ as follows,

$$\chi(0, v) \approx \frac{2}{1 - 2iv}, \qquad (4.148)$$

we find a saddle point at

$$v_{DLA}^* \approx -\frac{i}{2}\left(1 - \sqrt{\frac{2\bar{\alpha}_s Y}{\ln 1/(x_\perp Q_{s0})}}\right). \tag{4.149}$$

Using Eq. (4.148) in Eq. (4.147) and performing the integration over v in the saddle point approximation yields

$$N(x_\perp, Y)\Big|_{x_\perp Q_{s0} \ll 1} = (x_\perp Q_{s0})^2 C_{v_{DLA}^*} \frac{\sqrt{\pi}}{2} (2\bar{\alpha}_s Y)^{1/4} \ln^{-3/4} \frac{1}{x_\perp Q_{s0}}$$

$$\times \exp\left\{2\sqrt{2\bar{\alpha}_s Y \ln 1/(x_\perp Q_{s0})}\right\}. \tag{4.150}$$

We see that, as for the GGM multiple-rescatterings case (see Eq. (4.139)), the amplitude $N(x_\perp, Y)$ in the DLA regime grows quadratically with the dipole size $x_\perp$, though this rise receives a correction owing to the exponential in Eq. (4.150). On top of that, as is typical for the DLA case, the amplitude also grows with rapidity Y.

Extended geometric scaling region

Now let us study the region where the dipole size is still small, but not much smaller than the inverse saturation scale: $x_\perp Q_{s0} \lesssim 1$. In this region, evolution is still linear and one would still expect Eq. (4.147) to give us the solution. We begin by evaluating the v-integral in Eq. (4.147) in the saddle-point approximation. The location of the saddle point v_{sp} is determined by the standard condition

$$\bar{\alpha}_s \chi'(0, v_{sp})Y + 2i \ln x_\perp Q_{s0} = 0, \tag{4.151}$$

which gives the saddle point v_{sp} as a function of $x_\perp$ and Y: $v_{sp} = v_{sp}(x_\perp, Y)$. (The prime in Eq. (4.151) indicates a derivative with respect to v, $\chi'(0, v) = \partial\chi(0, v)/\partial v$.) Crudely approximating the v-integral in Eq. (4.147) by the value of the integrand at $v = v_{sp}$, we obtain

$$N(x_\perp, Y) \propto (x_\perp Q_{s0})^{1+2iv_{sp}} e^{\bar{\alpha}_s \chi(0, v_{sp})Y}. \tag{4.152}$$

The amplitude given by Eq. (4.152) grows with energy and with the dipole size $x_\perp$. When it becomes of order 1, say $N \approx 1/2$, the approximation in which this solution is derived breaks down and one has to go back to solving the nonlinear BK equation (4.142). Let us estimate where this breakdown of the linear regime occurs. We want to find a line in the $(x_\perp, Y)$-plane along which N is an order 1 constant: this will give us the saturation scale.

The saturation scale $Q_s(Y)$ (which now is a function of rapidity) is therefore defined by the condition

$$N(x_\perp = 1/Q_s(Y), Y) = \text{const}, \tag{4.153}$$

where the constant is of order 1. Using Eq. (4.152) in Eq. (4.153) yields

$$\bar{\alpha}_s \chi(0, v_0)Y + (1 + 2iv_0)\ln(Q_{s0}/Q_s(Y)) = 0, \tag{4.154}$$

where $\nu_0 \equiv \nu_{sp}(x_\perp = 1/Q_s(Y), Y)$. Taking the saddle point condition (4.151) along the saturation line we get

$$\bar{\alpha}_s \chi'(0, \nu_0)Y + 2i \ln(Q_{s0}/Q_s(Y)) = 0. \qquad (4.155)$$

Solving Eqs. (4.154) and (4.155) yields (Gribov, Levin, and Ryskin 1983, Iancu, Itakura, and McLerran 2002, Mueller and Triantafyllopoulos 2002)

$$Q_s(Y) = Q_{s0} \exp\left\{\bar{\alpha}_s \frac{\chi(0, \nu_0)}{1 + 2i\nu_0} Y\right\} \qquad (4.156)$$

with

$$\frac{\chi'(0, \nu_0)}{\chi(0, \nu_0)} = \frac{2i}{1 + 2i\nu_0}. \qquad (4.157)$$

It follows from Eq. (4.157), which one can solve numerically using Eq. (3.81), that ν_0 is simply a number (Gribov, Levin, and Ryskin 1983),

$$\nu_0 \approx -0.1275i. \qquad (4.158)$$

Using this value of ν_0 in Eq. (4.156) we get

$$Q_s(Y) \approx Q_{s0} e^{2.44\bar{\alpha}_s Y}. \qquad (4.159)$$

We have obtained a very important result: as follows from Eqs. (4.156) and (4.159), the saturation scale grows as an exponential of the rapidity. Since $Y = \ln 1/x$ this is indeed consistent with the power-of-$1/x$ growth in Eq. (3.135) obtained on general physical grounds in discussing GLR evolution. We now have the same qualitative result, with the exact exponent of the growth now specified by the slightly more detailed calculation that we have performed. Note that since $Q_{s0} \sim A^{1/6}$ (see Eq. (4.52)), we have $Q_s(Y) \sim A^{1/6}$ as well. This result can be understood as follows: the initial conditions for BK evolution (4.139) contain only one dimensionful scale Q_{s0} (we neglect the logarithm as a slowly varying function). The BK equation is conformally invariant; hence the scales resulting from this evolution, such as $Q_s(Y)$, should all be proportional to Q_{s0} and have the same A-scaling (see e.g. Kharzeev, Levin, and McLerran (2003)). It is also important to stress that the small-x evolution *does not* preserve the GGM formula (4.45) by simply including x- and A-dependence in the lowest-order nuclear gluon distribution, defined by $xG_A = AxG_N$; this would lead to a different scaling of $Q_s(Y)$ with A. In fact evolution corrections completely destroy the GGM form of N.

The region with momentum $Q < Q_s(Y)$ (corresponding to $x_\perp > 1/Q_s(Y)$), where the nonlinear term in the BK equation becomes important, is the *saturation region*.

Eliminating the rapidity dependence from Eq. (4.152), to absorb all the Y-dependence into $Q_s(Y)$, yields with the help of Eq. (4.156)

$$N(x_\perp, Y) \propto (x_\perp Q_{s0})^{1+2i\nu_{sp}} \left(\frac{Q_s(Y)}{Q_{s0}}\right)^{(1+2i\nu_0)\chi(0, \nu_{sp})/\chi(0, \nu_0)}. \qquad (4.160)$$

With the accuracy of our crude version of the saddle point approximation, we write $\nu_{sp} \approx \nu_0$ in the vicinity of the saturation scale. Substituting this into Eq. (4.160) we obtain (Iancu,

Itakura, and McLerran 2002, Mueller and Triantafyllopoulos 2002)

$$N(x_\perp, Y) \propto [x_\perp Q_s(Y)]^{1+2i\nu_0} . \qquad (4.161)$$

The dipole amplitude still grows with both $x_\perp$ and Y, just as in the DLA regime. However, the growth with $x_\perp$ is slower than the quadratic DLA scaling of Eq. (4.150). Conversely, the growth in N with rapidity appears to be stronger in Eq. (4.161) than in the DLA case (4.150).

We have another important result in Eq. (4.161): the dipole scattering amplitude $N(x_\perp, Y)$, which, in general, can be a function of two independent variables $x_\perp$ and Y, is here a function of a *single* variable, $x_\perp Q_s(Y)$. This result is known as *geometric scaling*. Geometric scaling has been demonstrated (and the term coined) in an analysis of the HERA DIS data by Stasto, Golec-Biernat, and Kwiecinski (2001) (see also Kwiecinski and Stasto 2002) presenting one of the strongest arguments for the observation of saturation phenomena at HERA (see Fig. 9.1). Theoretically it was first observed as a property of GLR-type equations by Bartels and Levin (1992). For the BK equation, geometric scaling was first demonstrated deep inside the saturation region by Levin and Tuchin (2000): this result will be derived below. As we will see shortly, the expression (4.161) that we have obtained is valid outside the saturation region, but not too far from the saturation boundary, i.e., for $x_\perp Q_s(Y) \lesssim 1$. The fact that geometric scaling is valid outside the saturation region was first observed by Iancu, Itakura, and McLerran (2002). This scaling phenomenon outside the saturation region is referred to as *extended geometric scaling*.

Note that the (absolute) value of ν_0 found in Eq. (4.158) is not very large. In fact, one can check explicitly that $\chi(0, \nu_0)$ is still well described by Eq. (3.84), which was used in the diffusion approximation presented in Sec. 3.3.4. The result (4.161) is valid as long as ν_{sp} is not too far from ν_0 (cf. (4.160)). If we decrease the dipole size $x_\perp$ then we would eventually end up in the DLA region, where the saddle point is close to $\nu = -i/2$ (cf. Eq. (4.149)). Clearly Eq. (4.150) cannot be written as a function of a single variable $x_\perp Q_s(Y)$ and thus violates geometric scaling. We conclude that the extended geometric scaling of Eq. (4.161) is valid only as long as $\chi(0, \nu_{sp})$ is described better by the diffusion formula (3.84) than by the DLA approximation (4.148). By equating the two approximations we see that the transition occurs near $\nu_{sp}^{geom} = -0.22i$, which, owing to Eq. (4.151), corresponds to

$$\ln \frac{1}{x_\perp^{geom} Q_{s0}} \approx 5.75 \, \bar{\alpha}_s Y, \qquad (4.162)$$

so that the border (upper limit) of the extended geometric scaling region is defined by the scale $k_{geom} = 1/x_\perp^{geom}$ given by (cf. Iancu, Itakura, and McLerran (2002))

$$k_{geom} = Q_{s0} e^{5.75 \bar{\alpha}_s Y} = Q_s(Y) \left(\frac{Q_s(Y)}{Q_{s0}} \right)^{1.35} . \qquad (4.163)$$

Therefore, the extended geometric scaling is valid up to

$$k_\perp = \frac{1}{x_\perp} \leq k_{geom}. \qquad (4.164)$$

Since $k_{geom} \gg Q_s(Y)$, the region of extended geometric scaling is parametrically broad and can be quite large at large Y. For $k_\perp > k_{geom}$ the solution maps back onto the DLA regime of Eq. (4.150).

Our current analytical knowledge of the saturation scale at high energy extends well beyond the approximation derived in Eq. (4.156). In fact we know that

$$Q_s(Y) = Q_{s0} \exp \left\{ \bar{\alpha}_s \frac{\chi(0, \nu_0)}{1 + 2i\nu_0} Y - \frac{3}{2(1 + 2i\nu_0)} \ln \bar{\alpha}_s Y + \text{const} \right.$$

$$\left. - \frac{6}{(1 + 2i\nu_0)^2} \sqrt{\frac{2\pi}{-\bar{\alpha}_s \chi''(0, \nu_0) Y}} + O\left(\frac{1}{Y}\right) \right\}. \tag{4.165}$$

One can show that the expression in the exponent of Eq. (4.165) is universal, in the sense that it is independent of the initial conditions for BK evolution with the exception of the constant term, which may depend on the initial conditions. As mentioned above, the first term in this expression was derived by Gribov, Levin, and Ryskin (1983) when analyzing GLR evolution and by Iancu, Itakura, and McLerran (2002) for the BK equation. The second term in the exponent of Eq. (4.165) was found by Mueller and Triantafyllopoulos (2002) and by Munier and Peschanski (2004a). The derivation of Mueller and Triantafyllopoulos (2002) is close to that presented above: however, they obtained the correct value of the second term on the right of Eq. (4.165) by modeling the saturation boundary as an absorptive barrier in the $(x_\perp, Y)$-plane. The derivation of Munier and Peschanski (2004a) employed a traveling wave solution of the BK equation. The third nontrivial ($O(1/\sqrt{Y})$) term in the exponent was also calculated by Munier and Peschanski (2004b). The traveling wave approach is very close in spirit and in letter to the method of characteristics used to solve differential equations: we will present both solutions below. For a comprehensive up-to-date summary of the results on the high energy behavior of the saturation scale we recommend a recent paper by Beuf (2010).

4.5.2 Solution inside the saturation region; geometric scaling

Let us now analyze the behavior of the solution of Eq. (4.138) deep inside the saturation region, where nonlinear effects are very important. Deep inside the saturation region, when the dipole size $x_\perp$ becomes large, $x_\perp \gg 1/Q_s(Y)$ (but we still have $x_\perp \ll 1/\Lambda_{QCD}$), the quasi-classical GGM amplitude from Eq. (4.51) approaches 1. As mentioned at the beginning of this section, analyzing Eq. (4.138) we can easily see that $N = 1$ is also a stationary solution of that equation. Therefore we conclude that, for large dipole sizes, BK evolution would not change the amplitude

$$N(\vec{x}_\perp, \vec{b}_\perp, Y) = 1, \quad x_\perp \gg 1/Q_s(Y), \tag{4.166}$$

which has reached the black-disk limit (BDL) (cf. Eq. (4.33)) and will remain there. Now let us determine the asymptotic approach to the black-disk limit (4.166). To do this we employ Eq. (4.137). As follows from Eq. (4.132), the S-matrix is small near the BDL, where $N \approx 1$. Keeping only terms linear in S in Eq. (4.137) yields

$$\frac{\partial S(\vec{x}_{10}, \vec{b}_\perp, Y)}{\partial Y} = -\frac{\alpha_s N_c}{2\pi^2} \int d^2 x_2 \frac{x_{10}^2}{x_{20}^2 x_{21}^2} S(\vec{x}_{10}, \vec{b}_\perp, Y), \qquad (4.167)$$

where the integral over dipole sizes goes over $x_{02}, x_{12} > 1/Q_s(Y)$. To perform the integral we replace the ultraviolet (UV) cutoff ρ in Eq. (A.20) (see also Eq. (4.64)) with $1/Q_s(Y)$ and use Eq. (A.29) to obtain

$$\frac{\partial S(\vec{x}_{10}, \vec{b}_\perp, Y)}{\partial Y} = -2\bar{\alpha}_s \ln [x_{10} Q_s(Y)] S(\vec{x}_{10}, \vec{b}_\perp, Y). \qquad (4.168)$$

Defining the *scaling* variable

$$\xi \equiv \ln \left[x_\perp^2 Q_s^2(Y) \right] \qquad (4.169)$$

with (cf. Eq. (4.156))

$$\frac{2\chi(0, \nu_0)}{1 + 2i\nu_0} \bar{\alpha}_s \equiv \frac{\partial \xi}{\partial Y} = \frac{\partial \ln \left[x_\perp^2 Q_s^2(Y) \right]}{\partial Y}, \qquad (4.170)$$

we can rewrite Eq. (4.168) as

$$\frac{\partial S}{\partial \xi} = -\frac{1 + 2i\nu_0}{2\chi(0, \nu_0)} \xi S. \qquad (4.171)$$

The solution of Eq. (4.171) can be written straightforwardly as (Levin and Tuchin 2000)

$$S(\xi) = S_0 \exp \left\{ -\frac{1 + 2i\nu_0}{2\chi(0, \nu_0)} \xi^2 \right\} \qquad (4.172)$$

with $S_0 < 1$ a constant. The corresponding dipole amplitude N is given by

$$N(\xi) \Big|_{x_\perp \gg 1/Q_s(Y)} = 1 - S_0 \exp \left\{ -\frac{1 + 2i\nu_0}{2\chi(0, \nu_0)} \xi^2 \right\} \qquad (4.173)$$

Equation (4.173) is known as the Levin–Tuchin formula (Levin and Tuchin 2000).

Note that the S-matrix and the amplitude N for dipole–nucleus scattering given by Eqs. (4.172) and (4.173) are functions of a single variable ξ, or, more precisely, of the combination $x_\perp Q_s(Y)$. This is indeed the *geometric scaling* found above: while before we obtained the scaling outside the saturation region, now we see that geometric scaling is also valid inside the saturation region.

Equations (4.173), (4.161), and (4.150) give us a good idea of the amplitude $N(x_\perp, Y)$ given by the solution of the BK equation as a function of rapidity Y and dipole size $x_\perp$. We see that $N(x_\perp, Y)$ grows with $x_\perp$ but at very large $x_\perp$ saturates to 1: thus the black-disk limit is not violated. Hence, at the qualitative level the overall shape of $N(x_\perp, Y)$ given by the GGM formula and shown in Fig. 4.11 is preserved. The amplitude $N(x_\perp, Y)$ also grows with rapidity Y though at larger $x_\perp$ the growth slows down, eventually stopping at the black-disk limit. The saturation scale increases with rapidity; this means that the GGM curve from Fig. 4.11 starts moving to the left on that plot. We will illustrate these conclusions with explicit plots when we discuss the numerical solution of BK evolution.

4.5.3 Semiclassical solution

We now present another powerful approach to solving the BK equation, which allows us to reproduce the results obtained above while providing new insight.

Defining (cf. Eq. (3.74))

$$\chi(\gamma) \equiv 2\psi(1) - \psi(\gamma) - \psi(1 - \gamma), \tag{4.174}$$

with γ related to ν via Eq. (3.79), we rewrite Eq. (4.144) as

$$\partial_Y \tilde{N}(\rho, Y) = \bar{\alpha}_s \chi(-\partial_\rho) \tilde{N}(\rho, Y) - \bar{\alpha}_s \tilde{N}^2(\rho, Y), \tag{4.175}$$

where $\partial_Y = \partial/\partial Y$ and

$$\rho = \ln \frac{k_\perp^2}{Q_{s0}^2}. \tag{4.176}$$

Let us now look for the solution of Eq. (4.175) using a semiclassical approximation. We write

$$\tilde{N}(\rho, Y) = e^{\Omega(\rho, Y)} \tag{4.177}$$

and assume that $\Omega(\rho, Y)$ is a slowly varying function of its arguments, such that $\Omega_{\rho Y} \ll \Omega_\rho \Omega_Y$, $\Omega_{\rho\rho} \ll \Omega_\rho^2$, $\Omega_{YY} \ll \Omega_Y^2$, with similar relations for the higher-order derivatives: the nth-order derivative is always much smaller than the nth power of the first derivative. (Here $\Omega_\rho = \partial\Omega/\partial\rho$, $\Omega_Y = \partial\Omega/\partial Y$, etc.)

Substituting Eq. (4.177) into Eq. (4.175) and employing the semiclassical approximations just outlined yields

$$\partial_Y \Omega = \bar{\alpha}_s \chi(-\partial_\rho \Omega) - \bar{\alpha}_s e^\Omega. \tag{4.178}$$

We will study Eq. (4.178) using the method of characteristics (see e.g. Courant and Hilbert 1953), following Gribov, Levin, and Ryskin (1983), Collins and Kwiecinski (1990), Bartels, Schuler, and Blumlein (1991), and Levin and Tuchin (2001). Defining partial derivatives

$$-\gamma \equiv \Omega_\rho, \quad \omega \equiv \Omega_Y, \tag{4.179}$$

we can rewrite Eq. (4.178) as

$$F \equiv \omega - \bar{\alpha}_s \chi(\gamma) + \bar{\alpha}_s e^\Omega = 0. \tag{4.180}$$

The characteristics of Eq. (4.178) can then be found by solving the following set of ordinary differential equations:

$$\frac{d\rho}{dt} = F_{-\gamma} = \bar{\alpha}_s \frac{d\chi(\gamma)}{d\gamma}, \tag{4.181a}$$

$$\frac{dY}{dt} = F_\omega = 1, \tag{4.181b}$$

$$\frac{d\gamma}{dt} = -F_\rho - (-\gamma)F_\Omega = \bar{\alpha}_s \gamma e^\Omega, \tag{4.181c}$$

$$\frac{d\omega}{dt} = -F_Y - \omega F_\Omega = -\bar{\alpha}_s \omega e^\Omega, \tag{4.181d}$$

$$\frac{d\Omega}{dt} = (-\gamma)F_{-\gamma} + \omega F_\omega = -\bar{\alpha}_s \gamma \frac{d\chi(\gamma)}{d\gamma} + \omega. \tag{4.181e}$$

Equation (4.181b) gives $Y = t$. In addition we can use Eq. (4.180) to eliminate ω from Eqs. (4.181). The remaining equations are

$$\frac{d\rho}{dY} = \bar{\alpha}_s \frac{d\chi(\gamma)}{d\gamma}, \tag{4.182a}$$

$$\frac{d\gamma}{dY} = \bar{\alpha}_s \gamma e^\Omega \tag{4.182b}$$

$$\frac{d\Omega}{dY} = \bar{\alpha}_s \left[\chi(\gamma) - \gamma \frac{d\chi(\gamma)}{d\gamma} - e^\Omega \right]. \tag{4.182c}$$

These equations are still difficult to solve in the general case. One may construct approximations of $\chi(\gamma)$ and solve the resulting equations exactly (see Levin and Tuchin (2001)). Instead of following this path we will keep $\chi(\gamma)$ exact and will again explore the linear regime. If $\tilde{N} = e^\Omega \ll 1$ then we can recast Eq. (4.182c) as

$$\frac{d\Omega}{dY} \approx \bar{\alpha}_s \left[\chi(\gamma) - \gamma \frac{d\chi(\gamma)}{d\gamma} \right]. \tag{4.183}$$

We see that there exists a *critical* characteristic trajectory of constant Ω (and hence $\tilde{N}$), defined by $d\Omega/dY = 0$, which leads to the following equation for $\gamma = \gamma_{cr}$ (Gribov, Levin, and Ryskin 1983):

$$\chi(\gamma_{cr}) = \gamma_{cr} \frac{d\chi(\gamma_{cr})}{d\gamma_{cr}}. \tag{4.184}$$

This is exactly equivalent to Eq. (4.157). Equation (4.184) gives $\gamma_{cr} \approx 0.6275$, which is consistent with Eq. (4.158) (see Eq. (3.79)). The critical line in the (ρ, Y)-plane follows from Eq. (4.182a):

$$\frac{d\rho_s}{dY} = \bar{\alpha}_s \frac{\chi(\gamma_{cr})}{\gamma_{cr}}. \tag{4.185}$$

Since we have defined the critical line as a line of constant $\tilde{N}$, the definition is analogous to that of Eq. (4.153) and therefore defines the saturation scale. Solving Eq. (4.185) with the initial condition $\rho_s(Y = 0) = 0$ yields

$$\rho_s(Y) = \bar{\alpha}_s \frac{\chi(\gamma_{cr})}{\gamma_{cr}} Y, \tag{4.186}$$

which is equivalent to Eq. (4.156) if we write $\rho_s(Y) = \ln Q_s^2(Y)/Q_{s0}^2$.

Note that one can verify the $e^{\Omega} \ll 1$ approximation made in arriving at Eq. (4.183): substituting $\gamma = \gamma_{cr}$ back into Eq. (4.182c) and using the relation (4.184) one can solve the resulting equation to show that the value of Ω along the critical trajectory (Ω_{cr}) is indeed large and negative, $\Omega_{cr} \approx -\ln \bar{\alpha}_s Y$.

Finally, near the critical (saturation) line we can expand (for fixed Y) as follows:

$$\Omega \approx \Omega_{cr} + \Omega_{\rho_s}(\rho - \rho_s) = \Omega_{cr} - \gamma_{cr}(\rho - \rho_s). \tag{4.187}$$

We have also used Eq. (4.179) in arriving at Eq. (4.187). Note that $\Omega_{cr} \approx -\ln \bar{\alpha}_s Y$ is independent of ρ and is a slowly varying function of Y: therefore, it carries little dynamical information and can be treated as a constant in our approximation. Employing Eq. (4.186) to define $Q_s^2(Y) = Q_{s0}^2 e^{\rho_s}$, we obtain

$$\tilde{N}(\rho, Y) = e^{\Omega} \propto e^{-\gamma_{cr}(\rho - \rho_s)} = \left(\frac{Q_s^2(Y)}{k_{\perp}^2} \right)^{\gamma_{cr}}, \tag{4.188}$$

which again is in perfect agreement with Eq. (4.161) (if we replace $x_{\perp}$ with $1/k_{\perp}$ in the latter). Thus we have rederived the extended geometric scaling behavior of the dipole–nucleus scattering amplitude, this time working in momentum space.

It is interesting to notice that the critical line has a very transparent physical meaning. The solution for Ω, given by Eq. (4.187), can be written as

$$\Omega \approx \Omega_{cr} - \gamma_{cr}\rho + \omega_{cr}Y, \tag{4.189}$$

with $\omega_{cr} = \bar{\alpha}_s \chi(\gamma_{cr})$. This is similar to the phase of a traveling wave packet moving along the ρ-axis with time Y, having wave number γ_{cr} and frequency ω_{cr}. (The profile of the wave packet would be determined by a prefactor to Eq. (4.188), which is not given by our approximate solution.) Such a wave packet has two characteristic velocities: the phase velocity v_{ph} (the velocity of a line with constant phase Ω) and the group velocity v_{gr} (the velocity of the maximum of the packet). Using Eq. (4.180) but dropping the e^{Ω} term, we can easily calculate these two velocities, obtaining

$$v_{gr} = \frac{\partial \omega}{\partial \gamma} = \bar{\alpha}_s \frac{d\chi(\gamma)}{d\gamma}, \tag{4.190}$$

$$v_{ph} = \frac{\omega}{\gamma} = \bar{\alpha}_s \frac{\chi(\gamma)}{\gamma}. \tag{4.191}$$

One can see that the critical line corresponds to the unique trajectory on which $v_{gr} = v_{ph}$ (Gribov, Levin, and Ryskin 1983, Munier and Peschanski 2004a).

The characteristics trajectories of BK evolution are shown in Fig. 4.28. They cannot cross each other, and the critical trajectory plays the role of a divider between two groups of trajectories, as shown in Fig. 4.28. This figure illustrates the special and essential role of

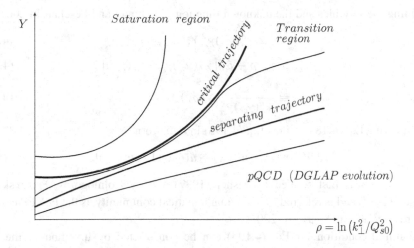

Fig. 4.28. The characteristics for Eq. (4.178) plotted in the (ρ, Y)-plane.

the critical (saturation) trajectory as a divider between pQCD (DGLAP) physics, possibly modified by small corrections due to the interactions of partons, and the saturation domain. The parton interactions are responsible for the characteristic phenomena in the latter region, in particular for the saturation of the parton density. The trajectories to the right of the critical line are very close to the trajectories of the linear (BFKL) evolution equation except in the region close to the critical line, when the effect of the critical (saturation) trajectory becomes important. This is illustrated by the "separating trajectory" also shown in Fig. 4.28, which separates the characteristics which are not affected by the saturation region, located to the right of that line, from those located to the left of the line. The unaffected characteristics are those of the DLA DGLAP. The trajectories to the left of the separating trajectory but to the right of the critical trajectory do not resemble the trajectories of the linear equation, and their behavior indicates that the linearized semiclassical approach is not applicable in this region.

4.5.4 Traveling wave solution

There is another elegant method of reproducing (and improving upon) the above results for geometric scaling and the critical anomalous dimension. We start with Eq. (4.175) and expand its kernel around γ_{cr}, defined in Eq. (4.184):

$$\chi(-\partial_\rho) = \chi(\gamma_{cr}) + (-\partial_\rho - \gamma_{cr})\chi'(\gamma_{cr}) + \tfrac{1}{2}(-\partial_\rho - \gamma_{cr})^2\chi''(\gamma_{cr}) + \cdots \qquad (4.192)$$

Truncating the expansion at the quadratic level of course limits the applicability of the approach we are about to develop. Certainly this approximation would not work in the DLA region. Equation (4.175) becomes

$$\partial_Y \tilde{N}(\rho, Y) = \bar{\alpha}_s \left[\chi(\gamma_{cr}) - \tilde{N}(\rho, Y)\right] \tilde{N}(\rho, Y) - \bar{\alpha}_s \chi'(\gamma_{cr})\left(\partial_\rho + \gamma_{cr}\right)\tilde{N}(\rho, Y)$$
$$+ \tfrac{1}{2}\bar{\alpha}_s \chi''(\gamma_{cr})(\partial_\rho + \gamma_{cr})^2 \tilde{N}(\rho, Y). \qquad (4.193)$$

Redefining the variables and the unknown function $\tilde{N}$ (Munier and Peschanski 2003),

$$t \equiv \tfrac{1}{2}\bar{\alpha}_s \chi''(\gamma_{cr})\gamma_{cr}^2 Y, \tag{4.194a}$$

$$x \equiv \gamma_{cr}\,\rho + \bar{\alpha}_s\left[\chi''(\gamma_{cr})\gamma_{cr}^2 - \chi(\gamma_{cr})\right]Y, \tag{4.194b}$$

$$u(t, x) \equiv \frac{2}{\chi''(\gamma_{cr})\gamma_{cr}^2}\tilde{N}(\rho, Y), \tag{4.194c}$$

and employing Eq. (4.184) brings Eq. (4.193) into the form

$$\partial_t u(t, x) = \partial_x^2 u(t, x) + u(t, x)[1 - u(t, x)]. \tag{4.195}$$

This equation was first studied by Fisher (1937) and by Kolmogorov, Petrovsky, and Piskunov (1937) and is referred to in the mathematical community as the F–KPP equation. (For a review see van Saarloos (2003).)

The initial condition for Eq. (4.195) can be constructed by first finding the initial condition for Eq. (4.144). Inverting Eq. (4.143) we write

$$\tilde{N}(k_\perp, Y) = \int \frac{d^2 x_\perp}{2\pi} e^{-i\vec{k}_\perp \cdot \vec{x}_\perp}\frac{N(x_\perp, Y)}{x_\perp^2}. \tag{4.196}$$

Dropping the b-dependence in Eq. (4.139) and using the result in Eq. (4.196) we obtain

$$\tilde{N}(k_\perp, Y = 0) = \int \frac{d^2 x_\perp}{2\pi} e^{-i\vec{k}_\perp \cdot \vec{x}_\perp}\frac{1}{x_\perp^2}\left[1 - \exp\left\{-\frac{x_\perp^2 Q_{s0}^2}{4}\ln\frac{1}{x_\perp \Lambda}\right\}\right]. \tag{4.197}$$

While the exact analytic integration in Eq. (4.197) does not lead to a compact answer, we can find the asymptotics of the initial conditions from it. At large $k_\perp$ (small $x_\perp$), expanding the exponential in Eq. (4.197) to the first nontrivial order and integrating using Eq. (A.9) yields

$$\left.\tilde{N}(k_\perp, Y = 0)\right|_{k_\perp/Q_{s0} \gg 1} \approx \frac{Q_{s0}^2}{4k_\perp^2}. \tag{4.198}$$

At small $k_\perp$ (large $x_\perp$), dropping the exponential and integrating over $x_\perp$ with $1/Q_{s0}$ as the IR cutoff we get

$$\left.\tilde{N}(k_\perp, Y = 0)\right|_{k_\perp/Q_{s0} \ll 1} \approx \ln\frac{Q_{s0}}{k_T}. \tag{4.199}$$

For $u(t = 0, x)$ these initial conditions imply

$$u(t = 0, x) \infty \begin{cases} \tfrac{1}{4}e^{-x/\gamma_{cr}}, & x \to +\infty, \\[2mm] -\dfrac{x}{2\gamma_{cr}}, & x \to -\infty. \end{cases} \tag{4.200}$$

It was proven that the F–KPP equation admits traveling wave solutions at late times t if the initial condition $u(0, x)$ decreases monotonically from 1 to 0 as x varies from $-\infty$ to $+\infty$, falling off exponentially with x as $x \to +\infty$ (Bramson 1983). While the initial condition giving $u(0, x)$ in Eq. (4.200) violates this condition at $x \to -\infty$, following Munier and Peschanski (2003) we assume that the high energy asymptotics of the solution outside the saturation region will not change significantly if we simply "freeze" $u(0, x)$ at $u = 1$ inside the saturation region. According to the theory of the F–KPP equation the asymptotic traveling wave solution depends on the speed of the exponential falloff $e^{-x/\gamma_{cr}}$ at $x \to +\infty$ in the initial condition for the equation: for $1/\gamma_{cr} \approx 1.5936 > 1$ the traveling wave solution is

$$u(t, x)\Big|_{t \to \infty} \sim f\left(x - 2t + \tfrac{3}{2}\ln t + O(1)\right), \tag{4.201}$$

for some function f. We see that $u(t, x)$ and, owing to Eq. (4.194), $\tilde{N}(\rho, Y)$ are functions of a single variable,

$$x - 2t + \tfrac{3}{2}\ln t = \gamma_{cr} \ln \frac{k_\perp^2}{Q_s(Y)^2} + \text{const}, \tag{4.202}$$

where

$$Q_s(Y)^2 = Q_{s0}^2 \exp\left\{\bar{\alpha}_s \frac{\chi(\gamma_{cr})}{\gamma_{cr}} Y - \frac{3}{2\gamma_{cr}} \ln \bar{\alpha}_s Y\right\}. \tag{4.203}$$

This is indeed geometric scaling. The saturation scale in Eq. (4.203) is identical to that in Eq. (4.165) up to the first two terms in the exponent.

Dropping the nonlinear term in Eq. (4.195) we see that $u(t, x) = e^{-x+2t}$ is clearly a solution of the resulting linearized equation, giving (Munier and Peschanski 2004a)

$$\tilde{N}(\rho, Y) \propto \left(\frac{Q_s^2(Y)}{k_\perp^2}\right)^{\gamma_{cr}}, \tag{4.204}$$

in agreement with Eqs. (4.161) and (4.188).

Owing to the approximations we have made, in expanding $\chi(\gamma)$ in order to arrive at Eq. (4.193) and in neglecting the fact that Eq. (4.200) violates the condition stated by Bramson (1983) for the existence of a traveling wave solution, we can conclude that the reduction of the BK equation to the F–KPP equation is valid only for $k_\perp$ values in the vicinity of the saturation scale. In particular, Eq. (4.193) does not give the solution (4.173) deep inside the saturation region. Interestingly, the traveling wave (geometric scaling) pattern itself appears to be more universal than the F–KPP reduction: for instance, Eq. (4.173) also has a traveling wave form. The traveling wave structure is also preserved in other models of the dipole BFKL kernel. For example, if we simplify the kernel of Eq. (4.138) by resumming only the transverse logarithms (such as $\ln x_\perp^2 Q_s^2(y)$ and $\ln x_\perp^2 \Lambda_{QCD}^2$), thus taking into account only the leading twist contributions to the full BFKL kernel, the BK equation can be reduced to a wave equation (Levin and Tuchin 2000, 2001) for which one also has a traveling wave solution (Polyanin and Zaitsev 2004, formula 3.4.1).

The existence of traveling wave solutions indicates that, at very high energy, $\tilde{N}(\rho, Y)$ behaves like a wave with a fixed coordinate (ρ) profile, which travels with increasing Y toward larger values of ρ without a change in profile. This is an important physical result from the traveling wave approach.

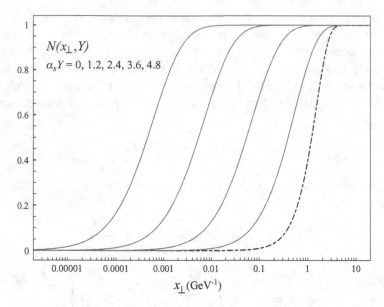

Fig. 4.29. Dipole amplitude $N(x_\perp, Y)$ plotted as a function of $x_\perp$; the curves right to left correspond to increasing values of $\alpha_s Y$, as shown in the legend. (We thank Javier Albacete for providing us with this figure.)

4.5.5 Numerical solutions

There are a number of numerical solutions of the BK equation. We are not going to give a comprehensive overview of these solutions here but will merely show some results.

A numerical solution of the BK equation (4.142) without impact parameter dependence, giving the amplitude $N(x_\perp, Y)$ as a function of $x_\perp$, is shown in Fig. 4.29 for several values of the rescaled rapidity $\alpha_s Y$. The initial condition is specified at $Y = 0$ by a slight modification of the GGM formula (4.51),

$$N(x_\perp, Y) = 1 - \exp\left\{-\frac{x_\perp^2 Q_{s0}^2}{4} \ln\left(\frac{1}{x_\perp \Lambda} + e\right)\right\} \tag{4.205}$$

with $Q_{s0} = 1$ GeV and $\Lambda = 0.2$ GeV, and is represented by the dashed line in Fig. 4.29. (Since the exponent of Eq. (4.51) is written in the $x_\perp \Lambda \ll 1$ approximation, e has been added in Eq. (4.205) to keep N positive for $x_\perp \Lambda > 1$.)

We see from Fig. 4.29 that the nonlinear small-x evolution pushes the dipole amplitude $N(x_\perp, Y)$ towards lower values of $x_\perp$ as Y increases. This is indeed in agreement with our analytical results: as $Q_s(Y)$ grows with rapidity, $1/Q_s(Y)$ decreases, moving the curve to the left along the $x_\perp$-axis. The growth in the saturation scale with rapidity Y is shown in Fig. 4.30. Here, the saturation scale is defined by requiring that $N(x_\perp = 1/Q_s(Y), Y) = 1/2$; it is plotted in Fig. 4.30 as a function of $\alpha_s Y$. Again we see qualitative agreement with the above analytical results: the saturation scale grows with rapidity. At large Y we see that $\ln Q_s(Y)$ grows linearly with $\alpha_s Y$, in agreement with Eq. (4.165); the slope of about

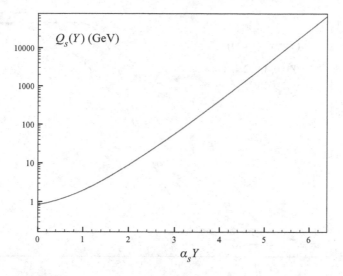

Fig. 4.30. The saturation scale $Q_s(Y)$ given by a numerical solution of the BK evolution equation, plotted as a function of the rescaled rapidity $\alpha_s Y$. (We thank Javier Albacete for providing us with this figure.)

2.1 is close to the analytical estimate, $2.44 N_c/\pi = 2.33$, in Eq. (4.159). (Note that for a more detailed comparison of Fig. 4.30 with the analytical results one also needs to take into account the logarithmic correction in the exponent of Eq. (4.165).)

The numerical solutions also exhibit the property of geometric scaling. This is demonstrated in Fig. 4.31, which shows the curves from Fig. 4.29 plotted as functions of the scaling variable

$$\tau = x_\perp Q_s(Y), \tag{4.206}$$

for the same set of $\alpha_s Y$ values as in Fig. 4.29. One can see the onset of the geometric scaling behavior both inside and outside the saturation region: as the rapidity Y increases, all the curves approach a universal scaling curve. (Indeed, at very large transverse momenta or very small $x_\perp$ the geometric scaling in Fig. 4.31 would be violated owing to the onset of the DLA DGLAP asymptotics; this is not shown in the figure because of its limited range in $x_\perp Q_s(Y)$.)

For another quantitative comparison of the analytic results and the numerical solutions we show in Fig. 4.32 a plot of the coordinate-space dipole amplitude $N(x_\perp, Y)$ as a function of the scaling variable τ over a broader range in τ, both for fixed-coupling (the dashed line) and running-coupling (the solid line) BK evolution (Albacete and Kovchegov 2007b). (The running-coupling BK evolution is given by the BK equation with running-coupling corrections included (rcBK). We will discuss rcBK in Chapter 6 (see Eq. (6.9) with kernels given either by Eq. (6.12) or Eq. (6.14)).) In the fixed-coupling case, comparing the power of 0.6 in Fig. 4.32 with Eq. (4.161) or Eq. (4.188) we see that it is consistent with the theoretical prediction of 0.6275 from, say, Eq. (4.184).

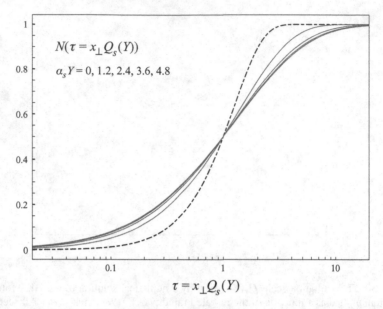

Fig. 4.31. The dipole amplitude $N(x_\perp, Y)$ as a function of the scaling variable $\tau = x_\perp Q_s(Y)$; the curves (in clockwise order in relation to the crossing point) correspond to the same values of $\alpha_s Y$ as in Fig. 4.29. (We thank Javier Albacete for providing us with this figure.) A color version of this figure is available online at www.cambridge.org/9780521112574.

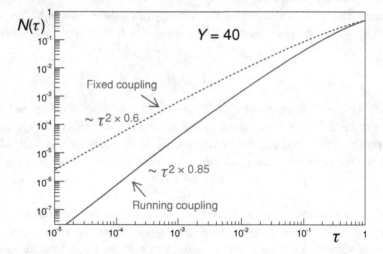

Fig. 4.32. Asymptotic solutions ($Y = 40$) of the evolution equation for running coupling (solid line) and fixed coupling with $\alpha_s = 0.2$ (dashed line). A fit to a power-law function $a\tau^{2\gamma}$ in the region $\tau \in [10^{-6}, 10^{-2}]$ yields $\gamma \approx 0.85$ for the running-coupling solution and $\gamma \approx 0.6$ for the fixed-coupling solution. (Reprinted with permission from Albacete and Kovchegov (2007b). Copyright 2007 by the American Physical Society.) A color version of this figure is available online at www.cambridge.org/9780521112574.

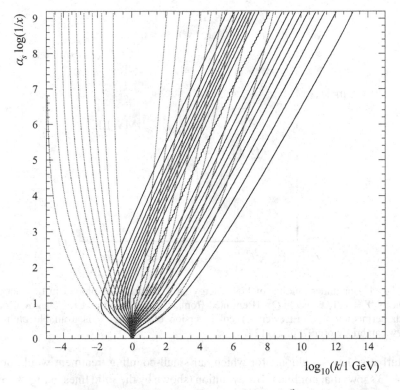

Fig. 4.33. A contour plot of the numerical solutions ($\sim k_\perp \tilde{N}(\vec{k}_\perp, Y)$) of the BFKL and BK evolution equations in momentum space, as functions of the transverse momentum $k_\perp$ and the rescaled rapidity $\alpha_s Y = \alpha_s \ln 1/x$. (We thank Anna Stasto for providing us with this figure.) A color version of this figure is available online at www.cambridge.org/9780521112574.

In Sec. 3.3.6 we discussed the two main problems of BFKL evolution unitarity violation and diffusion into the IR. From Fig. 4.29, along with the analytical calculations presented above in Sec. 4.5.2, we see that for the BK solution one always has $N \leq 1$, so that we can conclude that BK evolution does not violate the black-disk limit hence resolving this issue of the BFKL evolution. We see that a resolution of BDL violation still occurs in the perturbative domain, owing to the large value of saturation scale $Q_s(Y)$ there.

To answer the question regarding diffusion into the IR, represented by the Bartels cigar of Fig. 3.19, we will present one more result, from the numerical solution of the fixed-coupling BK equation of Golec-Biernat, Motyka, and Stasto (2002). Figure 4.33 depicts the lines of constant value for the numerical solution for $k_\perp \tilde{N}(k_\perp, Y)$ of the BFKL and BK equations in momentum space. Namely, Fig. 4.33 contains contour plots of $k_\perp \tilde{N}(k_\perp, Y)$ as a function of transverse momentum $k_\perp$ and rapidity $Y = \ln 1/x$. To illustrate the point, the initial conditions for both the BFKL and BK equations were chosen to be delta functions in the transverse momenta, $\delta(k_\perp - k_{0\perp})$ with $k_{0\perp} = 1$ GeV. One can see that the solutions of the BFKL equation (the dotted lines in Fig. 4.33) spread out as the rapidity increases. This is the diffusion discussed in Sec. 3.3.6, which is dangerous because it generates

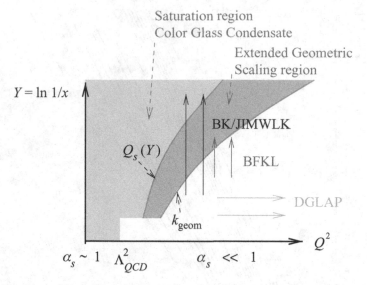

Fig. 4.34. Our understanding of high energy QCD interactions plotted in the plane of rapidity $Y = \ln 1/x$ and $\ln Q^2$. (Reprinted from Jalilian-Marian and Kovchegov (2006), with permission from Elsevier.) A color version of this figure is available online at www.cambridge.org/9780521112574.

nonperturbative low-$k_\perp$ gluons, for which our small-coupling treatment would not apply. Figure 4.33 shows that nonlinear BK evolution (shown by the solid lines in Fig. 4.33) avoids this problem. The nonlinear term in Eq. (4.138) leads to two main effects: (i) it drives the constant-value lines of the solution towards higher momenta, which is consistent with the increase in the saturation scale in Eq. (4.156), and (ii) it virtually eliminates the spread of the solution: as one can see from Fig. 4.33 the width of the $k_\perp$-distribution of the BK solution is roughly independent of rapidity. This solves the IR diffusion problem of the BFKL equation.

4.5.6 Map of high energy QCD

We summarize the results obtained in this Chapter in Fig. 4.34. This is a "map of high energy QCD", which may be compared with Fig. 4.28 and Fig. 3.22. Figure 4.34 represents the action of the evolution equations we have discussed, plotted in the $(Q^2, Y = \ln 1/x)$-plane. The region with $Q^2 \lesssim \Lambda^2_{QCD}$ is nonperturbative: there α_s is large and so we cannot use perturbation theory. The DGLAP evolution applies at large Q^2 and not very small x, as indicated by the horizontal arrows denoting evolution in Q^2. The BFKL equation is responsible for the evolution in x: it is represented by the short vertical arrows. At small enough x the linear BFKL evolution breaks down and nonlinear saturation effects set in. The transition to the saturation region is denoted by the saturation line $Q = Q_s(Y)$ (cf. the critical line in Fig. 4.28), and the saturation region is located above this line. The generalization of the BFKL evolution in x to include the saturation physics is accomplished by BK evolution in the large-N_c limit. Outside the large-N_c limit the nonlinear small-x evolution is described by the Jalilian–Marian–Iancu–McLerran–Weigert–Leonidov–Kovner

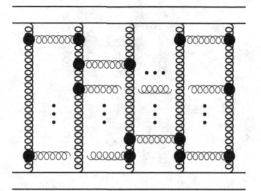

Fig. 4.35. Diagrams contributing to the BKP evolution. The vertical gluon lines are reggeized gluons, while the large solid circles denote Lipatov vertices.

(JIMWLK) evolution equation, which is a functional evolution equation to be presented in the next chapter (Jalilian-Marian *et al.* 1997b, 1999a, b, Iancu, Leonidov, and McLerran 2001a, b, Weigert 2002, Ferreiro *et al.* 2002). Both the BK and JIMWLK evolutions are shown by long vertical arrows. As shown above, geometric scaling works inside the saturation region. We have also indicated the lower boundary of the extended geometric scaling region by k_{geom} from Eq. (4.163).

The saturation region is also called the the color glass condensate, as indicated in Fig. 4.34: this term will be explained below. It is important to stress once more that all the nonlinear dynamics driving the saturation phenomena takes place for $Q_s \gtrsim Q \gg \Lambda_{QCD}$, i.e., in the perturbative region where the strong coupling constant is small and our perturbative calculations are justified.

4.6 The Bartels–Kwiecinski–Praszalowicz equation*

One may wonder whether the dipole evolution presented above should receive some potentially important subleading-N_c corrections. The problem is easier to address when formulated in terms of the standard BFKL approach. The BFKL equation of Sec. 3.3 gives the evolution for two reggeized gluons in the t-channel. Now imagine the case of an arbitrary number of t-channel reggeized gluons. Their small-x evolution is shown in Fig. 4.35.

To write down an evolution equation for n-reggeized gluon exchange, as in the BFKL case one has to define the Green function of the exchange, $G(\vec{k}_{1\perp}, \ldots, \vec{k}_{n\perp}; \vec{k}'_{1\perp}, \ldots, \vec{k}'_{n\perp}; Y)$, corresponding to a "rectangle" like that in Fig. 3.5, with n gluon legs attached to it from above and n more attached to it from below, as shown on the left of Fig. 4.36. Moreover, we only account for the discontinuities (the imaginary parts) of the Green function between all consecutive t-channel gluon exchanges, as shown by the cuts in Fig. 4.36. The scattering amplitude without such cuts can be reconstructed from the BKP Green function by the repeated use of dispersion relations, as in the case of gluon reggeization considered in Sec. 3.3.5.

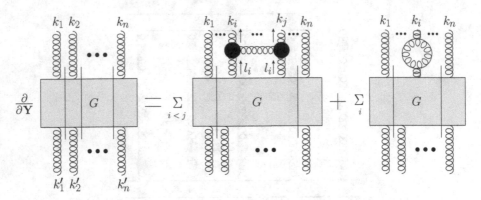

Fig. 4.36. Diagrammatic representation of the BKP evolution equation for the $2n$-point Green function. All momenta flow up in the diagrams and the vertical straight lines denote cuts.

One can write down an evolution equation for this Green function by applying the BFKL kernel to any pair of the reggeized gluons shown in Fig. 4.36. One obtains the Bartels–Kwiecinski–Praszalowicz (BKP) equation,

$$\partial_Y G(\vec{k}_{1\perp}, \ldots, \vec{k}_{n\perp}; \vec{k}'_{1\perp}, \ldots, \vec{k}'_{n\perp}; Y)$$

$$= \sum_{i=1}^{n} \omega_G(\vec{k}_{i\perp}) G(\vec{k}_{1\perp}, \ldots, \vec{k}_{n\perp}; \vec{k}'_{1\perp}, \ldots, \vec{k}'_{n\perp}; Y) + \lambda_{color} \sum_{i<j} \int d^2 l_i d^2 l_j$$

$$\times K^{ij}_{NF}(\vec{k}_{i\perp}, \vec{k}_{j\perp}; \vec{l}_{i\perp}, \vec{l}_{j\perp}) G(\vec{k}_{1\perp}, \ldots, \vec{l}_{i\perp}, \ldots, \vec{l}_{j\perp}, \ldots, \vec{k}_{n\perp}; \vec{k}'_{1\perp}, \ldots, \vec{k}'_{n\perp}; Y),$$

$$(4.207)$$

where $K^{ij}_{NF}(\vec{k}_{i\perp}, \vec{k}_{j\perp}; \vec{l}_{i\perp}, \vec{l}_{j\perp})$ is the nonforward BFKL kernel, given by (cf. Eq. (3.103))

$$K^{ij}_{NF}(\vec{k}_{i\perp}, \vec{k}_{j\perp}; \vec{l}_{i\perp}, \vec{l}_{j\perp}) = \frac{\alpha_s N_c}{2\pi^2} \delta^{(2)}(\vec{k}_{i\perp} + \vec{k}_{j\perp} - \vec{l}_{i\perp} - \vec{l}_{j\perp})$$

$$\times \left[\frac{k^2_{i\perp}}{l^2_{i\perp}(\vec{k}_{i\perp} - \vec{l}_{i\perp})} + \frac{k^2_{j\perp}}{l^2_{j\perp}(\vec{k}_{i\perp} - \vec{l}_{i\perp})} - \frac{(\vec{k}_{i\perp} + \vec{k}_{j\perp})^2}{l^2_{i\perp} l^2_{j\perp}} \right].$$

$$(4.208)$$

The coefficient λ_{color} depends on the color-SU(3) representation of the two gluons that interact, namely

$$\lambda_{singlet} = 1, \quad \lambda_{8_S} = \lambda_{8_A} = \tfrac{1}{2}, \quad \lambda_{10} = \lambda_{\overline{10}} = 0, \quad \lambda_{27} = -\tfrac{1}{3} \qquad (4.209)$$

where $8_S, 8_A, 10, \overline{10}, 27$ denote the representations of the SU(3) color group (the subscripts S, A denote the symmetric and antisymmetric representations). For $n = 2$ in the color-singlet case Eq. (4.207) reduces to the BFKL equation (3.58) and in the color-octet case Eq. (4.207) becomes the equation (3.102) leading to the bootstrap equation (3.107).

The general proof of Eq. (4.207) can be found in the papers of Bartels (1980) and Kwiecinski and Praszalowicz (1980) (see also Jaroszewicz 1980). A review of this approach

Table 4.1. *The intercepts of*
multi-reggeized gluon states as found in
Korchemsky, Kotanski, and Manashov
(2002)

Number of reggeized gluons	Intercept
$n = 2$	$2.77\bar{\alpha}_s$
$n = 4$	$0.67\bar{\alpha}_s$
$n = 6$	$0.39\bar{\alpha}_s$

was given by Lipatov (1999). (See also the papers of Korchemsky, Kotanski, and Manashov (2004) and Lipatov (2009) and references therein for recent developments.)

The contribution of multi-reggeized gluon states to the onium–onium scattering cross section is N_c-suppressed compared with that of single-BFKL-pomeron exchange. To see this suppression for a four-reggeized gluon state (with all four gluons in a net color-singlet state) one has to first subtract from it the color configurations corresponding to single- and double-BFKL-ladder exchanges (which are included in the four-gluon evolution, owing to the bootstrap property of Sec. 3.3.5), along with the three-reggeized gluon configurations. The remaining quantity would contribute to the onium–onium cross section at order $1/N_c^4$ (for $\alpha_s N_c$ fixed); this is suppressed in comparison with single-BFKL-ladder exchange, which is of order $1/N_c^2$, (4.102), and is comparable with double-BFKL-pomeron exchange, also of order $1/N_c^4$. The dipole model presented above does not contain such contributions: in fact one has to augment the dipole model with color quadrupoles to reproduce the evolution of the four-gluon BKP state (Chen and Mueller 1995).

While the coupling of the four-gluon BKP state, and of the states with higher numbers of gluons, to the onia is N_c^2-suppressed, their evolution (4.207) clearly contains a leading-N_c contribution.[8] One may, therefore, be concerned that the solutions of Eq. (4.207) could lead to these multi-reggeized gluon states giving contributions to the cross section that grow with rapidity faster than the multi-BFKL-pomeron exchanges at the same order in N_c^2-suppression and thus become order-1 corrections to the dipole model (or even dominating the cross section). For instance, the four-gluon state, after all subtractions, might grow with energy faster than the two-BFKL-pomeron exchange contribution. Such worries have been put to rest by an explicit solution of the large-N_c version of the BKP equation, performed by Korchemsky, Kotanski, and Manashov (2002), whose results (for even n) are summarized in Table 4.1. One can see, for instance, that the intercept of the $n = 4$ state, equal to $0.67\bar{\alpha}_s$, is significantly smaller than the effective intercept due to two BFKL ladder exchanges, which is $2(\alpha_P - 1) = 8\bar{\alpha}_s \ln 2 \approx 5.55\bar{\alpha}_s$ and can be safely neglected. More importantly, it follows from Table 4.1 that the higher-n states actually have intercepts that *decrease* with n, thus becoming progressively less important than the n-pomeron contribution with intercept $n(\alpha_P - 1)$, growing linearly with n. States with odd n are also unimportant for the total

---

[8] For a general SU(N_c) group the decomposition of $N_c \otimes N_c$ contains another representation, denoted $\mathcal{R}_7$, with $\lambda_{\mathcal{R}_7} = 1/N_c$. The evolution of two gluons in this representation, along with that in representation 27 ($\lambda_{27} = -1/N_c$) is N_c-suppressed. (See Kovner and Lublinsky (2007) for a detailed presentation of group factors.)

cross section: their intercepts are less than or equal to zero. In the next section we will work out the case of an odderon, which corresponds to three gluons in the t-channel.

4.7 The odderon[*]

Consider a scattering amplitude mediated by a t-channel exchange of a C-odd object, all of whose other quantum numbers are those of a pomeron (or QCD vacuum). Such a process is important phenomenologically for the exclusive production of some C-odd vector mesons, such as pions, J/ψ, η_C, etc. The exchanged object with vacuum quantum numbers except for $C = -1$ is called the *odderon* (by analogy with the definition of the pomeron in Sec. 3.2). Its existence in QCD was first suggested by Lukaszuk and Nicolescu (1973).

Let us determine the odderon contribution to onium–onium scattering. We start by considering the general scattering amplitude for an onium on a target. The charge conjugation operation interchanges the quark and the antiquark:

$$C: \qquad \vec{x}_{1\perp} \leftrightarrow \vec{x}_{0\perp}, \qquad z \leftrightarrow 1 - z, \tag{4.210}$$

where as usual the quark in the onium is located at $\vec{x}_{1\perp}$ while the antiquark is at $\vec{x}_{0\perp}$, and z is the longitudinal momentum fraction carried by the quark. In the GGM or LLA approximations the z-dependence can be neglected: we will discard it here. The odderon-exchange amplitude, by definition, corresponds to an elastic amplitude that is anti-symmetric under the operation (4.210):

$$\mathcal{O}(\vec{x}_{1\perp}, \vec{x}_{0\perp}, Y) = -\mathcal{O}(\vec{x}_{0\perp}, \vec{x}_{1\perp}, Y). \tag{4.211}$$

We can relate the elastic odderon amplitude $\mathcal{O}$ to the S-matrix $S(\vec{x}_{1\perp}, \vec{x}_{0\perp}, Y)$ (in the notation of Eq. (4.140)) by

$$\mathcal{O}(\vec{x}_{1\perp}, \vec{x}_{0\perp}, Y) = \frac{1}{2i}\left[S(\vec{x}_{1\perp}, \vec{x}_{0\perp}, Y) - S(\vec{x}_{0\perp}, \vec{x}_{1\perp}, Y)\right]. \tag{4.212}$$

In the eikonal and LLA approximations we have $S(\vec{x}_{0\perp}, \vec{x}_{1\perp}) = S^\dagger(\vec{x}_{1\perp}, \vec{x}_{0\perp})$, so that

$$\mathcal{O}(\vec{x}_{1\perp}, \vec{x}_{0\perp}, Y) = \frac{1}{2i}\left[S(\vec{x}_{1\perp}, \vec{x}_{0\perp}, Y) - S^\dagger(\vec{x}_{1\perp}, \vec{x}_{0\perp}, Y)\right]$$

$$= \operatorname{Im} S(\vec{x}_{1\perp}, \vec{x}_{0\perp}, Y) \tag{4.213}$$

(Hatta *et al.* 2005a). We see that the odderon amplitude is equal to the imaginary part of the S-matrix and hence, to the real part of the T-matrix. We can thus generalize Eq. (4.132) to

$$S(\vec{x}_{1\perp}, \vec{x}_{0\perp}, Y) = 1 - N(\vec{x}_{1\perp}, \vec{x}_{0\perp}, Y) + i\mathcal{O}(\vec{x}_{1\perp}, \vec{x}_{0\perp}, Y). \tag{4.214}$$

Now let us return to onium–onium scattering. The lowest-order C-odd onium–onium scattering amplitude is given by the three-gluon exchange diagrams depicted in Fig. 4.37. The gluons in Fig. 4.37 couple only to the quarks and antiquarks in the onia: contributions with gluons coupling to each other are either C-even or zero, by color algebra considerations. In general, three gluons can be either in the f^{abc} or d^{abc} color configuration, where $d^{abc} = 2\operatorname{Tr}[t^a\{t^b, t^c\}]$ is an absolutely symmetric object, the braces denoting an anticommutator. We are interested in the part of the diagram in Fig. 4.37 (in the eikonal approximation)

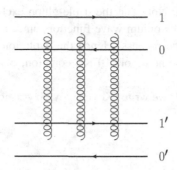

Fig. 4.37. The lowest order odderon exchange diagram in onium–onium scattering. As usual, the disconnected gluon lines imply sums over all couplings to the quark and antiquark lines.

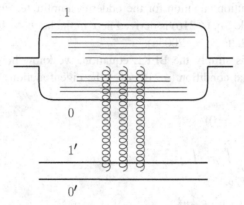

Fig. 4.38. Odderon evolution in the large-N_c approximation: a dipole cascade convoluted with triple-gluon exchange.

contributing to the odderon onium–onium amplitude $\mathcal{O}(\vec{x}_{1\perp}, \vec{x}_{0\perp}; \vec{x}'_{1\perp}, \vec{x}'_{0\perp}, Y)$. One can show, using Eq. (4.212), that only the d^{abc} configuration contributes to $\mathcal{O}$. By analogy with Eq. (4.117) we write, switching to the complex-variable notation of Sec. 4.3.3,

$$\mathcal{O}(\rho_1, \rho_0; \rho_{1'}, \rho_{0'}; Y = 0) = c_0\, \alpha_s^3 \ln^3 \left| \frac{\rho_{11'}\rho_{00'}}{\rho_{10'}\rho_{01'}} \right| \tag{4.215}$$

with c_0 some constant (dependent on N_c), whose exact value is not important for us. Clearly the exchange in Eq. (4.215) is C-odd, as it changes sign under either the $1 \leftrightarrow 0$ or the $1' \leftrightarrow 0'$ interchange.

Equation (4.215) can be used as the initial condition of the LLA small-x evolution in the large-N_c limit, which we would like to construct for the odderon amplitude $\mathcal{O}(\vec{x}_{1\perp}, \vec{x}_{0\perp}, Y)$. Above, in constructing the dipole model cascade we did not make any assumptions about the C-parity of the scattering amplitude: hence the dipole evolution should apply to the odderon case. Working in the rest frame of one of the colliding onia we present the odderon evolution in Fig. 4.38, constructed by analogy with Fig. 4.22. The incoming onium develops a dipole cascade, with one dipole exchanging three t-channel gluons with the lower onium at rest. Note that the triple-gluon exchange of Fig. 4.37 can only couple to a single dipole in the onium wave function, since a single gluon cannot couple to a color dipole in an elastic process. Hence the evolution for the odderon amplitude should be described by the same dipole BFKL equation, with different initial condition (4.215).

By analogy with Eq. (4.87) we write (Kovchegov, Szymanowski, and Wallon (2004))

$$\partial_Y \mathcal{O}(\vec{x}_{1\perp}, \vec{x}_{0\perp}; \vec{x}'_{1\perp}, \vec{x}'_{0\perp}; Y)$$

$$= \frac{\alpha_s N_c}{2\pi^2} \int d^2 x_2 \frac{x_{10}^2}{x_{20}^2 x_{21}^2}$$

$$\times \left[\mathcal{O}(\vec{x}_{1\perp}, \vec{x}_{2\perp}; \vec{x}'_{1\perp}, \vec{x}'_{0\perp}; Y) + \mathcal{O}(\vec{x}_{2\perp}, \vec{x}_{0\perp}; \vec{x}'_{1\perp}, \vec{x}'_{0\perp}; Y) - \mathcal{O}(\vec{x}_{1\perp}, \vec{x}_{0\perp}; \vec{x}'_{1\perp}, \vec{x}'_{0\perp}; Y) \right].$$

$$(4.216)$$

This is the linear evolution equation for the odderon amplitude; Eq. (4.215) provides its initial conditions. While Eq. (4.216) is derived here in the large-N_c limit it is also valid for any N_c, as shown by Hatta *et al.* (2005a).

Since Eq. (4.216) is simply the BFKL equation, we know how to find its solution. Decomposing the initial condition (4.215) over the eigenfunctions of the Möbius group, we write

$$\mathcal{O}(\rho_1, \rho_2; \rho_{1'}, \rho_{2'}; Y = 0)$$

$$= c_0 \alpha_s^3 \frac{6}{\pi^2} \sum_{\text{odd } n} \int\limits_{-\infty}^{\infty} dv \int d^2\rho_a$$

$$\times \frac{v^2 + \frac{1}{4}n^2}{\left[v^2 + \frac{1}{4}(n+1)^2\right]\left[v^2 + \frac{1}{4}(n-1)^2\right]} \chi(n, v) E^{n,v}(\rho_{1a}, \rho_{2a}) E^{n,v*}(\rho_{1'a}, \rho_{2'a})$$

$$(4.217)$$

and, using Eq. (4.115), we obtain the general solution of Eq. (4.216) for the odderon amplitude:

$$\mathcal{O}(\rho_1, \rho_2; \rho_{1'}, \rho_{2'}; Y)$$

$$= c_0 \alpha_s^3 \frac{6}{\pi^2} \sum_{\text{odd } n} \int\limits_{-\infty}^{\infty} dv \int d^2\rho_a \, e^{\bar{\alpha}_s \chi(n,v)Y}$$

$$\times \frac{v^2 + \frac{1}{4}n^2}{\left[v^2 + \frac{1}{4}(n+1)^2\right]\left[v^2 + \frac{1}{4}(n-1)^2\right]} \chi(n, v) E^{n,v}(\rho_{1a}, \rho_{2a}) E^{n,v*}(\rho_{1'a}, \rho_{2'a}).$$

$$(4.218)$$

Owing to the property (4.120) of the functions $E^{n,v}$ this amplitude is indeed C-odd. Note that the solution (4.125) of the BFKL equation for the forward amplitude is C-even as it contains a sum over even n, while the odderon solution (4.218) has a sum over odd n and is therefore C-odd. We see from Eq. (4.216) that in dipole language the odderon and the pomeron evolutions are given by the same dipole BFKL equation; the different initial conditions project out different contributions.

The solution (4.218), constructed in this form by Kovchegov, Szymanowski, and Wallon (2004), is analogous to that found in momentum space earlier by Bartels, Lipatov, and Vacca (2000), which is known as the BLV solution. The latter was obtained by solving the BKP equation (4.207) for three gluons in the d^{abc} color state. We can now find the intercept of the three-gluon BKP d^{abc}-state. The leading high energy asymptotics of the BFKL solution (4.125) is given by the $n = 0$ term in the series, since it carries the largest intercept. In the case of Eq. (4.218) the $n = 0$ term is no longer included in the sum, and the largest intercept comes from the $n = \pm 1$ terms, giving, in the saddle-point approximation around $v = 0$ (Bartels, Lipatov, and Vacca 2000),

$$\alpha_{odd} - 1 = \bar{\alpha}_s \chi(n = \pm 1, v = 0) = 0. \tag{4.219}$$

We see that the odderon amplitude does not grow with energy, even when small-x evolution is included! This is an interesting result, which may be the reason for the lack of experimental observation of the odderon.

The odderon amplitude also receives saturation corrections due to nonlinear evolution. Consider dipole–nucleus scattering. The nonlinear evolution equation for the C-odd amplitude $\mathcal{O}(\vec{x}_{1\perp}, \vec{x}_{0\perp}, Y)$ in the large-N_c approximation can be found by inserting Eq. (4.214) into Eq. (4.137) and taking the imaginary part of the resulting expression, keeping in mind that both N and $\mathcal{O}$ are real quantities. This gives (Kovchegov, Szymanowski, and Wallon 2004, Hatta *et al.* 2005a)

$$\partial_Y \mathcal{O}(\vec{x}_{1\perp}, \vec{x}_{0\perp}, Y) = \frac{\alpha_s N_c}{2\pi^2} \int d^2 x_2 \frac{x_{01}^2}{x_{20}^2 x_{21}^2}$$

$$\times \left[\mathcal{O}(\vec{x}_{1\perp}, \vec{x}_{2\perp}, Y) + \mathcal{O}(\vec{x}_{2\perp}, \vec{x}_{0\perp}, Y) - \mathcal{O}(\vec{x}_{1\perp}, \vec{x}_{0\perp}, Y) \right]$$

$$- \frac{\alpha_s N_c}{2\pi^2} \int d^2 x_2 \frac{x_{01}^2}{x_{20}^2 x_{21}^2}$$

$$\times \left[\mathcal{O}(\vec{x}_{1\perp}, \vec{x}_{2\perp}, Y) N(\vec{x}_{2\perp}, \vec{x}_{0\perp}, Y) + N(\vec{x}_{1\perp}, \vec{x}_{2\perp}, Y) \mathcal{O}(\vec{x}_{2\perp}, \vec{x}_{0\perp}, Y) \right]. \tag{4.220}$$

We conclude that saturation simply suppresses the odderon amplitude $\mathcal{O}$ further, by making it decrease with energy. This can be readily seen from Eq. (4.220) by, for instance, substituting $N = 1$ in it for $x_{20}, x_{21} > 1/Q_s(Y)$, corresponding to the saturated total dipole–nucleus cross section. One would then get the S-matrix version of the Levin–Tuchin formula, Eq. (4.172), but now for the odderon amplitude $\mathcal{O}$, indicating that it falls off steeply with increasing energy or rapidity in the presence of saturation.

Further reading

More details on some aspects of the GGM multiple-rescattering formula and on BK evolution, with its solution, can be found in the reviews by Iancu and Venugopalan (2003),

Weigert (2005), and Jalilian-Marian and Kovchegov (2006). Further information on the semiclassical approximation for the solution of the BK equation can be found in Levin and Tuchin (2000, 2001) (see also Diaz Saez and Levin (2011) and references therein). A comprehensive summary of the current status and achievements of the traveling wave approach can be found in Beuf (2010). The most recent status of the saturation boundary approaches to solving the BK equation at higher orders is discussed in Avsar *et al.* (2011) and references therein. The consequences of the conformal symmetry of the fixed-coupling BK evolution for its solution have been recently explored by Gubser (2011).

The solution of the BKP equation, especially for the odderon case, has been widely discussed, and we consider that the review and paper of Lipatov (1999, 2009) together with the paper of Korchemsky, Kotanski, and Manashov (2004) can bring the reader to the current understanding of this problem. A comprehensive review of the theory of odderon evolution and the status of experimental searches for the odderon was given by Ewerz (2003).

Exercises

4.1 By summing all possible connections of the t-channel gluons to the dipole in Fig. 4.10, derive Eq. (4.40) explicitly.

4.2** Find the virtual correction to the onium wave function in Eq. (4.64) by an explicit calculation of diagrams. One may directly sum the LCPT diagrams in Fig. 4.14 (Chen and Mueller 1995). Alternatively, one may start with a momentum-space expression for each distinct contributing Feynman diagram, Fourier-transform it into coordinate space in x^-, regulate the x^--integrals, and integrate over the x^--coordinates of the quark–gluon vertices from $-\infty$ to 0. Fourier-transforming the obtained expression into transverse coordinate space should yield (4.64).

4.3 Follow the steps outlined in the text to find the eigenvalues of the kernel of Eq. (4.90). Namely, starting with Eq. (4.95) reduce it to Eq. (4.97).

4.4 Starting with Eq. (4.90) use the substitution (4.98) to obtain the BFKL equation (3.58) for the function f. You may find Eqs. (A.9) and (A.10) handy.

4.5 (a) Solve the following zero-transverse-dimensional equation for the generating functional $Z(Y, u)$ (cf. Eq. (4.77)):

$$\partial_Y Z = \alpha_s (Z^2 - Z) \qquad (4.221)$$

with initial condition $Z(Y = 0, u) = 0$. Using Eq. (4.81) find the number of dipoles in this "wave function".

(b) Perform a similar exercise for a toy model of the BK equation (4.138): solve

$$\partial_Y N = \alpha_s N - \alpha_s N^2, \qquad (4.222)$$

with $N(Y = 0) = N_0 \ll 1$ as the initial condition; $N_0 > 0$ is a constant. Show that $N(Y) \to 1$ as $Y \to \infty$.

4.6 Suppose that the dipole–nucleus scattering amplitude in the linear regime outside the saturation region is given by the following approximation of the DLA formula (4.150) (for $x_\perp Q_{s0} \ll 1$)

$$N(x_\perp, Y) = (x_\perp Q_{s0})^2 \, \exp\left\{2\sqrt{\bar{\alpha}_s Y \ln[1/(x_\perp Q_{s0})^2]}\right\}. \qquad (4.223)$$

(a) Find the energy-dependent saturation scale $Q_s(Y)$ by requiring that

$$N(x_\perp = 1/Q_s(Y), Y) = 1. \qquad (4.224)$$

(b) Show that for $1/Q_s(Y) \gg x_\perp \gg 1/k_{geom}$, with $k_{geom} = Q_s^2(y)/Q_{s0}$, (cf. Eq. (4.163)), Eq. (4.223) can be rewritten as

$$N(x_\perp, Y) \approx x_\perp Q_s(Y), \qquad (4.225)$$

i.e., as a function of a single variable $x_\perp Q_s(Y)$ instead of the two variables $x_\perp$ and Y (cf. Eq. (4.161)). This is a simplified derivation of the extended geometric scaling (Iancu, Itakura, and McLerran 2002).

4.7* Derive Eq. (4.217) with $\mathcal{O}(\rho_1, \rho_2; \rho_{1'}, \rho_{2'}; Y = 0)$ as given in Eq. (4.215).

4.8 Determine the high energy asymptotics of the F_2 structure function (and $\sigma_{tot}^{\gamma^* A}$). At very small x we have $Q_s \gg Q$. Argue that in such a case the $x_\perp$-integral in (4.12) is dominated by $1/Q_s \ll x_\perp \ll 1/Q$. Approximating the dipole–nucleus interaction by a black disk of radius R, so that $N(x_\perp, b_\perp, Y) \approx \theta(R - b_\perp)$, and expanding the Bessel functions in Eqs. (4.18) and (4.21), show using Eq. (4.10a) that

$$F_2 \sim \sigma_{tot}^{\gamma^* A} \sim R^2 \ln \hat{s} \sim \ln^3 \hat{s}; \qquad (4.226)$$

the last conclusion results from the substitution $R = R_0 + a \ln \hat{s}$, reflecting the diffusion of the black-disk radius (3.115). Equation (4.226) sets an upper limit on $\sigma_{tot}^{\gamma^* A}$ known as the *Gribov bound* (Gribov 1970).

5

Classical gluon fields and the color glass condensate

In the previous chapter we developed a two-step approach to DIS: one first sums the multiple rescatterings, leading to the GGM formula resumming powers of $\alpha_s^2 A^{1/3}$, and then one includes the small-x evolution effects, which enter via s-channel gluon emissions and absorptions, by resumming powers of $\alpha_s Y$. Here we generalize this two-step approach, making it applicable to other high energy scattering processes. We show that the GGM approximation is equivalent to treating the gluon field in the nucleus classically, according to the prescription of the McLerran–Venugopalan (MV) model. Quantum evolution corrections to the MV model come in through the Jalilian-Marian–Iancu–McLerran–Weigert–Leonidov–Kovner (JIMWLK) evolution equation, which, in particular, provides an all-N_c generalization of the dipole approach. The color glass condensate (CGC) is introduced.

5.1 Strong classical gluon fields: the McLerran–Venugopalan model

5.1.1 The key idea of the approach

Let us consider a large ultrarelativistic nucleus in the infinite-momentum frame. The nucleus is taken as being described by the Glauber model of Sec. 4.2. We are interested in the small-x tail of the gluon wave function in the nucleus. As follows from Eq. (2.56), in the rest frame of the nucleus the small-x gluons have a coherence length of order

$$l_{coh} \sim \frac{1}{m_N x}, \tag{5.1}$$

where m_N is the mass of a nucleon. If the Bjorken-x variable is sufficiently small then the coherence length may become very large, much larger than the size of the nucleus. Such small-x gluons would be produced by the whole nucleus coherently in the longitudinal direction. An example of this interaction is shown in the left-hand panel of Fig. 5.1. There the small-x gluon (denoted by the wavy line) interacts coherently with several Lorentz-contracted nucleons. Indeed the nucleons, and the nucleus as a whole, are color-neutral and one might think that a coherent gluon would simply not "see" them. However, the gluon is coherent only in the longitudinal direction: in the transverse direction it is localized on the scale $x_\perp \sim 1/k_T$, with $k_T \equiv k_\perp$ the transverse momentum of the gluon. If $k_T \gg \Lambda_{QCD}$, which is a necessary condition for using gluon degrees of freedom, the transverse extent of the gluon is much smaller than the sizes of the nucleons. Because of this the gluon interacts

198

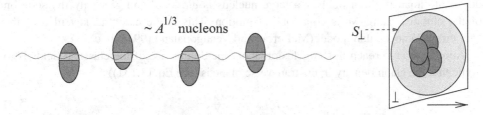

Fig. 5.1. Left-hand panel: a small-x gluon sees the whole nucleus coherently in the longitudinal direction and interacts with several different nucleons in it. Right-hand panel: the effective color charge seen by the gluon in the IMF as a result of a random walk in color space. (Reprinted from Jalilian-Marian and Kovchegov (2006), with permission from Elsevier.) A color version of this figure is available online at www.cambridge.org/9780521112574.

with only part of each nucleon in the transverse direction, as shown in the left-hand panel of Fig. 5.1. The color charge in the segment of a nucleon that the gluon is traversing does not have to be zero: the gluon may run into, say, a single valence quark. As a result of such interactions, the gluon "feels" some effective color charge in all the nucleons' segments that it traverses. In our Glauber approximation we assume that all the nucleons are independent, so that interactions with parts of different nucleons are similar to a random walk in color space. If each individual nucleon's segment has a typical color charge g, then, owing to the random walk nature of the process, the total color charge seen by the gluon at a fixed impact parameter is $g\sqrt{n}$, where $n \sim A^{1/3}$ is the number of nucleons at a fixed transverse coordinate of the gluon.

In the infinite-momentum frame, owing to Lorentz contraction all the nucleons appear to be squeezed into a thin "pancake" of Lorentz-contracted nucleus, as shown at the right in Fig. 5.1. One may then define the effective color charge density seen by a gluon in the transverse plane of the nucleus (McLerran and Venugopalan 1994a, b, c). The typical magnitude of these color charge density fluctuations is given by the color charge squared divided by the transverse area of the nucleus, $(g\sqrt{n})^2/S_\perp = g^2 n/S_\perp$. The number of color charge sources in the whole nucleus is proportional to the number of nucleons in the nucleus, $n \sim A$. The typical color charge density fluctuations are, therefore, characterized by the momentum scale

$$\mu^2 \sim \frac{g^2 A}{S_\perp} \sim \Lambda_{QCD}^2 A^{1/3}. \tag{5.2}$$

It is important to notice that the momentum scale in Eq. (5.2) grows with A as $A^{1/3}$, similarly to the saturation scale in the GGM model (4.50) (see also (4.52)). The important conclusion we can draw from Eq. (5.2) is that for sufficiently large nuclei their small-x wave functions are characterized by a hard momentum scale μ that is much larger than Λ_{QCD}. It is likely that the large scale μ determines the running of the strong-coupling constant, $\alpha_s = \alpha_s(\mu^2)$, allowing for a small-coupling α_s description of the process. Field theories with small coupling are usually dominated by classical fields, with the quantum corrections suppressed by extra powers of the small coupling constant α_s. Therefore the dominant small-x gluon field of a large nucleus is *classical* and given by the solution of the classical Yang–Mills equations of motion. This is the essential key idea of the McLerran–Venugopalan model (McLerran and Venugopalan (1994a, b, c)).

Another way to reach this conclusion about the dominance of the classical fields is to argue that the gluon density in the transverse plane is (see Eq. (3.131))

$$\rho_{glue} = \frac{x G_A}{S_\perp}.$$

(5.3)

For a dilute nucleus $x G_A = A x G_N \sim A$, so that $\rho_{glue} \sim A^{1/3}$ and is therefore large for a nucleus with $A \gg 1$, resulting in the high occupation number of gluons. Such a high occupation number implies the dominance of the classical physics: hence the gluon field should be classical (McLerran and Venugopalan 1994a). Moreover, ρ_{glue} has the dimensions of mass squared, giving us a new momentum scale $\mu^2 \sim \rho_{glue}$, which is consistent with that in Eq. (5.2). The strongest gluon field possible in the QCD Lagrangian (1.1) at small coupling g is of order some momentum scale times $1/g$, as can be inferred by equating the linear and nonlinear terms in the field strength tensor (1.4). Hence the resulting strong gluon field should be of order $A_\mu \sim 1/g$ (cf. Eq. (3.137)), which is characteristic of classical gluon fields (e.g. instanton fields).

We see that the MV model is based on the observation that the larger-x partons (such as the valence quarks in the nucleons) in a large nucleus serve as classical sources for the smaller-x gluons. We now are going to find this classical gluon field.

5.1.2 Classical gluon field of a single nucleus

According to the prescription of the MV model, we need to solve the classical Yang–Mills equations

$$\mathcal{D}_\mu F^{\mu\nu} = J^\nu,$$

(5.4)

with an ultrarelativistic nucleus providing the source current J^ν, so that in the infinite-momentum frame

$$J^\nu = \delta^{\nu+} \rho(x^-, \vec{x}_\perp),$$

(5.5)

where $\rho(x^-, \vec{x}_\perp)$ is the color charge density.[1] The adjoint covariant derivative is defined by

$$\mathcal{D}_\mu F^{\mu\nu} \equiv \partial_\mu F^{\mu\nu} - ig \left[A_\mu, F^{\mu\nu} \right]$$

(5.6)

in the standard convention.

The classical gluon field of a nucleus is easier to find in the covariant $\partial_\mu A^\mu = 0$ gauge. To do this we will assume, for simplicity, that all the relevant large-x color charge in the nucleus is carried by the valence quarks. Furthermore, we will specifically choose to consider a nucleus with "mesonic" nucleons made out of $q\bar{q}$ pairs instead of three valence

---

[1] Unlike in the previous chapter, where the nucleus was either at rest or moving along the x^- light cone, in this chapter we take the nucleus to be moving along the x^+ light cone direction, in order for the notation to agree with the majority of the literature on the subjects discussed here. A simple $+ \leftrightarrow -$ substitution relates the results of the two chapters.

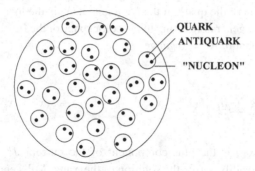

Fig. 5.2. The model for a nucleus where the "nucleons" are quark–antiquark pairs. (Reprinted with permission from Kovchegov (1996). Copyright 1996 by the American Physical Society.)

quarks (Kovchegov 1996). (This latter assumption merely simplifies the calculations; the conclusions are easy to generalize to the case of real nuclei.) Our model of the nucleus is depicted in Fig. 5.2. Considering the nucleus to be moving ultrarelativistically in the light cone plus direction, we label "valence" quark and antiquark coordinates by $\vec{x}_{i\perp}, x_i^-$ and $\vec{x}'_{i\perp}, x_i'^-$ in accordance with their position along the x^--axis, so that

$$x_1^-, x_1'^- < x_2^-, x_2'^- < \cdots < x_A^-, x_A'^-. \tag{5.7}$$

In the recoilless eikonal approximation considered here, neither coordinate in a quark-antiquark pair changes due to the emission of gluon fields.

In our theoretical nucleus a nucleon consists of a $q\bar{q}$ pair, where the quark and antiquark move as free particles inside the nucleon but are not able to leave the nucleon due to confinement. Similarly, in the Glauber model of the nucleus the nucleons can be anywhere within the nucleus with equal probability. As we will see shortly, in the IMF and in the covariant $\partial_\mu A^\mu = 0$ gauge the gluon field of, say, quark i located at x_i^- is proportional to $\delta(x^- - x_i^-)$. Since the quarks (and anti-quarks) in the model have different x_i^--coordinates, the fields of the individual quarks (anti-quarks) cannot overlap and we can construct the gluon field of the nucleus as a sum of the fields of the quarks and anti-quarks. We first will find this sum in the covariant gauge, and after that we will transform the total field to the $A^+ = 0$ light cone gauge, which is more convenient for calculating the gluon distribution function.

Starting from a nucleus at rest in the covariant gauge, we write the color charge density as

$$\rho_{cov}(\vec{x}) = \sum_{a-1}^{N_c^2-1} t^a \rho_{cov}^a(\vec{x}), \tag{5.8}$$

with

$$\rho_{cov}^a(\vec{x}) = g \sum_{i=1}^{A} (t_i^a)[\delta^3(\vec{x} - \vec{x}_i) - \delta^3(\vec{x} - \vec{x}_i')] \tag{5.9}$$

where $\vec{x}_i$ is the location of the quark in the ith nucleon, $\vec{x}_i'$ is the location of the antiquark, and the (t_i^a) are SU(N_c) generators acting in the color space of the ith nucleon. The subscript *cov* denotes the covariant $\partial_\mu A^\mu = 0$ gauge.

Boosting into the IMF we obtain

$$\rho_{cov}^a(x^-, \vec{x}_\perp) = 2g \sum_{i=1}^A (t_i^a)[\delta(x^- - x_i^-)\delta^2(\vec{x}_\perp - \vec{x}_{i\perp}) - \delta(x^- - x_i'^-)\delta^2(\vec{x}_\perp - \vec{x}_{i\perp}')],$$
(5.10)

where now $\rho_{cov}(x^-, \vec{x}_\perp)$ is the plus component of the current J^μ, in accordance with Eq. (5.5). As one can readily verify, the solution of the Yang–Mills equations (5.4) with the source given in (5.5), (5.10) is

$$A_{cov}^+ = -\frac{g}{\pi} \sum_{a=1}^{N_c^2-1} \sum_{i=1}^A t^a(t_i^a) \left[\delta(x^- - x_i^-)\ln(|\vec{x}_\perp - \vec{x}_{i\perp}|\Lambda) - \delta(x^- - x_i'^-)\ln(|\vec{x}_\perp - \vec{x}_{i\perp}'|\Lambda)\right],$$

$$A_{cov}^- = 0, \quad \vec{A}_{cov}^\perp = 0,$$
(5.11)

where Λ is some infrared cutoff. The only nonzero component of the field strength in the covariant gauge is then

$$F_{cov}^{\perp+} = \frac{g}{\pi} \sum_{a=1}^{N_c^2-1} \sum_{i=1}^A t^a(t_i^a) \left[\delta(x^- - x_i^-)\frac{\vec{x}_\perp - \vec{x}_{i\perp}}{|\vec{x}_\perp - \vec{x}_{i\perp}|^2} - \delta(x^- - x_i'^-)\frac{\vec{x}_\perp - \vec{x}_{i\perp}'}{|\vec{x}_\perp - \vec{x}_{i\perp}'|^2}\right].$$
(5.12)

The gluon field in Eq. (5.11) is itself a solution of the classical Yang–Mills equations. However, as we mentioned before, the field in the $A^+ = 0$ light cone gauge is needed to find the gluon distribution resulting from classical physics. We have to gauge-transform the field from Eq. (5.11) into the light cone gauge. The field in the new gauge is

$$A_\mu^{LC} = S A_\mu^{cov} S^{-1} - \frac{i}{g}(\partial_\mu S)S^{-1}.$$
(5.13)

Requiring the new gauge to be the light cone gauge, $A_{LC}^+ = 0$, we solve for S to obtain[2]

$$S(x^-, \vec{x}_\perp) = \mathrm{P}\exp\left\{\frac{ig}{2}\int_{x^-}^{-\infty} dx'^- A_{cov}^+(x'^-, \vec{x}_\perp)\right\},$$
(5.14)

where, as usual, the symbol P denotes path-ordering of the operators in the integral. The matrix of the gauge transformation is given by a Wilson line (Wilson 1974) along the x^- light cone. (The choice of the contour of the Wilson line in Eq. (5.14) is not unique: the freedom to choose the contour is directly related to the residual gauge freedom within the

---

[2] The factor $1/2$ in Eq. (5.14) is due to our definition of the light cone components in Sec. 1.3.

$A^+ = 0$ gauge.) The gluon field in the $A^+ = 0$ light cone gauge is

$$\vec{A}_{LC}^{\perp}(x^-, \vec{x}_\perp) = \frac{1}{2} \int\limits_{-\infty}^{x^-} dx'^- F_{LC}^{+\perp}(x'^-, \vec{x}_\perp)$$

$$= \frac{1}{2} \int\limits_{-\infty}^{x^-} dx'^- S(x'^-, \vec{x}_\perp) F_{cov}^{+\perp}(x'^-, \vec{x}_\perp) S^{-1}(x'^-, \vec{x}_\perp), \tag{5.15}$$

with $A_{LC}^- = 0$. Since the fields do not depend on x^+ we have suppressed x^+ in all the arguments.

Substituting Eq. (5.12) into Eq. (5.15) we obtain the classical gluon field for an ultrarelativistic nucleus in its light cone gauge (Kovchegov 1996, Jalilian-Marian *et al.* 1997a):

$$\vec{A}_\perp^{LC}(x^-, \vec{x}_\perp) = \frac{g}{2\pi} \sum_{a=1}^{N_c^2-1} \sum_{i=1}^{A} (t_i^a) \left[S(x_i^-, \vec{x}_\perp) t^a S^{-1}(x_i^-, \vec{x}_\perp) \frac{\vec{x}_\perp - \vec{x}_{i\perp}}{|\vec{x}_\perp - \vec{x}_{i\perp}|^2} \theta(x^- - x_i^-) \right.$$

$$\left. - S(x_i'^-, \vec{x}_\perp) t^a S^{-1}(x_i'^-, \vec{x}_\perp) \frac{\vec{x}_\perp - \vec{x}_{i\perp}'}{|\vec{x}_\perp - \vec{x}_{i\perp}'|^2} \theta(x^- - x_i'^-) \right]. \tag{5.16}$$

An explicit expression for $S(x^-, \vec{x}_\perp)$ can be obtained by substituting the covariant-gauge field (5.11) into Eq. (5.14) and integrating over the delta-functions. This yields

$$S(x^-, \vec{x}_\perp) = \prod_{i=1}^{A} \exp\left\{ \frac{ig^2}{2\pi} \sum_{a=1}^{N_c^2-1} t^a (t_i^a) \ln \frac{|\vec{x}_\perp - \vec{x}_{i\perp}|}{|\vec{x}_\perp - \vec{x}_{i\perp}'|} \theta(x^- - x_i^-) \right\}, \tag{5.17}$$

where the terms in the product are ordered from left to right with increasing index i. In arriving at Eq. (5.17) we have coarse-grained our treatment of the nucleus, assuming that the coordinate x^- is either larger or smaller than the position of the nucleon on the light cone, taken now to be approximately equal to x_i^-. Hence we do not have situations where only one quark in a nucleon contributes to $S(x^-, \vec{x}_\perp)$. Individual nucleon contributions are suppressed by powers of A, hence neglecting one of them is justified in our Glauber, $A \gg 1$, approximation for the nucleus.

The calculation of the Wilson line (5.14), which led to Eq. (5.17), also allows us to determine the region of applicability of the classical approximation used in the MV model. Note that the covariant-gauge field (5.11) is of order g; hence, in terms of the Feynman diagrams it corresponds to the emission of a gluon by the valence (anti)quarks (see also Exercise 5.1). The Wilson line (5.14) is then given by gluon exchanges between valence quarks and the path of the Wilson line, as shown in Fig. 5.3A. In fact the product in Eq. (5.17) consists of one-gluon exchanges in the exponents, each term corresponding to a given nucleon. It seems that if we expand the exponentials in the product (5.17) we can have as many gluon exchanges with each nucleon as we like. Formally, this is indeed the case: nonetheless, we claim that, to keep the classical approximation under control we

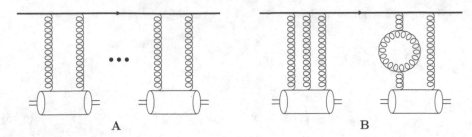

Fig. 5.3. Diagrams contributing to the Wilson line (5.14) in the validity domain of the classical approximation (A) and beyond (B).

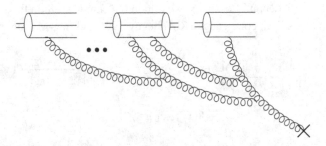

Fig. 5.4. Diagrammatic representation of the non-Abelian Weizsäcker–Williams field of a nucleus. The cross denotes the location x^μ where the field is measured.

cannot exceed more than two gluons per nucleon, as shown in Fig. 5.3A (which means expanding each exponential in Eq. (5.17) up to order g^4). Indeed, one would be tempted to go beyond this limit and include a three-gluon exchange diagram by expanding the exponentials to order g^6, as depicted on the left of Fig. 5.3B. However, at order g^6, in calculating the Wilson line in the full perturbative QCD theory we should also include the diagram on the right of Fig. 5.3B. Such a diagram contains a gluon self-energy correction and is essentially nonclassical, as it cannot be evaluated by classical methods. Therefore we would lose control over the diagram calculation if we tried to use classical methods at order g^6. Hence the classical approximation is only valid in QCD as long as we do not exceed the two-gluon per nucleon limit (Kovchegov 1997). This conclusion is similar to what we saw in the GGM approximation: the resulting resummation parameter for the classical MV approach is again $\alpha_s^2 A^{1/3}$, just as in the GGM case (4.46). Even the diagram in Fig. 5.3A is similar to the GGM diagram in Fig. 4.5. We see that the GGM formula can be thought of as having been obtained in the classical approximation.

Equation (5.16) gives the solution of the classical equations of motion for a given configuration of valence quarks (and antiquarks in our model) inside the nucleons and of nucleons inside the nucleus. We will refer to the field in Eq. (5.16) as the non-Abelian Weizsäcker–Williams field, since this is a non-Abelian analogue of the well-known Weizsäcker–Williams field in electrodynamics. A diagram corresponding to the non-Abelian Weizsäcker–Williams field of a nucleus is shown in Fig. 5.4 (Kovchegov

1997). Diagrams corresponding to the classical gluon field are tree level, in accordance with the conventional understanding of classical dynamics. (The apparent loop in Fig. 5.4 is not a quantum loop, as, together with a diagram in which the gluon couplings to the quark line are interchanged, it contributes as if the intermediate quark line were on mass shell (cf. Fig. 4.7) and thus is equivalent to two independent quark–gluon interactions.)

The classical gluon field (5.16) in the MV model can alternatively be found using a description of the valence quark distribution in the nucleus by a continuous light cone color charge density $\rho_{LC}(x^-, \vec{x}_\perp)$ related to the covariant-gauge density (5.10) by a gauge rotation:

$$\rho_{LC}(x^-, \vec{x}_\perp) = S(x^-, \vec{x}_\perp)\rho_{cov}(x^-, \vec{x}_\perp)S^{-1}(x^-, \vec{x}_\perp) \tag{5.18}$$

(McLerran and Venugopalan 1994a, b, c, Jalilian-Marian *et al.* 1997a). In such a description one does not have to specify a model for the point valence charges, such as that in Fig. 5.2, though the dilute nucleus approximation is employed.

In the point-charge approach presented above, in order to calculate a physically observable quantity one has to average over all possible positions of quarks and anti-quarks in the nucleons and of nucleons in the nucleus, which, in this classical approximation, would correspond to averaging over many scattering events. In the continuous density approach this would correspond to integrating the observable over all charge densities $\rho_{LC}(x^-, \vec{x}_\perp)$ with some weight functional $W[\rho_{LC}]$. The weight functional for a group of independent valence quarks has to be Gaussian, so that the integral would be of the form (McLerran and Venugopalan 1994a, Jalilian-Marian *et al.* 1997a)

$$\int \mathcal{D}\rho_{LC}W[\rho_{LC}] \equiv \int \mathcal{D}\rho_{LC} \exp\left\{-\int d^2x_\perp \int\limits_{-\infty}^{\infty} dx^- \frac{\mathrm{tr}\left[\rho_{LC}^2(x^-, \vec{x}_\perp)\right]}{\mu^2(x^-, \vec{x}_\perp)}\right\}. \tag{5.19}$$

Here $\mu^2(x^-, \vec{x}_\perp)$ is some function of the coordinates: it is a measure of the color-charge fluctuations and is a generalization of μ from Eq. (5.2). (The Gaussian form of Eq. (5.19) can be verified explicitly and $\mu^2(x^-, \vec{x}_\perp)$ can be found in the point-charge approach presented above: this was done in Kovchegov (1997).) The expectation value of some density-dependent operator $\hat{O}_\rho$ would then be given by

$$\langle \hat{O}_\rho \rangle = \frac{\int \mathcal{D}\rho_{LC}\hat{O}_\rho W[\rho_{LC}]}{\int \mathcal{D}\rho_{LC}W[\rho_{LC}]}. \tag{5.20}$$

5.1.3 Classical gluon distribution

Using Eq. (5.16) we can derive a formula for the distribution of gluons in the nucleus. First we need to derive an expression for the gluon distribution as a function of the gluon field operator. Working in the $A^+ = 0$ light cone gauge, we expand the gluon field operator in terms of creation and annihilation operators in the form (see e.g. Lepage and

Brodsky 1980)

$$\vec{A}^a_{LC\perp}(x^+ = 0, x^-, \vec{x}_\perp) = \int\limits_{k^+>0} \frac{d^2k_\perp dk^+}{(2\pi)^3 2k^+} \sum_{\lambda=\pm 1} \left\{ \hat{a}^a_\lambda(\vec{k}_\perp, k^+)\vec{\epsilon}^\lambda_\perp e^{-ik\cdot x} + \hat{a}^{a\dagger}_\lambda(\vec{k}_\perp, k^+)\vec{\epsilon}^{\lambda*}_\perp e^{ik\cdot x} \right\},$$

$$(5.21)$$

where

$$[\hat{a}^a_\lambda(\vec{k}_\perp, k^+), \hat{a}^{b\dagger}_{\lambda'}(\vec{k}'_\perp, k'^+)] = 2k^+(2\pi)^3\delta(k^+ - k'^+)\delta^2(\vec{k}_\perp - \vec{k}'_\perp)\delta_{\lambda\lambda'}\delta^{ab}. \quad (5.22)$$

Using these creation and annihilation operators we can write the number of gluons with transverse momentum $k_\perp$ and light cone momentum k^+ (per unit transverse momentum squared $dk_\perp^2$ and per unit rapidity dk^+/k^+) as the *Weizsäcker–Williams distribution function*,

$$\phi^{WW}(x, k_\perp^2) = \frac{\pi}{2(2\pi)^3} \sum_{\lambda=\pm 1} \sum_{a=1}^{N_c^2-1} \langle A|\hat{a}^{a\dagger}_\lambda(\vec{k}_\perp, k^+)\hat{a}^a_\lambda(\vec{k}_\perp, k^+)|A\rangle, \quad (5.23)$$

where $|A\rangle$ is a state of the nucleus and, as usual, $x = k^+/p^+$ with p^+ the large light cone momentum of the nucleons in the nucleus. We have implicitly assumed that the gluon distribution does not depend on the direction of the gluon transverse momentum and have replaced $d^2k_\perp$ by $\pi dk_\perp^2$. The quantity $\phi^{WW}(x, k_\perp^2)$ is the unintegrated gluon distribution of the nucleus (cf. Eq. (3.92)). The standard (integrated) gluon distribution is related to $\phi^{WW}(x, k_\perp^2)$ by Eq. (3.93).

Solving Eq. (5.21) for $\hat{a}^a_\lambda$ and $\hat{a}^{a\dagger}_\lambda$ and using the result in Eq. (5.23) yields

$$\phi^{WW}(x, k_\perp^2) = \frac{(k^+)^2}{8\pi^2} \int d^2x_\perp d^2y_\perp e^{i\vec{k}_\perp\cdot(\vec{x}_\perp - \vec{y}_\perp)} \int\limits_{-\infty}^{\infty} dx^- dy^- e^{-ik^+(x^- - y^-)/2}$$

$$\times \left\langle A \left| \mathrm{tr}\left[\vec{A}^{LC}_\perp(0, x^-, \vec{x}_\perp) \cdot \vec{A}^{LC}_\perp(0, y^-, \vec{y}_\perp) \right] \right| A \right\rangle. \quad (5.24)$$

To perform the Fourier transformations over x^- and y^- note that the non-Abelian WW field of Eq. (5.16) is essentially a theta function in x^-, i.e., $\theta(x^-)$, since the x^--extent of the ultrarelativistic nucleus moving in the x^+-direction is negligibly small. Writing

$$\vec{A}^{LC}_\perp(0, x^-, \vec{x}_\perp) \approx \theta(x^-)\vec{A}^{LC}_\perp(0, x^- = +\infty, \vec{x}_\perp) \equiv \theta(x^-)\vec{A}^{LC}_\perp(\vec{x}_\perp), \quad (5.25)$$

we reduce Eq. (5.24) to

$$\phi^{WW}(x, k_\perp^2) = \frac{1}{2\pi^2} \int d^2x_\perp d^2y_\perp e^{i\vec{k}_\perp\cdot(\vec{x}_\perp - \vec{y}_\perp)} \left\langle \mathrm{tr}\left[\vec{A}^{LC}_\perp(\vec{x}_\perp) \cdot \vec{A}^{LC}_\perp(\vec{y}_\perp) \right] \right\rangle, \quad (5.26)$$

where, for brevity, we denote the averaging in the state $|A\rangle$ simply by angle brackets. For the classical gluon field (5.16), averaging in the state $|A\rangle$ implies averaging over the positions of the valence quarks in the nucleons and of the nucleons in the nucleus, along with averaging over the quark colors. For the field found as a function of the charge density $\rho_{LC}(x^-, \vec{x}_\perp)$, the averaging is the same as that defined in Eq. (5.20). One can also show that the definition of the unintegrated gluon distribution (5.26), after integration over $\vec{k}_\perp$, can be recast into a form consistent with the standard definition of the integrated gluon distribution, which can be found in Sterman (1993).

Substituting the classical gluon field (5.16) into the expression for the unintegrated gluon distribution (5.26) we obtain

$$\phi^{WW}\left(x, k_\perp^2\right) = \frac{\alpha_s}{2\pi^3} \int d^2 x_\perp d^2 y_\perp e^{i\vec{k}_\perp \cdot (\vec{x}_\perp - \vec{y}_\perp)} \sum_{i,j=1}^{A}$$

$$\times \Bigg\langle (t_i^a)(t_j^b) \, \mathrm{tr} \left[S(x_i^-, \vec{x}_\perp) t^a S^{-1}(x_i^-, \vec{x}_\perp) S(x_j^-, \vec{y}_\perp) t^b S^{-1}(x_j^-, \vec{y}_\perp) \right]$$

$$\times \frac{\vec{x}_\perp - \vec{x}_{i\perp}}{|\vec{x}_\perp - \vec{x}_{i\perp}|^2} \cdot \frac{\vec{y}_\perp - \vec{x}_{j\perp}}{|\vec{y}_\perp - \vec{x}_{j\perp}|^2} + \text{a.c.} \Bigg\rangle, \tag{5.27}$$

where summation over repeated color indices is implied and a.c., the antiquark contributions, stands for three more terms, involving antiquarks.

In the spirit of the Glauber large-nucleus approximation, we assume that the contribution of the ith nucleon is not contained in $S(x_i^-, \vec{x}_\perp)$ (the same for the jth nucleon in $S(x_j^-, \vec{y}_\perp)$): this means that averaging over the color space of the quarks in the ith and the jth nucleons can be carried out separately, giving $(1/N_c)\mathrm{tr}_i[(t_i^a)] = 0$ and $(1/N_c)\mathrm{tr}_j[(t_j^b)] = 0$ unless $i = j$, in which case we get $(1/N_c)\mathrm{tr}_i[(t_i^a)(t_i^b)] = [1/(2N_c)]\delta^{ab}$. This simplifies Eq. (5.27) to

$$\phi^{WW}\left(x, k_\perp^2\right) = \frac{\alpha_s}{4\pi^3 N_c} \int d^2 x_\perp d^2 y_\perp e^{i\vec{k}_\perp \cdot (\vec{x}_\perp - \vec{y}_\perp)} \sum_{i=1}^{A}$$

$$\times \left\langle \mathrm{tr} \left[S(x_i^-, \vec{x}_\perp) t^a S^{-1}(x_i^-, \vec{x}_\perp) S(x_i^-, \vec{y}_\perp) t^a S^{-1}(x_i^-, \vec{y}_\perp) \right] \right\rangle$$

$$\times \left[\int d^2 x_i \frac{T(\vec{x}_{i\perp})}{A} \frac{\vec{x}_\perp - \vec{x}_{i\perp}}{|\vec{x}_\perp - \vec{x}_{i\perp}|^2} \cdot \frac{\vec{y}_\perp - \vec{x}_{i\perp}}{|\vec{y}_\perp - \vec{x}_{i\perp}|^2} + \text{a.c.} \right]. \tag{5.28}$$

We have now written out the averaging over $\vec{x}_{i\perp}$ explicitly, but neglecting the difference between the location of a nucleon and the location of a quark in the nucleon. We have also neglected the difference between x_i^- and $x_i'^-$ in the arguments of S, since, as we have assumed, the ith nucleon does not contribute to S. The nuclear profile function $T(\vec{b}_\perp)$ was defined in Eq. (4.31): the ratio $T(\vec{b}_\perp)/A$ is the transverse-plane probability density for finding a nucleon at impact parameter $\vec{b}_\perp$.

To simplify Eq. (5.28) further we will use the following group theory identity, which we will formulate in general terms for future use. Define a fundamental Wilson line along an arbitrary (not necessarily closed) contour C by

$$V \equiv \mathrm{P} \exp \left\{ ig \int_C dx \cdot A \right\}, \tag{5.29}$$

where, as usual, $A_\mu = \sum_a t^a A_\mu^a$ and the t^a are the SU(N_c) generators in the fundamental representation. Similarly, define the adjoint Wilson line along the same contour C by

$$U \equiv \mathrm{P} \exp\left(ig \int_C dx \cdot \mathcal{A} \right) \tag{5.30}$$

where now $\mathcal{A}_\mu = \sum_a T^a A_\mu^a$ with $(T^a)_{bc} = -if^{abc}$ the SU(N_c) generators in the adjoint representation. As can be verified explicitly, the following identity relates these two Wilson lines:

$$U_{ab} t^b = V^\dagger t^a V. \tag{5.31}$$

This relation also leads to another useful formula:

$$U_{ab} = 2\mathrm{tr}\left[t^b V^\dagger t^a V \right]. \tag{5.32}$$

Note also that, since the adjoint SU(N_c) generators T^a are purely imaginary,

$$U_{ab} = U_{ab}^* = U_{ba}^\dagger. \tag{5.33}$$

Using Eq. (5.31) with $V = S^{-1} = S^\dagger$ we write

$$S(x_i^-, \vec{x}_\perp) t^a S^{-1}(x_i^-, \vec{x}_\perp) = U_{ab}^\dagger(x_i^-, \vec{x}_\perp) t^b \tag{5.34}$$

where (cf. Eqs. (5.14) and (5.17))

$$U(x^-, \vec{x}_\perp) = \mathrm{P} \exp\left\{ \frac{ig}{2} \int_{x^-}^{-\infty} dx'^- \mathcal{A}_{cov}^+(x'^-, \vec{x}_\perp) \right\}$$

$$= \prod_{i=1}^{A} \exp\left\{ \frac{ig^2}{2\pi} T^a(t_i^a) \ln \frac{|\vec{x}_\perp - \vec{x}_{i\perp}|}{|\vec{x}_\perp - \vec{x}_{i\perp}'|} \theta(x^- - x_i^-) \right\}. \tag{5.35}$$

We now can rewrite the term in the second line of Eq. (5.28) as

$$\langle \mathrm{tr}\left[S(x_i^-, \vec{x}_\perp) t^a S^{-1}(x_i^-, \vec{x}_\perp) S(x_i^-, \vec{y}_\perp) t^a S^{-1}(x_i^-, \vec{y}_\perp) \right] \rangle = \frac{1}{2} \langle \mathrm{Tr}\left[U^\dagger(x_i^-, \vec{x}_\perp) U(x_i^-, \vec{y}_\perp) \right] \rangle \tag{5.36}$$

where the trace Tr is over the adjoint indices.

Employing Eq. (5.35) and expanding the contribution of the $(i-1)$th nucleon up to order g^4, in accordance with the two-gluons-per-nucleon limitation of the classical approach, we get

$$
\mathrm{Tr}\left[U(x_i^-, \vec{y}_\perp)U^\dagger(x_i^-, \vec{x}_\perp)\right]
$$

$$
= \mathrm{Tr}\left\{U(x_{i-1}^-, \vec{y}_\perp)\left[1 + \frac{ig^2}{2\pi}T^c(t_{i-1}^c)\ln\frac{|\vec{y}_\perp - \vec{x}_{i-1\perp}||\vec{x}_\perp - \vec{x}_{i-1\perp}'|}{|\vec{y}_\perp - \vec{x}_{i-1\perp}'||\vec{x}_\perp - \vec{x}_{i-1\perp}|}\right.\right.
$$

$$
\left.\left. - \frac{g^4}{2(2\pi)^2}T^cT^d(t_{i-1}^c)(t_{i-1}^d)\ln^2\frac{|\vec{y}_\perp - \vec{x}_{i-1\perp}||\vec{x}_\perp - \vec{x}_{i-1\perp}'|}{|\vec{y}_\perp - \vec{x}_{i-1\perp}'||\vec{x}_\perp - \vec{x}_{i-1\perp}|} + O(g^6)\right]U^\dagger(x_{i-1}^-, \vec{x}_\perp)\right\}.
$$

$$(5.37)$$

Averaging over the color space of the $(i-1)$th nucleon we obtain

$$
\left\langle\mathrm{Tr}\left[U(x_i^-, \vec{y}_\perp)U^\dagger(x_i^-, \vec{x}_\perp)\right]\right\rangle = \left\langle\mathrm{Tr}\left[U(x_{i-1}^-, \vec{y}_\perp)U^\dagger(x_{i-1}^-, \vec{x}_\perp)\right]\right\rangle
$$

$$
\times\left[1 - \alpha_s^2\left\langle\ln^2\frac{|\vec{y}_\perp - \vec{x}_{i-1\perp}|}{|\vec{y}_\perp - \vec{x}_{i-1\perp}'|}\frac{|\vec{x}_\perp - \vec{x}_{i-1\perp}'|}{|\vec{x}_\perp - \vec{x}_{i-1\perp}|}\right\rangle\right]. \quad (5.38)
$$

The logarithm in the second line of Eq. (5.38) looks like that arising from the two-gluon-exchange high energy interaction of an onium $\vec{x}_\perp, \vec{y}_\perp$ with an onium $\vec{x}_{i-1\perp}, \vec{x}_{i-1\perp}'$, as can be seen from comparing Eq. (5.38) with Eq. (3.139). Indeed this is natural, since the result arises from the expansion of up to two gluons per nucleon shown in Fig. 5.3 (except that here we have two adjoint Wilson lines instead of the single fundamental Wilson line in Fig. 5.3). The result of averaging this term over the impact parameter and over angular orientations of the nucleon can be obtained by comparing Eq. (3.139) with its averaged version (3.25). We are assuming that our nucleus is very large; hence, averaging over all impact parameter values up to infinity is applicable here.

We now assume that $\vec{x}_\perp$ and $\vec{y}_\perp$ are perturbatively close to each other, so that $|\vec{x}_\perp - \vec{y}_\perp| \ll 1/\Lambda_{QCD}$ and is much smaller than the nucleon size. In the nucleus, when averaging the logarithm-squared term in Eq. (5.38) we also have to multiply the transverse integral by the probability density for finding the nucleon at $\vec{b}_\perp$, i.e., by $T(\vec{b}_\perp)/A$ (cf. Eq. (5.28)). In our coarse-grained picture of the nucleus we will assume that both coordinates are located at the same impact parameter $\vec{b}_\perp = (\vec{x}_\perp + \vec{y}_\perp)/2$ as far as the nuclear profile function $T(\vec{b}_\perp)$ is concerned. Then we can rewrite Eq. (5.38) as

$$
\left\langle\mathrm{Tr}\left[U(x_i^-, \vec{y}_\perp)U^\dagger(x_i^-, \vec{x}_\perp)\right]\right\rangle = \left\langle\mathrm{Tr}\left[U(x_{i-1}^-, \vec{y}_\perp)U^\dagger(x_{i-1}^-, \vec{x}_\perp)\right]\right\rangle
$$

$$
\times\left[1 - 2\pi\alpha_s^2\frac{T(\vec{b}_\perp)}{A}(\vec{x}_\perp - \vec{y}_\perp)^2\ln\frac{1}{|\vec{x}_\perp - \vec{y}_\perp|\Lambda}\right], \quad (5.39)
$$

where we have neglected the term 1 in comparison with the logarithm in Eq. (3.25), since $|\vec{x}_\perp - \vec{y}_\perp| \ll 1/\Lambda$. As usual $\Lambda \sim \Lambda_{QCD}$ is an IR cutoff, with $1/\Lambda$ approximately the nucleon size. Equation (5.39) has the contribution of the $(i-1)$th nucleon factorized from the rest of the expression.

Iterating the above steps for all the other nucleons we end up with

$$\langle \text{Tr}[U(x_i^-, \vec{y}_\perp)U^\dagger(x_i^-, \vec{x}_\perp)] \rangle$$

$$= (N_c^2 - 1)\left[1 - \frac{Q_{sG}^2(\vec{b}_\perp)}{4A}(\vec{x}_\perp - \vec{y}_\perp)^2 \ln \frac{1}{|\vec{x}_\perp - \vec{y}_\perp|\Lambda}\right]^{i-1}$$

$$\approx (N_c^2 - 1)\exp\left\{-\frac{i-1}{A}\frac{Q_{sG}^2(\vec{b}_\perp)}{4}(\vec{x}_\perp - \vec{y}_\perp)^2 \ln \frac{1}{|\vec{x}_\perp - \vec{y}_\perp|\Lambda}\right\}, \qquad (5.40)$$

where in the last step we have used the fact that $A \gg 1$. The *gluon saturation scale*,

$$Q_{sG}^2(\vec{b}_\perp) = 8\pi\alpha_s^2 T(\vec{b}_\perp), \qquad (5.41)$$

can be obtained from the quark saturation scale Eq. (4.50) if one replaces C_F by N_c in the latter and multiplies the result by 2. This factor 2 is due to the fact that in arriving at Eq. (4.50) we modeled each nucleon by a quark, while now nucleons are modeled as quarkonia.

Before we continue, let us pause to stress the importance of the result obtained in Eq. (5.40).

On Wilson lines and the S-matrix

Equation (5.40), which is necessary for our calculation of the WW gluon distribution, is in fact a very important result in itself. As the nucleons are ordered along the x^--axis we can make the replacement

$$\frac{i-1}{A} \to \frac{x_i^-}{L}, \qquad (5.42)$$

with L the net x^--extent of the nucleus as defined in Sec. 4.2 (up to a $+ \leftrightarrow -$ interchange). The exponent in Eq. (5.40) then becomes equivalent to Eq. (4.43) if in the latter we note that ρ_A is independent of the longitudinal coordinate (inside the nucleus), use $\sigma^{q\bar{q}N}$ from Eq. (4.25), replace C_F by N_c in Eq. (4.43), and interchange the $+$ and $-$ coordinates in order to work in the same coordinate frame. The only real difference, C_F versus N_c, is due to quark degrees of freedom versus gluon degrees of freedom. We see that, in the covariant gauge, the S-matrix of a dipole scattering on a nucleus is equivalent to the correlator of the two Wilson lines. Namely, $U(x^-, \vec{y}_\perp)$ describes a gluon propagating from x^- to $-\infty$ along the x^--axis with the transverse coordinate fixed at $\vec{y}_\perp$. Similarly, $U^\dagger(x^-, \vec{x}_\perp)$ describes a gluon at $\vec{x}_\perp$ propagating along the x^--axis from $-\infty$ to x^-. The fact that the transverse coordinates of the gluons are invariant is the same property of eikonal scattering as we saw in the GGM and dipole models. (In the classical field correlator (5.28), no actual gluon propagates: it just so happens that the correlator is related to an average of two adjoint Wilson lines, which, in turn, is equivalent to a gluon dipole scattering matrix.)

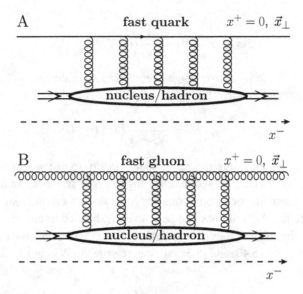

Fig. 5.5. The Wilson lines for (A) a fast quark and (B) a fast gluon scattering in the field of the target nucleus.

Generalizing this conclusion, we see that the propagator of an eikonal quark moving along the light cone x^--axis can be replaced by the Wilson line

$$V_{\vec{x}_\perp} = P \exp \left\{ \frac{ig}{2} \int\limits_{-\infty}^{\infty} dx^- A^+(x^+ = 0, x^-, \vec{x}_\perp) \right\} \qquad (5.43)$$

and the propagator of the eikonal antiquark can be replaced by the conjugate Wilson line $V_{\vec{x}_\perp}^\dagger$. The propagator of an eikonal gluon moving along the light cone x^--axis can be replaced by the adjoint Wilson line

$$U_{\vec{x}_\perp} = P \exp \left\{ \frac{ig}{2} \int\limits_{-\infty}^{\infty} dx^- \mathcal{A}^+(x^+ = 0, x^-, \vec{x}_\perp) \right\}. \qquad (5.44)$$

The Wilson lines defined in Eqs. (5.43) and (5.44) are illustrated schematically in Figs. 5.5A and B, respectively, as propagators of the eikonal quark and gluon moving along the x^--axis and interacting with the gluon field of the nucleus.

The Wilson line correlator (5.40) and the correspondence between such a correlator and the S-matrix were derived in the $\partial_\mu A^\mu = 0$ covariant gauge: the same results are true in the light cone gauge of the projectile, $A^- = 0$. For the light cone gauge of the nucleus, $A^\perp = 0$, the Wilson line correlator has to be augmented by gauge links at $x^- = \pm\infty$, making it a closed gauge-invariant Wilson loop: the links do not contribute in the $A^- = 0$ and $\partial_\mu A^\mu = 0$ gauges but are important in the $A^+ = 0$ gauge.

The S-matrix for a quark dipole scattering on a nuclear target, defined in Eq. (4.38) (in the notation of Eq. (4.140) and/or Eq. (4.214)) can be rewritten in terms of the

Wilson lines as

$$S(\vec{x}_{1\perp}, \vec{x}_{0\perp}, Y) = \frac{1}{N_c} \left\langle \text{tr} \left[V_{\vec{x}_{1\perp}} V_{\vec{x}_{0\perp}}^\dagger \right] \right\rangle, \tag{5.45}$$

with the factor of $1/N_c$ inserted to average over the colors of the quarks (up to the $+ \leftrightarrow -$ convention difference). Similarly, for a gluon dipole the S-matrix is

$$S_G(\vec{x}_{1\perp}, \vec{x}_{0\perp}, Y) = \frac{1}{N_c^2 - 1} \left\langle \text{Tr} \left[U_{\vec{x}_{1\perp}} U_{\vec{x}_{0\perp}}^\dagger \right] \right\rangle. \tag{5.46}$$

As we have just observed, using the result (5.40) in Eq. (5.46) would lead to the gluon S-matrix in the GGM model. We see that, for high energy scattering in the covariant and $A^- = 0$ gauges, diagrammatic calculations are equivalent to calculations of Wilson lines. Below we will see that Wilson lines can be conveniently used to construct S-matrices for the scattering of other objects, more complicated than a dipole, on a nuclear target.

With the help of Eqs. (5.40) and (5.36) we can rewrite the WW gluon distribution (5.28) as

$$\phi^{WW}(x, k_\perp^2) = \frac{\alpha_s C_F}{4\pi^3} \int d^2 b_\perp d^2 r_\perp e^{i\vec{k}_\perp \cdot \vec{r}_\perp} \sum_{i=1}^A \exp \left\{ -\frac{i-1}{A} \frac{r_\perp^2 Q_{sG}^2(\vec{b}_\perp)}{4} \ln \frac{1}{r_\perp \Lambda} \right\}$$

$$\times \frac{T(\vec{b}_\perp)}{A} \left[\int d^2 x_i \frac{\vec{x}_\perp - \vec{x}_{i\perp}}{|\vec{x}_\perp - \vec{x}_{i\perp}|^2} \cdot \frac{\vec{y}_\perp - \vec{x}_{i\perp}}{|\vec{y}_\perp - \vec{x}_{i\perp}|^2} + \text{a.c.} \right], \tag{5.47}$$

where

$$\vec{r}_\perp = \vec{x}_\perp - \vec{y}_\perp, \quad \vec{b}_\perp = \frac{\vec{x}_\perp + \vec{y}_\perp}{2}, \tag{5.48}$$

and we have assumed that for a large nucleus $T(\vec{x}_{i\perp}) \approx T(\vec{b}_\perp)$. The integration over $\vec{x}_{i\perp}$ in Eq. (5.47) can now be carried out using the Fourier decomposition from Eq. (A.10) and employing Eq. (A.9) and is left as an exercise for the reader. It yields

$$\int d^2 x_i \frac{\vec{x}_\perp - \vec{x}_{i\perp}}{|\vec{x}_\perp - \vec{x}_{i\perp}|^2} \cdot \frac{\vec{y}_\perp - \vec{x}_{i\perp}}{|\vec{y}_\perp - \vec{x}_{i\perp}|^2} = 2\pi \ln \frac{1}{|\vec{x}_\perp - \vec{y}_\perp| \Lambda}. \tag{5.49}$$

The antiquark contribution in Eq. (5.47) contains a term depending simply on $x_{i\perp}'$, which simply doubles the contribution in Eq. (5.49), while the terms depending on both $x_{i\perp}$ and $x_{i\perp}'$ simply modify the IR cutoff in Eq. (5.49) by a multiplicative constant that we can neglect. In the end the contents of the square brackets in the last line of Eq. (5.47) give us only twice the contribution in Eq. (5.49).

Summing over the index i in Eq. (5.47) and remembering yet again that $A \gg 1$ we at last obtain the non-Abelian WW gluon distribution for a large nucleus (Jalilian-Marian *et al.* 1997a)

$$\phi^{WW}(x, k_\perp^2) = \frac{C_F}{2\pi^3 \alpha_s} \int d^2 b_\perp d^2 r_\perp e^{i\vec{k}_\perp \cdot \vec{r}_\perp} \frac{1}{r_\perp^2} N_G(\vec{r}_\perp, \vec{b}_\perp, Y = 0), \tag{5.50}$$

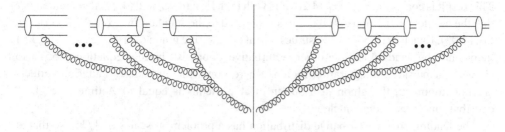

Fig. 5.6. Diagrammatic representation of the non-Abelian Weizsäcker–Williams distribution function ϕ^{WW} from Eq. (5.50).

where, by analogy with Eq. (4.51), we have defined the forward amplitude for a gluon dipole scattering on a nucleus

$$N_G(\vec{r}_\perp, \vec{b}_\perp, Y = 0) = 1 - \exp\left\{ -\frac{r_\perp^2 Q_{sG}^2(\vec{b}_\perp)}{4} \ln \frac{1}{r_\perp \Lambda} \right\}. \tag{5.51}$$

As expected,

$$N_G(\vec{r}_\perp, \vec{b}_\perp, Y) = 1 - S_G(\vec{r}_\perp, \vec{b}_\perp, Y); \tag{5.52}$$

the gluon S-matrix S_G was defined in Eq. (5.46).

The result in Eqs. (5.50) and (5.51) is insensitive to the details of the nuclear model and can be obtained using a continuous color-charge density description (Jalilian-Marian *et al.* 1997a). To put it in line with the GGM result for the saturation scale (4.50), we note that for a model in which nucleons are replaced by single valence quarks we have

$$Q_{sG}^2(\vec{b}_\perp) = 4\pi\alpha_s^2 T(\vec{b}_\perp). \tag{5.53}$$

The only difference between (5.53) and (4.50) is the Casimir operator replacement $C_F \to N_c$ associated with going from the quark to the gluon degrees of freedom.

Equation (5.50) is the central result of the McLerran–Venugopalan model for a single nucleus. It is represented diagrammatically in Fig. 5.6 in analogy with the gluon distribution in Fig. 2.12. Let us now describe its main properties. While exact analytic integration over $r_\perp$ in Eq. (5.50) appears to be a rather unwieldy task, we can still study the limiting cases of large and small $k_\perp$ analytically.

For $k_\perp \gg Q_{sG}$ we expand the exponential in Eq. (5.51) to the lowest nontrivial order and integrate over $r_\perp$, obtaining

$$\phi^{WW}\left(x, k_\perp^2\right)\Big|_{k_\perp \gg Q_{sG}} \approx \frac{C_F}{4\pi^2\alpha_s} \frac{1}{k_\perp^2} \int d^2b_\perp Q_{sG}^2(\vec{b}_\perp). \tag{5.54}$$

For a nucleus with "nucleons" each consisting of a single valence quark we use Eq. (5.53) along with Eq. (4.31) to derive

$$\phi^{WW}\left(x, k_\perp^2\right)\Big|_{k_\perp \gg Q_{sG}} \approx A\frac{\alpha_s C_F}{\pi} \frac{1}{k_\perp^2}. \tag{5.55}$$

This result is consistent with Eq. (4.26) and with both Eqs. (4.48) and (4.27) if we remember that the unintegrated gluon distribution is connected to the standard integrated one, xG, via Eq. (3.93). Equation (5.55) demonstrates that at large $k_\perp$ the gluon distribution $\phi^{WW}(x, k_\perp^2)$ maps onto the standard leading-order perturbative gluon distribution. Equation (5.55) also shows that outside the saturation region, where nonlinear multiple-rescatterings effects are not important, the gluon distribution of A nucleons is equal to A times the gluon distributions of individual nucleons.

The leading-order perturbative distribution has a problem: it scales as $1/k_\perp^2$, so that at low $k_\perp$ it will diverge, leading to an infinite number of gluons. Moreover, the corresponding integrated gluon distribution xG, obtained by integrating ϕ^{WW} over $k_\perp^2$, is also IR divergent; thus, in the absence of a cutoff, the net number of gluons would still be infinite.

The full distribution ϕ^{WW} in Eq. (5.50) is actually free of such a problem, as can be seen by studying the opposite limit, deep inside the saturation region, where $k_\perp \ll Q_{sG}$. There we see that $r_\perp \sim 1/k_\perp \gg 1/Q_{sG}$, so that we can neglect the exponential in Eq. (5.51). Putting $N_G = 1$ in Eq. (5.50) and integrating over $r_\perp > 1/Q_{sG}$ yields

$$\phi^{WW}(x, k_\perp^2)\Big|_{k_\perp \ll Q_{sG}} \approx \frac{C_F}{\alpha_s \pi^2} \int d^2 b_\perp \ln \frac{Q_{sG}^2(\vec{b}_\perp)}{k_\perp^2}. \qquad (5.56)$$

We see that the power-law divergence of Eq. (5.55) is softened down to a logarithmic divergence. While some IR divergence still remains, when Eq. (5.56) is integrated over $k_\perp$ the number of gluons xG is now finite. We conclude that the effect of saturation in the MV model is to soften the IR divergence, resulting in a finite net number of gluons.

Note also that in Eq. (5.56), deep inside the saturation region, $\phi^{WW} \sim 1/\alpha_s$. Remembering the relation between the unintegrated gluon distribution and the classical gluon fields in Eq. (5.26) we see that

$$A_\mu^{LC} \sim \frac{1}{g}, \qquad (5.57)$$

as expected for classical gluon fields. This is as strong as a gluon field can be at weak coupling g: we see that the occupation numbers of the classical gluons in the nuclear wave function are very high, on the one hand justifying the classical approximation while on the other hand demonstrating an interesting phenomenon, that the virtual gluons in the small-x wave function form a very dense system.

The unintegrated gluon distribution ϕ^{WW} multiplied by the two-dimensional phase-space factor $k_\perp$ is plotted schematically in Fig. 5.7 as a function of transverse momentum $k_T = k_\perp$. (In the plot we have assumed for simplicity that the nucleus is a cylinder with its axis along the z-axis, so that Q_{sG} does not depend on $\vec{b}_\perp$ and the $\vec{b}_\perp$-integral in Eq. (5.50) can be carried out simply by multiplying the integrand by the transverse area.) The quantity $k_T \phi^{WW}$ is the number of gluons with a given k_T (as opposed to $\phi^{WW}(x, k_T^2)$, which counts the gluons with a given k_T^2). The dashed curve in Fig. 5.7 represents the leading-order result (5.55) with $k_T \phi^{WW} \sim 1/k_T$, which indeed is IR divergent. The solid curve represents the full result: one can see from Eq. (5.56) that $k_T \phi^{WW}$ in fact goes to zero as $k_T \to 0$. The distribution $k_T \phi^{WW}$ peaks around the saturation scale, which means that most gluons in

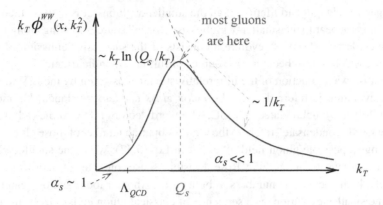

Fig. 5.7. Phase-space distribution of gluons in transverse momentum space. (Reprinted from Jalilian-Marian and Kovchegov (2006), with permission from Elsevier.) A color version of this figure is available online at www.cambridge.org/9780521112574.

the WW wave functions have $k_T \approx Q_{sG}$, and the wave function is indeed describable by perturbative small-coupling methods.

5.2 The Jalilian-Marian–Iancu–McLerran–Weigert–Leonidov–Kovner evolution equation

5.2.1 The color glass condensate (CGC)

Let us now find the quantum corrections to the classical MV model due to nonlinear small-x evolution in the LLA. Small-x evolution can be included either in the wave function of a projectile, as in Chapter 4, or in the wave function of the target. The Jalilian-Marian–Iancu–McLerran–Weigert–Leonidov–Kovner (JIMWLK) evolution equation, which we will derive here, accomplishes the latter. It generalizes the Gaussian weight functional $W[\rho_{LC}]$ from Eq. (5.19) to a rapidity-dependent functional $W_Y[\rho_{LC}]$, which is no longer of Gaussian form and instead has to be determined from the JIMWLK functional equation for evolution in the rapidity Y. The averaging prescription (5.20) still holds, generating rapidity dependence for the expectation values of operators:

$$\langle \hat{O}_\rho \rangle_Y = \frac{\int \mathcal{D}\rho_{LC} \, \hat{O}_\rho \, W_Y[\rho_{LC}]}{\int \mathcal{D}\rho_{LC} \, W_Y[\rho_{LC}]}. \tag{5.58}$$

The original JIMWLK equation was derived by including quantum corrections in the classical MV wave function of a large nucleus (Jalilian-Marian *et al.* 1997b, 1999a, b, Iancu *et al.* 2001a, b). The main principle of the JIMWLK derivation comes from the MV model: one has to separate the partons into those with large x and those with small x; the large-x partons serve as classical sources for the small-x partons. As we build up small-x evolution and go to lower x by making steps in rapidity $Y \to Y + dY$, the gluons at rapidity Y become large-x gluons, and are incorporated into a source of classical fields. Clearly, as we have already seen in Mueller's dipole model, the larger-x gluons have a much longer wavelength and lifetime than the smaller-x gluons: it is natural, then, that the larger-x gluons appear to the smaller-x gluons as "frozen" sources moving along light cone straight lines. Hence JIMWLK evolution consists of the successive emission of classical gluon fields, which in turn become the sources of further gluon fields, etc.

The small-x wave function of the ultrarelativistic nucleus given by the MV model with JIMWLK evolution is referred to as the *color glass condensate* (Iancu, Leonidov, and McLerran 2001a, b), abbreviated as CGC. The word "color" refers to the (adjoint) gluon colors; the word "condensate" refers to the high occupation number of those gluons, leading to the strongest possible gluon field, like that in Eq. (5.57): while the small-x gluons do not form a condensate in, say, the Bose–Einstein sense, one can draw a loose analogy based on the high occupation numbers in both cases. Another loose analogy can be drawn between the small-x evolution, as a sequence of classical gluon emissions from stationary sources, and spin glasses, which also have a separation of degrees of freedom according to a multitude of time scales; this is the origin of the word "glass" in CGC.

Here we will rederive the JIMWLK equation following Mueller (2001). The main idea for deriving the JIMWLK equation suggested by Mueller is to treat the small increase in energy (or rapidity) in two different, but equivalent, ways. In the first, one incorporates the modifications due to the increase in energy into the nuclear wave function (the CGC), which will then change (evolve); this was done in the original JIMWLK derivation. In the second, which we have already seen in Mueller's dipole model, this energy increase is incorporated into the projectile wave function. Then the projectile will emit one gluon per step of LLA evolution, and such an emission can be treated perturbatively in a rather simple manner. Equating these two ways of including high energy corrections, one obtains the JIMWLK evolution equation for the CGC nuclear wave function.

5.2.2 Derivation of JIMWLK evolution

Just as in the rest of this chapter we will work in the frame where the nucleus is moving along the x^+-axis while the projectile is moving along the x^--axis. We will use the $A^- = 0$ light cone gauge of the projectile. One can show that for the nucleus this gauge is equivalent to the covariant gauge: clearly the field (5.11) both solves the Yang–Mills equations (5.4) and satisfies the $A^- = 0$ gauge condition. To make our notation more compact, we define

$$\alpha(x^-, \vec{x}_\perp) \equiv A^+(x^+ = 0, x^-, \vec{x}_\perp), \tag{5.59}$$

with A^+ the fundamental-representation gluon field in the $A^- = 0$ gauge. The Yang–Mills equations give

$$\Box \alpha(x^-, \vec{x}_\perp) = \rho(x^-, \vec{x}_\perp), \tag{5.60}$$

where ρ is also taken in the $A^- = 0$ gauge and $\Box = \partial_\mu \partial^\mu$. We see that the two functions $\alpha(x^-, \vec{x}_\perp)$ and $\rho(x^-, \vec{x}_\perp)$ are straightforwardly connected, with the latter also related to ρ_{LC} (see Eq. (5.18)): therefore we can replace the integration over ρ_{LC} in Eq. (5.58) by integration over α. Defining a weight functional $W_Y[\alpha]$ we can rewrite the averaged values of operators as (cf. (5.58))

$$\langle \hat{O}_\alpha \rangle_Y = \int \mathcal{D}\alpha \, \hat{O}_\alpha W_Y[\alpha]. \tag{5.61}$$

where we agree that the normalization of $W_Y[\alpha]$ is such that

$$\int \mathcal{D}\alpha \, W_Y[\alpha] = 1. \tag{5.62}$$

Indeed, the functional $W_Y[\alpha]$ is formally different from $W_Y[\rho_{LC}]$, though the two are of course related: we use the same letter W for both only to simplify the notation. Since in this section we will be working solely with the field $\alpha(x^-, \vec{x}_\perp)$ this recycling of symbols should not cause confusion.

Our goal is to construct an evolution equation for $W_Y[\alpha]$. Our strategy is first to derive an evolution equation for the expectation value of some (arbitrary chosen) test operator $\hat{O}_\alpha$, obtaining on the one hand

$$\partial_Y \langle \hat{O}_\alpha \rangle_Y = \langle \mathcal{K}_\alpha \otimes \hat{O}_\alpha \rangle_Y = \int \mathcal{D}\alpha \, (\mathcal{K}_\alpha \otimes \hat{O}_\alpha) W_Y[\alpha], \tag{5.63}$$

where $\mathcal{K}_\alpha$ is the kernel of the equation and may be a function of the field $\alpha(x^-, \vec{x}_\perp)$; the symbol $\otimes$ denotes its action. The rightmost expression in Eq. (5.63) was obtained using the definition of averaging in Eq. (5.61). On the other hand, differentiating Eq. (5.61) with respect to Y we get

$$\partial_Y \langle \hat{O}_\alpha \rangle_Y = \int \mathcal{D}\alpha \, \hat{O}_\alpha \partial_Y W_Y[\alpha]. \tag{5.64}$$

Equating the right-hand sides of Eqs. (5.63) and (5.64), and arranging for the kernel in Eq. (5.63) to act on $W_Y[\alpha]$ (by employing integration by parts), we arrive at an evolution equation for $W_Y[\alpha]$.

To construct the test operator we define the Wilson lines in accordance with Eqs. (5.43) and (5.44). The fundamental Wilson line is defined by

$$V_{\vec{x}_\perp} = \mathrm{P} \exp \left\{ \frac{ig}{2} \int\limits_{-\infty}^{\infty} dx^- t^a \alpha^a(x^-, \vec{x}_\perp) \right\}, \tag{5.65}$$

while the adjoint Wilson line is

$$U_{\vec{x}_\perp} = \mathrm{P} \exp \left\{ \frac{ig}{2} \int\limits_{-\infty}^{\infty} dx^- T^a \alpha^a(x^-, \vec{x}_\perp) \right\}. \tag{5.66}$$

Following Mueller (2001) we choose the trial operator to be

$$\hat{O}_{\vec{x}_{1\perp}, \vec{x}_{0\perp}} = V_{\vec{x}_{1\perp}} \otimes V_{\vec{x}_{0\perp}}^\dagger. \tag{5.67}$$

This is almost the dipole S-matrix of Eq. (5.45): the operator $\hat{O}_{\vec{x}_{1\perp}, \vec{x}_{0\perp}}$ consists of the quark propagator (Wilson line) $V_{\vec{x}_{1\perp}}$ at $\vec{x}_{1\perp}$ and the antiquark propagator $V_{\vec{x}_{0\perp}}^\dagger$ at $\vec{x}_{0\perp}$. What is

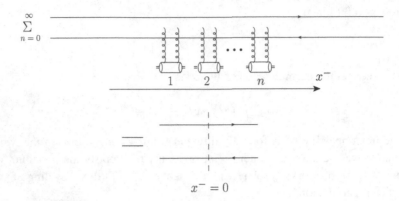

Fig. 5.8. Abbreviated notation for the GGM multiple-rescatterings interaction of a projectile with a nuclear target.

missing is the trace and the average over colors: the symbol $\otimes$ in Eq. (5.67) underscores the fact that the color indices of V and $V^\dagger$ are fixed (and not summed over):

$$V_{\vec{x}_{1\perp}} \otimes V_{\vec{x}_{0\perp}}^\dagger = \left(V_{\vec{x}_{1\perp}}\right)_{ij} \left(V_{\vec{x}_{0\perp}}^\dagger\right)_{kl}. \tag{5.68}$$

We want to derive an evolution equation for $\hat{O}_{\vec{x}_{1\perp},\vec{x}_{0\perp}}$. Its construction is analogous to that of the BK equation. The evolution is given by the long-lived s-channel gluons, which interact with the target over a relatively short period of time. To represent it diagrammatically we first define an abbreviated notation, in Fig. 5.8. As discussed in Sec. 4.4, the lifetime of the s-channel gluons, which in our coordinates is $x_{coh}^- = k^-/k_\perp^2$, is much longer than the duration of the GGM multiple-rescatterings interaction of the gluon system with the nucleus, which is of order $1/p^+$, with p^+ the large light cone momentum of the nucleons. This should be clear from Fig. 4.23. We now employ this result to define the abbreviated notation in Fig. 5.8. Since the GGM multiple rescatterings occur over a relatively short time (compared with the time needed for the development of quantum evolution), we can, for the purpose of the evolution calculations, include them all in one "instantaneous" interaction at $x^- = 0$, denoted by the vertical dashed line on the right in Fig. 5.8. Interactions with the target are summed over for *any* gluon or quark line crossing the dashed line. We also include the no-interaction contribution in the sum (the $n = 0$ term in Fig. 5.8). Note that below we will sometimes use this dashed-line notation to include successive evolution emissions as well: owing to the ordering of the s-channel gluons in k^- (in the LLA), the lifetimes of the smaller-k^- gluons are shorter and hence they may also appear as instantaneous events to the larger-k^- gluons, which are emitted much earlier and absorbed much later.

Using the notation introduced in Fig. 5.8 we can draw diagrams generating one step of the evolution of the operator in Eq. (5.67), as shown in Figs. 5.9 and 5.10. Note again that the dashed line denotes the interaction with the target for any propagator line that it crosses. Hence diagrams A, B, H, and K in Figs. 5.9 and 5.10 are real, in the sense that in them the gluon interacts with the target, while the rest of the diagrams are virtual corrections.

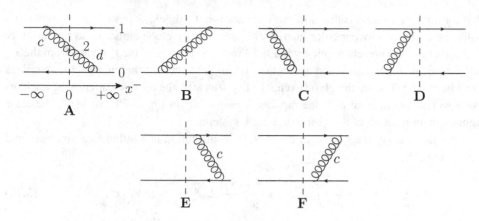

Fig. 5.9. One step of a small-x evolution for the operator $V_{\vec{x}_{1\perp}} \otimes V_{\vec{x}_{0\perp}}^{\dagger}$ with the s-channel gluon interacting both with the quark and the antiquark Wilson lines.

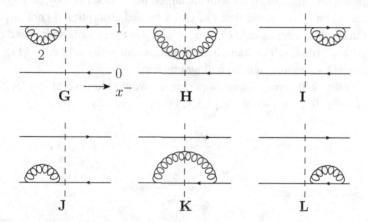

Fig. 5.10. One step of a small-x evolution for the operator $V_{\vec{x}_{1\perp}} \otimes V_{\vec{x}_{0\perp}}^{\dagger}$; here an s-channel gluon is emitted and absorbed solely by either the quark or the antiquark Wilson lines.

The diagrams are grouped into those where the gluon is emitted by the quark and absorbed by the antiquark or vice versa (Fig. 5.9), and those where the gluon is both emitted and absorbed only by the quark or only by the antiquark (Fig. 5.10).

We start by analyzing diagrams E and F in Fig. 5.9. The gluon in these graphs does not interact with the target: hence it has the same color throughout its propagation. The contribution of these diagrams can be obtained using LCPT methods, similarly to how we performed the calculations for Mueller's dipole model. We get

$$E + F = \frac{\alpha_s}{\pi^2} \int d^2 x_2 \, dY \, \frac{\vec{x}_{21} \cdot \vec{x}_{20}}{x_{21}^2 x_{20}^2} \, t^c V_{\vec{x}_{1\perp}} \otimes V_{\vec{x}_{0\perp}}^{\dagger} t^c. \tag{5.69}$$

The minus sign due to the coupling to the antiquark, is canceled by the minus sign arising because the graph is virtual (cf. Eq. (4.64)). The main difference between this result and that for dipole evolution is that now our operator is not a dipole, it is $V_{\vec{x}_{1\perp}} \otimes V_{\vec{x}_{0\perp}}^{\dagger}$ without a color trace: this is why our color matrices t^c generated by gluon emission do not multiply each other and are present explicitly in Eq. (5.69). The placing of the t^c comes from the x^- ordering of gluon emission and absorption, taking into account that quark–gluon vertices must be at $x^- > 0$ since the gluon exchange happens after the quarks scatter at the nucleus. We also have an integral over the rapidity $Y = p'^-/k^-$ in Eq. (5.69), with p'^- the large light cone momentum of the quark–antiquark system.

Next let us study diagram A in Fig. 5.9. Its contribution is a little more involved, and can be written as

$$A = -\frac{\alpha_s}{\pi^2} \int d^2 x_2 \, dY \, \frac{\vec{x}_{21} \cdot \vec{x}_{20}}{x_{21}^2 x_{20}^2} \, U_{\vec{x}_{2\perp}}^{dc} \, V_{\vec{x}_{1\perp}} t^c \otimes V_{\vec{x}_{0\perp}}^{\dagger} t^d. \tag{5.70}$$

The expression in Eq. (5.70) consists of the same emission kernel as in Eq. (5.69) and can be obtained using LCPT. The main difference between (5.70) and (5.69) is due to the fact that in diagram A the gluon interacts with the target: this is described by the adjoint Wilson line $U_{\vec{x}_{2\perp}}^{dc}$. The gluon colors at the times of emission and absorption do not have to be the same in this diagram and are labeled c and d, bringing in the color factors t^c and t^d. Finally, diagram A is "real" (that is, it contains a gluon interacting with the target, of Fig. 4.13) and hence is different by a minus sign from diagrams E and F.

For reasons that will soon become apparent we would like to cast Eq. (5.70) into the form of Eq. (5.69). To do so, we use Eq. (5.31) with Eq. (5.33) to write

$$V t^a V^{\dagger} = U_{ab}^{\dagger} t^b, \tag{5.71}$$

so that

$$V_{\vec{x}_{1\perp}} t^c = U_{\vec{x}_{1\perp}}^{\dagger ca} t^a V_{\vec{x}_{1\perp}}. \tag{5.72}$$

This, along with Eq. (5.33), allows us to rewrite Eq. (5.70) as

$$A = -\frac{\alpha_s}{\pi^2} \int d^2 x_2 \, dY \, \frac{\vec{x}_{21} \cdot \vec{x}_{20}}{x_{21}^2 x_{20}^2} \left[U_{\vec{x}_{1\perp}} U_{\vec{x}_{2\perp}}^{\dagger} \right]_{ab} t^a V_{\vec{x}_{1\perp}} \otimes V_{\vec{x}_{0\perp}}^{\dagger} t^b, \tag{5.73}$$

where the subscript or superscript positions of the color indices a, b are chosen for convenience only; however, the ordering of the indices is important.

The rest of the calculation is now clear: using Eqs. (5.31), (5.71), and (5.33) we can write down expressions for the remaining diagrams in Fig. 5.9:

$$B = -\frac{\alpha_s}{\pi^2} \int d^2 x_2 \, dY \, \frac{\vec{x}_{21} \cdot \vec{x}_{20}}{x_{21}^2 x_{20}^2} \left[U_{\vec{x}_{2\perp}} U_{\vec{x}_{0\perp}}^{\dagger} \right]_{ab} t^a V_{\vec{x}_{1\perp}} \otimes V_{\vec{x}_{0\perp}}^{\dagger} t^b, \tag{5.74a}$$

$$C + D = \frac{\alpha_s}{\pi^2} \int d^2 x_2 \, dY \, \frac{\vec{x}_{21} \cdot \vec{x}_{20}}{x_{21}^2 x_{20}^2} \left[U_{\vec{x}_{1\perp}} U_{\vec{x}_{0\perp}}^{\dagger} \right]_{ab} t^a V_{\vec{x}_{1\perp}} \otimes V_{\vec{x}_{0\perp}}^{\dagger} t^b. \tag{5.74b}$$

The sum of all the graphs in Fig. 5.9 is

$$
A + B + C + D + E + F
$$

$$
= \frac{\alpha_s}{\pi^2} \int d^2 x_2 \, dY \frac{\vec{x}_{21} \cdot \vec{x}_{20}}{x_{21}^2 x_{20}^2}
$$

$$
\times \left[1 - U_{\vec{x}_{1\perp}} U_{\vec{x}_{2\perp}}^\dagger - U_{\vec{x}_{2\perp}} U_{\vec{x}_{0\perp}}^\dagger + U_{\vec{x}_{1\perp}} U_{\vec{x}_{0\perp}}^\dagger \right]_{ab} t^a V_{\vec{x}_{1\perp}} \otimes V_{\vec{x}_{0\perp}}^\dagger t^b, \qquad (5.75)
$$

where $\mathbf{1}_{ab} = \delta_{ab}$.

Now we turn our attention to the diagrams in Fig. 5.10. Using the same group-theoretical identities, (5.31), (5.71), and (5.33), we obtain

$$
G + H + I = \frac{\alpha_s}{\pi^2} \int \frac{d^2 x_2}{x_{21}^2} \, dY \left[U_{\vec{x}_{1\perp}} U_{\vec{x}_{2\perp}}^\dagger - 1 \right]_{ab} t^b t^a V_{\vec{x}_{1\perp}} \otimes V_{\vec{x}_{0\perp}}^\dagger, \qquad (5.76a)
$$

$$
J + K + L = \frac{\alpha_s}{\pi^2} \int \frac{d^2 x_2}{x_{20}^2} \, dY \left[U_{\vec{x}_{2\perp}} U_{\vec{x}_{0\perp}}^\dagger - 1 \right]_{ab} V_{\vec{x}_{1\perp}} \otimes V_{\vec{x}_{0\perp}}^\dagger t^b t^a. \qquad (5.76b)
$$

To cast our results into a more compact form suitable for deriving JIMWLK evolution we need to introduce the derivative with respect to the function $\alpha^a(x^-, \vec{x}_\perp)$, with $\alpha = t^a \alpha^a$. We note that

$$
\frac{\delta}{\delta \alpha^a(y^-, \vec{y}_\perp)} V_{\vec{x}_\perp} = \frac{ig}{2} \delta^2 (\vec{x}_\perp - \vec{y}_\perp) U_{\vec{y}_\perp}^{\dagger ab}[\infty, y^-] t^b V_{\vec{x}_\perp} \qquad (5.77)
$$

where

$$
U_{\vec{y}_\perp}[\infty, y^-] = \mathrm{P} \exp \left\{ \frac{ig}{2} \int_{y^-}^{\infty} dx^- T^a \alpha^a(x^-, \vec{y}_\perp) \right\}, \qquad (5.78)
$$

so that $U_{\vec{y}_\perp} = U_{\vec{y}_\perp}[\infty, -\infty]$. The Wilson lines in our setup are only nontrivial because of interactions with the target at $x^- = 0$, as shown in Fig. 5.8 in the GGM approximation. As already mentioned, the same is true for successive small-x evolution, which generates gluons with much shorter lifetimes than those of the gluon that we are considering at this evolution step. Hence, if $y^- > 0$ then $U_{\vec{y}_\perp}[\infty, y^-] = \mathbf{1}$ and

$$
\frac{\delta}{\delta \alpha^a(y^-, \vec{y}_\perp)} V_{\vec{x}_\perp} = \frac{ig}{2} \delta^2 (\vec{x}_\perp - \vec{y}_\perp) t^a V_{\vec{x}_\perp}, \quad y^- > 0. \qquad (5.79)
$$

Taking the hermitian conjugate of this result we obtain

$$
\frac{\delta}{\delta \alpha^a(y^-, \vec{y}_\perp)} V_{\vec{x}_\perp}^\dagger = -\frac{ig}{2} \delta^2 (\vec{x}_\perp - \vec{y}_\perp) V_{\vec{x}_\perp}^\dagger t^a, \quad y^- > 0. \qquad (5.80)
$$

Using Eqs. (5.79) and (5.80) we can rewrite Eq. (5.75) as

$$
A + \cdots + F = \frac{\alpha_s}{2} \int d^2 x_\perp d^2 y_\perp dY \eta_{\vec{x}_\perp \vec{y}_\perp}^{ab} \frac{\delta^2 \left(V_{\vec{x}_{1\perp}} \otimes V_{\vec{x}_{0\perp}}^\dagger \right)}{\delta \alpha^a(x^-, \vec{x}_\perp) \delta \alpha^b(y^-, \vec{y}_\perp)}, \qquad (5.81)
$$

with

$$\eta^{ab}_{\vec{x}_{1\perp}\vec{x}_{0\perp}} = \frac{4}{g^2\pi^2}\int d^2x_2 \frac{\vec{x}_{21}\cdot\vec{x}_{20}}{x^2_{21}x^2_{20}}\left[1 - U_{\vec{x}_{1\perp}}U^\dagger_{\vec{x}_{2\perp}} - U_{\vec{x}_{2\perp}}U^\dagger_{\vec{x}_{0\perp}} + U_{\vec{x}_{1\perp}}U^\dagger_{\vec{x}_{0\perp}}\right]^{ab} \qquad (5.82)$$

for $x^-, y^- > 0$. Note that one of the two functional derivatives on the right-hand side of Eq. (5.81) acts on V and the other acts on $V^\dagger$. Naively one might expect that to obtain diagrams G through L one has to generalize Eq. (5.81) by allowing that both derivatives can act on V or that both can act on $V^\dagger$. This is almost correct. However, performing a detailed calculation one gets

$$\frac{\alpha_s}{2}\int d^2x_\perp d^2y_\perp dY \eta^{ab}_{\vec{x}_\perp\vec{y}_\perp}\left[\frac{\delta^2 V_{\vec{x}_{1\perp}}}{\delta\alpha^a(x^-,\vec{x}_\perp)\delta\alpha^b(y^-,\vec{y}_\perp)}\right]\otimes V^\dagger_{\vec{x}_{0\perp}}$$

$$= \frac{\alpha_s}{\pi^2}\int \frac{d^2x_2}{x^2_{21}}dY\left[\tfrac{1}{2}U_{\vec{x}_{1\perp}}U^\dagger_{\vec{x}_{2\perp}} + \tfrac{1}{2}U_{\vec{x}_{2\perp}}U^\dagger_{\vec{x}_{1\perp}} - 1\right]_{ab}t^b t^a V_{\vec{x}_{1\perp}}\otimes V^\dagger_{\vec{x}_{0\perp}}, \qquad (5.83)$$

which is different from Eq. (5.76a) by

$$\frac{\alpha_s}{2\pi^2}\int \frac{d^2x_2}{x^2_{21}}dY\left[U_{\vec{x}_{1\perp}}U^\dagger_{\vec{x}_{2\perp}} - U_{\vec{x}_{2\perp}}U^\dagger_{\vec{x}_{1\perp}}\right]_{ab}t^b t^a V_{\vec{x}_{1\perp}}\otimes V^\dagger_{\vec{x}_{0\perp}}$$

$$= -\frac{\alpha_s}{2\pi^2}\int \frac{d^2x_2}{x^2_{21}}dY\text{Tr}\left[T^a U_{\vec{x}_{1\perp}}U^\dagger_{\vec{x}_{2\perp}}\right]t^a V_{\vec{x}_{1\perp}}\otimes V^\dagger_{\vec{x}_{0\perp}}, \qquad (5.84)$$

where we have used Eq. (5.33) along with the definition of the adjoint generators T^a.

Defining

$$v^a_{\vec{x}_{1\perp}} = \frac{i}{g\pi^2}\int \frac{d^2x_2}{x^2_{21}}\text{Tr}\left[T^a U_{\vec{x}_{1\perp}}U^\dagger_{\vec{x}_{2\perp}}\right], \qquad (5.85)$$

we can write the contribution of diagrams G, H, and I in Eq. (5.76a) as

$$G+H+I = \frac{\alpha_s}{2}\int d^2x_\perp d^2y_\perp dY\eta^{ab}_{\vec{x}_\perp\vec{y}_\perp}\left[\frac{\delta^2 V_{\vec{x}_{1\perp}}}{\delta\alpha^a(x^-,\vec{x}_\perp)\delta\alpha^b(y^-,\vec{y}_\perp)}\right]\otimes V^\dagger_{\vec{x}_{0\perp}}$$

$$+ \alpha_s\int d^2x_\perp dY v^a_{\vec{x}_\perp}\left[\frac{\delta V_{\vec{x}_{1\perp}}}{\delta\alpha^a(x^-,\vec{x}_\perp)}\right]\otimes V^\dagger_{\vec{x}_{0\perp}}. \qquad (5.86)$$

Similarly, diagrams J, K, and L from Eq. (5.76b) give

$$J+K+L = \frac{\alpha_s}{2}\int d^2x_\perp d^2y_\perp dY\eta^{ab}_{\vec{x}_\perp\vec{y}_\perp}V_{\vec{x}_{1\perp}}\otimes\left[\frac{\delta^2 V^\dagger_{\vec{x}_{0\perp}}}{\delta\alpha^a(x^-,\vec{x}_\perp)\delta\alpha^b(y^-,\vec{y}_\perp)}\right]$$

$$+ \alpha_s\int d^2x_\perp dY v^a_{\vec{x}_\perp} V_{\vec{x}_{1\perp}}\otimes\left[\frac{\delta V^\dagger_{\vec{x}_{0\perp}}}{\delta\alpha^a(x^-,\vec{x}_\perp)}\right], \qquad (5.87)$$

with $x^-, y^- > 0$ throughout.

Combining Eqs. (5.81), (5.86), and (5.87) we derive an evolution equation for the operator (5.67) in the LLA:

$$\partial_Y \langle \hat{O}_{\vec{x}_{1\perp}, \vec{x}_{0\perp}} \rangle_Y = \frac{\alpha_s}{2} \int d^2 x_\perp d^2 y_\perp \left\langle \eta^{ab}_{\vec{x}_\perp \vec{y}_\perp} \frac{\delta^2 \hat{O}_{\vec{x}_{1\perp}, \vec{x}_{0\perp}}}{\delta \alpha^a(x^-, \vec{x}_\perp) \delta \alpha^b(y^-, \vec{y}_\perp)} \right\rangle_Y$$

$$+ \alpha_s \int d^2 x_\perp \left\langle \nu^a_{\vec{x}_\perp} \frac{\delta \hat{O}_{\vec{x}_{1\perp}, \vec{x}_{0\perp}}}{\delta \alpha^a(x^-, \vec{x}_\perp)} \right\rangle_Y, \tag{5.88}$$

where $\eta^{ab}_{\vec{x}_{1\perp} \vec{x}_{0\perp}}$ and $\nu^a_{\vec{x}_{1\perp}}$ are given by Eqs. (5.82) and (5.85) respectively.

We have obtained Eq. (5.63) in an explicit form for the test operator $\hat{O}_{\vec{x}_{1\perp}, \vec{x}_{0\perp}}$. Using Eq. (5.64) and integrating by parts, we can recast Eq. (5.88) as

$$\int \mathcal{D}\alpha \, \hat{O}_{\vec{x}_{1\perp}, \vec{x}_{0\perp}} \partial_Y W_Y[\alpha]$$

$$= \int \mathcal{D}\alpha \, \hat{O}_{\vec{x}_{1\perp}, \vec{x}_{0\perp}} \left\{ \frac{\alpha_s}{2} \int d^2 x_\perp d^2 y_\perp \frac{\delta^2}{\delta \alpha^a(x^-, \vec{x}_\perp) \delta \alpha^b(y^-, \vec{y}_\perp)} \left(\eta^{ab}_{\vec{x}_\perp \vec{y}_\perp} W_Y[\alpha] \right) \right.$$

$$\left. - \alpha_s \int d^2 x_\perp \frac{\delta}{\delta \alpha^a(x^-, \vec{x}_\perp)} \left(\nu^a_{\vec{x}_\perp} W_Y[\alpha] \right) \right\}. \tag{5.89}$$

Equation (5.89) is valid for any operator $\hat{O}_{\vec{x}_{1\perp}, \vec{x}_{0\perp}}$ with arbitrary transverse positions $\vec{x}_{1\perp}, \vec{x}_{0\perp}$ and for any quark colors. Following the above steps, one may derive the same equation for an operator constructed from two adjoint Wilson lines. This derivation can be repeated for an operator constructed from an arbitrary number of fundamental and adjoint Wilson lines, resulting again in Eq. (5.89). We see that the equation is valid for a broad class of test operators. We can therefore equate the integrands on both sides of (5.89) to obtain the Jalilian-Marian–Iancu–McLerran–Weigert–Leonidov–Kovner (JIMWLK) evolution equation (Jalilian-Marian *et al.* 1997b, 1999a, b, Iancu, Leonidov, and McLerran 2001a, b, Weigert 2002, Ferreiro *et al.* 2002):

$$\partial_Y W_Y[\alpha] = \frac{\alpha_s}{2} \int d^2 x_\perp d^2 y_\perp \frac{\delta^2}{\delta \alpha^a(x^-, \vec{x}_\perp) \delta \alpha^b(y^-, \vec{y}_\perp)} \left(\eta^{ab}_{\vec{x}_\perp \vec{y}_\perp} W_Y[\alpha] \right)$$

$$- \alpha_s \int d^2 x_\perp \frac{\delta}{\delta \alpha^a(x^-, \vec{x}_\perp)} \left(\nu^a_{\vec{x}_\perp} W_Y[\alpha] \right). \tag{5.90}$$

This is a differential equation for the weight functional $W_Y[\alpha]$, the Gaussian form of the functional (5.19) serving as its initial condition. This equation resums all powers of $\alpha_s Y$, and the Gaussian initial condition resums all classical physic effects (powers of $\alpha_s^2 A^{1/3}$). (As before we have $x^-, y^- > 0$.)

Owing to its complexity, no analytic solution of the JIMWLK equation exists. Its solution has been obtained only numerically, using lattice gauge theory methods (Rummukainen and Weigert 2004).

Returning to operators $\hat{O}$ constructed from the fundamental and/or adjoint Wilson lines (5.65) and (5.66), we see that the JIMWLK evolution for the expectation value of any such operator reduces to Eq. (5.88):

$$\partial_Y \langle \hat{O} \rangle_Y = \frac{\alpha_s}{2} \int d^2 x_\perp d^2 y_\perp \left\langle \eta^{ab}_{\vec{x}_\perp \vec{y}_\perp} \frac{\delta^2 \hat{O}}{\delta \alpha^a (x^-, \vec{x}_\perp) \delta \alpha^b (y^-, \vec{y}_\perp)} \right\rangle_Y$$

$$+ \alpha_s \int d^2 x_\perp \left\langle v^a_{\vec{x}_\perp} \frac{\delta \hat{O}}{\delta \alpha^a (x^-, \vec{x}_\perp)} \right\rangle_Y. \tag{5.91}$$

The diagrammatic representation of the JIMWLK operator evolution (5.91) is again given by diagrams of the types shown in Figs. 5.9 and 5.10, with s-channel gluon emissions from all the Wilson lines involved; thus we see that the JIMWLK evolution is driven by the same physics as the dipole BK evolution but provides an all-N_c generalization of the large-N_c BK equation. We will show how to obtain BK from JIMWLK in the next section.

Equation (5.91) can be recast in a more compact form if one notices that

$$\frac{1}{2} \frac{\delta}{\delta \alpha^a (x^-, \vec{x}_\perp)} \eta^{ab}_{\vec{x}_\perp \vec{y}_\perp} = \delta^2 (\vec{x}_\perp - \vec{y}_\perp) v^b_{\vec{x}_\perp}, \tag{5.92}$$

which reduces Eq. (5.91) to the Fokker–Planck form (Weigert 2002)

$$\partial_Y \langle \hat{O} \rangle_Y = \frac{\alpha_s}{2} \int d^2 x_\perp d^2 y_\perp \left\langle \frac{\delta}{\delta \alpha^a (x^-, \vec{x}_\perp)} \eta^{ab}_{\vec{x}_\perp \vec{y}_\perp} \frac{\delta}{\delta \alpha^b (y^-, \vec{y}_\perp)} \hat{O} \right\rangle_Y. \tag{5.93}$$

The JIMWLK equation for operators, in the form (5.91) or (5.93), allows one to construct the usual integro-differential evolution equation for any operator consisting of Wilson lines. This is a great strength of the JIMWLK approach: one can construct small-x evolution equations, bypassing diagrammatic analysis, and simply differentiate the operators with respect to α^a.

5.2.3 Obtaining BK from JIMWLK and the Balitsky hierarchy

In this section we are going to show that the Balitsky–Kovchegov equation is obtained by the CGC (JIMWLK) approach in the limit of a large number of colors ($N_c \gg 1$). As demonstrated above (see Eq. (5.45)), the S-matrix for dipole–nucleus scattering is closely related to the operator $\hat{O}_{\vec{x}_{1\perp}, \vec{x}_{0\perp}}$ of Eq. (5.67). Define the S-matrix operator

$$\hat{S}_{\vec{x}_{1\perp}, \vec{x}_{0\perp}} = \frac{1}{N_c} \text{tr} \left[V_{\vec{x}_{1\perp}} V^\dagger_{\vec{x}_{0\perp}} \right] \tag{5.94}$$

with V defined in Eq. (5.65). The S-matrix (5.45) is then given by

$$S(\vec{x}_{1\perp}, \vec{x}_{0\perp}, Y) = \langle \hat{S}_{\vec{x}_{1\perp}, \vec{x}_{0\perp}} \rangle_Y. \tag{5.95}$$

Substituting the operator (5.94) into Eq. (5.91) involves a considerable amount of algebra, which can be navigated by employing Eqs. (5.31) and (5.71), along with the Fierz identities

$$(t^a)_{ij} (t^a)_{kl} = \frac{1}{2} \left(\delta_{il} \delta_{jk} - \frac{1}{N_c} \delta_{ij} \delta_{kl} \right), \tag{5.96}$$

which imply that

$$\text{tr}\left[t^a M_1 t^a M_2\right] = \frac{1}{2}\,\text{tr}\,M_1\,\text{tr}\,M_2 - \frac{1}{2N_c}\,\text{tr}[M_1 M_2]\,, \tag{5.97a}$$

$$\text{tr}\left[t^a M_1\right]\text{tr}\left[t^a M_2\right] = \frac{1}{2}\,\text{tr}[M_1 M_2] - \frac{1}{2N_c}\,\text{tr}\,M_1\text{tr}\,M_2\,, \tag{5.97b}$$

for any $N_c \times N_c$ matrices M_1, M_2. In the end one obtains

$$\partial_Y\langle\hat{S}_{\vec{x}_{1\perp},\vec{x}_{0\perp}}\rangle_Y = \frac{\bar{\alpha}_s}{2\pi}\int d^2 x_2\,\frac{x_{10}^2}{x_{20}^2 x_{21}^2}\left[\langle\hat{S}_{\vec{x}_{1\perp},\vec{x}_{2\perp}}\hat{S}_{\vec{x}_{2\perp},\vec{x}_{0\perp}}\rangle_Y - \langle\hat{S}_{\vec{x}_{1\perp},\vec{x}_{0\perp}}\rangle_Y\right], \tag{5.98}$$

which looks very similar to the BK equation (4.137). The difference is in the first (nonlinear) term on the right-hand side of Eq. (5.98): to transform Eq. (5.98) into Eq. (4.137) one has to make the replacement

$$\langle\hat{S}_{\vec{x}_{1\perp},\vec{x}_{2\perp}}\hat{S}_{\vec{x}_{2\perp},\vec{x}_{0\perp}}\rangle_Y \longrightarrow \langle\hat{S}_{\vec{x}_{1\perp},\vec{x}_{2\perp}}\rangle_Y\langle\hat{S}_{\vec{x}_{2\perp},\vec{x}_{0\perp}}\rangle_Y. \tag{5.99}$$

Such a replacement is only justified in the large-N_c limit: clearly each $\hat{S}$ is a single-trace operator and corresponds to a quark loop (a dipole). Cross talk between the loops (dipoles) corresponds to nonplanar diagrams and, therefore, is N_c-suppressed at large N_c. Hence, for large-N_c, Eq. (5.98) reduces to the BK equation (4.137) (Weigert 2002, Kovner, Milhano, and Weigert 2000).

Since in the linearized regime the BK equation reduces to the BFKL equation, we can also conclude that BFKL evolution is obtained from JIMWLK in the linear regime outside the saturation region.

Outside the large-N_c limit Eq. (5.98) is not a closed equation, i.e., its right-hand side contains a quantity $\langle\hat{S}_{\vec{x}_{1\perp},\vec{x}_{2\perp}}\hat{S}_{\vec{x}_{2\perp},\vec{x}_{0\perp}}\rangle_Y$ and we do not know how to express this in terms of $\langle\hat{S}_{\vec{x}_{1\perp},\vec{x}_{0\perp}}\rangle_Y$. This quantity $\langle\hat{S}_{\vec{x}_{1\perp},\vec{x}_{2\perp}}\hat{S}_{\vec{x}_{2\perp},\vec{x}_{0\perp}}\rangle_Y$ is a new four-Wilson-line operator, for which one has to write down a separate evolution equation, again using Eq. (5.91). This evolution equation in turn contains on its right-hand side an operator with six fundamental Wilson lines, which would require its own evolution equation, etc. The result of applying the JIMWLK evolution (5.91) to all these operators would be an infinite set of evolution equations, in each of which the evolution of the n-Wilson-line operator would be driven by an $(n + 2)$-Wilson-line operator. This infinite system of equations is called the Balitsky hierarchy (Balitsky 1996, 1999a, b). The large-N_c limit truncates the Balitsky hierarchy at the lowest order, making Eq. (5.98) a closed (BK) equation. Other, perhaps less parametrically justified, truncations have been proposed (see Weigert 2005). While, just as for JIMWLK, no analytical solution of the Balitsky hierarchy of equations exists, numerical studies of JIMWLK in principle allow one to determine the evolution of these multi-Wilson-line operators with rapidity.

An interesting question concerns the importance of the $1/N_c$ corrections to BK evolution. Their size can be found by comparing the expectation value of the S-matrix operator $\langle\hat{S}_{\vec{x}_{1\perp},\vec{x}_{0\perp}}\rangle_Y$ obtained from the numerical solution of the full JIMWLK equation with

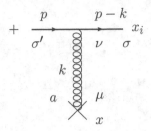

Fig. 5.11. The gluon field due to one ultrarelativistic quark.

that for the S-matrix resulting from solving the BK equation for the same initial conditions. We know that for gluon-driven dynamics the $1/N_c$ corrections are usually of order $1/N_c^2 \approx 11\%$. However, saturation effects tend to play an important role in suppressing the $1/N_c$ corrections. It has been shown by explicit numerical solution of JIMWLK that the corrections to $\langle \hat{S}_{\vec{x}_{1\perp}, \vec{x}_{0\perp}} \rangle_Y$ as compared with those for the BK case are actually close to 0.1% (Rummukainen and Weigert 2004, Kovchegov *et al.* 2009), which is two orders of magnitude smaller than the naive estimate above.

We wish to finish this chapter with a general remark: the color glass condensate gives us a beautiful example of how one can develop an effective theory starting from only a handful of physical assumptions. This theory is rather complex but it leads to new fundamental insights about our microscopic theory, QCD, in high energy scattering.

Further reading

In our presentation in this chapter we have tried to give the simplest possible derivations of the main results of the CGC formalism. We hope that the reader who wants to learn more on this subject will be able to read the original papers after reading this chapter. Many aspects of both CGC physics and the relevant derivations have been discussed in the reviews by McLerran (2005, 2008, 2009b), Iancu and Venugopalan (2003), Weigert (2005), Jalilian-Marian and Kovchegov (2006), and Gelis *et al.* (2010). In these reviews the theoretical topics are discussed together with practical applications and some challenges for further thinking are given. The relationship between JIMWLK evolution written in terms of derivatives with respect to the field α as opposed to the originally used color charge density ρ was explored by Kovner and Milhano (2000). For extended versions of the CGC formalism we recommend four papers of Kovner and Lublinsky (2005a–d) and the paper of Hatta *et al.* (2006).

Exercises

5.1 (a) Construct diagrammatically the gluon field of a single ultrarelativistic quark in the $\partial_\mu A^\mu = 0$ covariant gauge, which contributed to Eq. (5.11). Begin with the diagram in Fig. 5.11, where the gluon line is off mass shell. Show that the field in momentum space is[3]

$$A_\mu^a(k) = -ig(t_i^a)\frac{-ig_{\mu\nu}}{k^2 + i\epsilon}\bar{u}_\sigma(p-k)\gamma^\nu u_{\sigma'}(p)(2\pi)\delta\left((p-k)^2\right), \qquad (5.100)$$

where the delta function insures that the outgoing quark is on mass shell (the quark is assumed to be massless). Simplify Eq. (5.100) using the fact that p^+ is very large and employing Table A.1.

(b) Fourier-transform the result of part (a) into coordinate space using

$$A_\mu^a(x) = \int \frac{d^4k}{(2\pi)^4} e^{-ik\cdot(x-x_i)} A_\mu^a(k). \qquad (5.101)$$

You should obtain (suppressing the quark polarization indices)

$$A_{cov}^{a+} = -\frac{g}{\pi}(t_i^a)\delta(x^- - x_i^-)\ln\left(|\vec{x}_\perp - \vec{x}_{i\perp}|\Lambda\right) \qquad (5.102)$$

as the only nonzero field component. (You may find Eq. (A.9) useful.)

(c) Repeat the calculation from parts (a) and (b) in the $A^+ = 0$ light cone gauge.

5.2 Prove Eq. (5.31).

5.3 Using Eqs. (A.10) and (A.9) prove Eq. (5.49).

5.4 Neglecting the logarithm in the exponent of Eq. (5.51), integrate Eq. (5.50) over $\vec{r}_\perp$ exactly to obtain an approximate expression for the unintegrated WW gluon distribution ϕ^{WW}. Simplify the answer further by assuming that the nucleus is a cylinder oriented along the z-axis, so that $Q_{sG}(\vec{b}_\perp) = Q_{sG}\theta(R-b)$ and the $\vec{b}_\perp$-integration is trivial. Plot the expression obtained for $k_T\phi^{WW}$ as a function of k_T/Q_{sG} and compare the curve with Fig. 5.7.

5.5 Prove Eq. (5.92) by direct differentiation.

5.6 Substitute Eq. (5.94) into Eq. (5.91) and take the functional derivatives to show explicitly that one arrives at Eq. (5.98).

---

[3] The extra minus sign is due to the fact that the current in Eq. (5.4) is given by $J_\mu^a = -g\bar{\psi}\gamma_\mu t^a\psi$, which can be seen by comparing it with the QCD Lagrangian (1.1).

6

Corrections to nonlinear evolution equations

In this chapter we describe developments at the very forefront of research on nonlinear evolution equations. We first outline the calculation of running-coupling corrections to the BFKL, BK, and JIMWLK evolution equations. Such corrections slow down the growth of the saturation scale with energy, putting the predictions of saturation physics more in line with the experimental data. We then discuss the next-to-leading order (NLO) corrections to the BFKL and BK evolutions, which resum the subleading logarithms of energy, i.e., powers of $\alpha_s^2 Y$. The NLO BFKL corrections are rather large numerically; we present a proposal for resumming these large corrections to all orders that results in a reduction in their net effect on the LO calculation. Owing to the highly technical nature of many of the results presented, in most topics considered in this chapter we will merely outline the main points of the derivation. Interested readers can find the calculational details in the references supplied.

6.1 Why we need higher-order corrections

There are several reasons to study higher-order corrections to the BFKL, BK, and JIMWLK evolution equations presented in the previous chapters. Some reasons are theoretical, some are phenomenological, and some are both.

On the phenomenological side, the LO BFKL approach encounters a very simple problem. The BFKL pomeron intercept given by Eq. (3.86) is

$$\alpha_P - 1 \approx 2.77 \bar{\alpha}_s, \tag{6.1}$$

which, for a phenomenologically reasonable value of the strong coupling α_s, say 0.3, gives $\alpha_P - 1 \approx 0.79$, which is too large to describe any existing data in DIS, proton–proton, or nuclear collisions. One would therefore hope that higher-order corrections would lower this result, pushing the theory closer to the data.

On the more theoretical side we note that the BFKL, BK, and JIMWLK equations were derived in earlier chapters for fixed coupling constant. A question arises concerning the value of the coupling constant that should be used; this is important, since the validity of the whole saturation approach depends on whether the coupling is small. Theoretically we cannot answer this question from fixed-coupling calculations; one has to perform higher-order calculations to fix the scale of the running-coupling constant. This question about the scale of the coupling also has phenomenological importance, since one has to know which values of the coupling to use in comparing the small-x evolution with experiment.

The BK equation derived earlier contains powers of $\alpha_s N_c Y$ resummed through large-N_c LLA evolution along with powers of $\alpha_s^2 A^{1/3}$ resummed by the GGM initial conditions. Generalizing BK to JIMWLK relaxes the large-N_c approximation: the JIMWLK equation resums powers of $\alpha_s Y$ and $\alpha_s^2 A^{1/3}$. Both the LO BK and LO JIMWLK evolutions are valid as long as the NLO corrections are small, i.e., for $\alpha_s^2 Y \ll 1$, which means $Y \ll 1/\alpha_s^2$. Therefore, the problem of calculating the NLO correction to the BFKL, BK, and JIMWLK kernels is very important for understanding the region of applicability of the high density QCD theory in the form that has been developed above and for further extension of this region. Corrections to the initial conditions for the evolution equations (for instance, terms containing powers of $\alpha_s^4 A^{1/3}$) are also important, both theoretically and phenomenologically; however, attempts to calculate those have not reached the level required for coherent presentation in a book and will not be described here.

From a purely theoretical standpoint it is also important to understand whether the expansion in logarithms of $1/x$ is stable, that is, whether one can calculate corrections to the LO result and whether such corrections are finite (after all the standard field-theoretical divergences have been taken into consideration). Again, this question is, in the end, related to the first, purely phenomenological, one: what are the size and the sign of the NLO corrections?

The presentation below attempts to answer many of the above questions.

6.2 Running-coupling corrections to the BFKL, BK, and JIMWLK evolutions

We begin by calculating the scale of the running-coupling constant in the BFKL, BK, and JIMWLK evolution equations. The running-coupling corrections to small-x evolution are calculated following the Brodsky–Lepage–Mackenzie (BLM) scale-setting procedure (Brodsky, Lepage, and Mackenzie 1983). Working in the setting we used for the derivation of the JIMWLK and BK evolutions, below we will first resum the contributions of all quark-loop corrections to the LLA kernel. Each quark-loop correction brings in a power of $\alpha_\mu N_f$, with N_f the number of quark flavors (see Sec. 1.5) and $\alpha_\mu = \alpha_s(\mu^2)$ the physical coupling at some arbitrary renormalization scale μ. Inspired by Abelian gauge theories, Brodsky, Lepage, and Mackenzie (1983) argued that the powers of $\alpha_\mu N_f$ come mainly from the powers of the one-loop QCD beta function, that is, from the powers of $\alpha_\mu \beta_2$, where β_2 is given in Eq. (1.89). Following the BLM prescription, we will then complete N_f to the full coefficient of the one-loop beta function by means of the replacement

$$N_f \to -6\pi \beta_2 \qquad (6.2)$$

in the expression obtained by including quark-loop corrections in the BK and JIMWLK kernels. After this, the powers of $\alpha_\mu \beta_2$ should combine to give the physical running coupling $\alpha_s(Q^2)$ defined in Eq. (1.88) at the various momentum scales Q that would follow from this calculation.

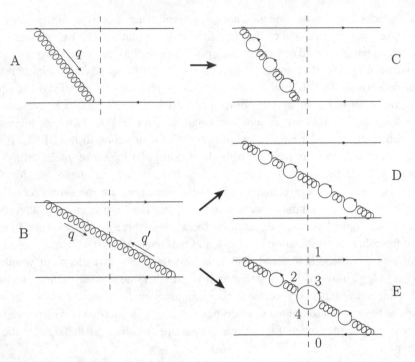

Fig. 6.1. Diagrams with quark-loop corrections to the BK and JIMWLK evolution kernels. The iteration of quark-loop insertions to all orders is implied in each graph on the right.

The original BLM prescription did not address the diagrams with gluon–gluon interactions. Here we will assume that the prescription is still valid for diagrams with triple-gluon vertices. This statement has not been rigorously proven, though in small-x physics it was explicitly verified in the NLO BK calculation by Balitsky and Chirilli (2008). A complementary way of thinking about a running-coupling calculation is by defining it as a resummation of $\alpha_s N_f$-corrections to the LO BFKL, BK, and JIMWLK kernels, the completion of N_f to the full beta function using (6.2) being an intelligent guess at the size of the gluon contribution, explicitly confirmed at NLO.

6.2.1 An outline of the running-coupling calculation

The main types of diagrams containing quark-loop corrections to the LO BK and JIMWLK evolution kernels are shown in Fig. 6.1 using the notation of Figs. 5.8, 5.9, and 5.10. The vertical dashed line again denotes the interaction with the target (or the subsequent evolution along with the interaction with the target). On the left of Fig. 6.1 we show one virtual (A) and one real (B) diagram contributing to one step of the LO BK or JIMWLK evolutions (cf. Fig. 5.9). All other real and virtual diagrams in the evolution kernel generated by connecting the gluon line to the quark and antiquark lines in all possible ways (see Figs. 5.9 and 5.10) should be included in the calculation; they are not shown explicitly. For LCPT diagrams, instantaneous terms like those shown in Fig. 4.14 need to be included as well.

The running-coupling corrections for the BK and JIMWLK kernels are obtained by inserting all-order quark bubbles into the gluon propagator in all possible ways. On the right of Fig. 6.1 we show the quark-loop-corrected diagrams corresponding to the fixed-coupling graphs on the left. The virtual correction (in the upper-left panel, labeled A) gives rise only to one class of diagrams, with quark loops iterated on the gluon propagator to all orders, shown in panel C. Working in momentum space, it is clear that the quark bubbles in this case form a geometric series, which is resummed to give

$$\alpha_s(q_\perp^2) = \frac{\alpha_\mu}{1 + \alpha_\mu \beta_2 \ln(q_\perp^2/\mu^2)}, \qquad (6.3)$$

where we have used the replacement (6.2) to complete N_f to the full beta function, and the factor α_μ in the numerator comes from the coupling of the gluon to the parent dipole. We see how physical running coupling emerges for the virtual diagrams. The coupling runs with the transverse momentum of the gluon line $q_\perp$; this can be found by calculating the diagram in panel C of Fig. 6.1 in, say, LCPT (Kovchegov and Weigert 2007a).

The real-emission diagram B generates two classes of quark-loop corrections, as shown in the lower two panels on the right of Fig. 6.1, labeled D and E. The first class of corrections, shown in panel D, corresponds to the case when it is the gluon that interacts with the target. This is to be compared to the other class of corrections, where the gluon fluctuates into a $q\bar{q}$ pair, which is still in the wave function at the time it enters the nucleus, so that now the quark and antiquark in the pair interact with the nucleons, as depicted in panel E of Fig. 6.1.

The momenta of the gluon line to the left and to the right of the interaction with the target are different in general: we label them q and q' respectively, as shown in panel B of Fig. 6.1. Note that the running-coupling corrections to the interaction of the gluon (and now $q\bar{q}$) cascade with the target factorize from the running-coupling corrections to the small-x evolution and are included separately (Balitsky 2007, Kovchegov and Weigert 2007b). Concentrating on the evolution, we see that the quark bubbles in the diagrams like that in panel D of Fig. 6.1 give us two separate geometric series, one to the left and one to the right of the interaction with the target. We thus get

$$\frac{\alpha_\mu}{\left[1 + \alpha_\mu \beta_2 \ln(q_\perp^2/\mu^2)\right]\left[1 + \alpha_\mu \beta_2 \ln(q_\perp'^2/\mu^2)\right]} = \frac{\alpha_s(q_\perp^2)\,\alpha_s(q_\perp'^2)}{\alpha_\mu}, \qquad (6.4)$$

where again we have used Eq. (6.2) to complete N_f to the full beta function and the factor α_μ stems from the coupling of the gluon to the dipole. We can see a problem with Eq. (6.4): using Eq. (1.88) we cannot rewrite it as a product of powers of the running coupling only, as we did in Eq. (6.3). One factor α_μ would still remain, as shown on the right of Eq. (6.4). Hence diagrams in the class represented by panel D cannot be expressed in terms of the running couplings only.

To resolve the issue we have to include the diagram in panel E as well. At first glance, the diagrams in this class, just as in the panel D class, would seem to have two geometric series but now with a factor of α_μ^2 in the numerator; the extra coupling arises from the coupling of gluons to the quark bubble that interacts with the target (which is slightly larger than the other bubbles in panel E). This would give

$$\frac{\alpha_\mu^2}{\left[1 + \alpha_\mu \beta_2 \ln(q_\perp^2/\mu^2)\right]\left[1 + \alpha_\mu \beta_2 \ln(q_\perp'^2/\mu^2)\right]} = \alpha_s(q_\perp^2)\alpha_s(q_\perp'^2). \tag{6.5}$$

However, this cannot be the complete answer. For one thing, it seems absurd that one power of the fixed coupling, corresponding to the gluon emission and absorption, has been replaced by two powers of the running coupling, making the evolution kernel contribution of order α_s^2. Analyzing the matter further, one realizes that the quark loop that interacts with the target also brings in a factor N_f that should be completed to β_2 and, more importantly, that the integration over momentum in the loop leads to a UV divergence, i.e., generates a $\ln \mu^2$ term. Keeping this logarithmically divergent term, we write the contribution of diagram E in Fig. 6.1 to the running of the coupling as

$$\frac{\alpha_\mu^2 \beta_2 \ln(Q^2/\mu^2)}{\left[1 + \alpha_\mu \beta_2 \ln(q_\perp^2/\mu^2)\right]\left[1 + \alpha_\mu \beta_2 \ln(q_\perp'^2/\mu^2)\right]}; \tag{6.6}$$

the scale Q is determined by an explicit calculation. Adding Eqs. (6.4) and (6.6) we see that diagrams D and E combine to give

$$\frac{\alpha_\mu \left[1 + \alpha_\mu \beta_2 \ln(Q^2/\mu^2)\right]}{\left[1 + \alpha_\mu \beta_2 \ln(q_\perp^2/\mu^2)\right]\left[1 + \alpha_\mu \beta_2 \ln(q_\perp'^2/\mu^2)\right]} = \frac{\alpha_s(q_\perp^2)\alpha_s(q_\perp'^2)}{\alpha_s(Q^2)}. \tag{6.7}$$

We see that now the answer for the real graphs is expressible in term of factors of the running coupling only. Note the unexpected structure of the result (6.7): in the BK and JIMWLK evolution kernels, one factor of the fixed coupling α_s in the LO evolution kernel is replaced by three running couplings, two in the numerator and one in the denominator,

$$\alpha_\mu \longrightarrow \frac{\alpha_s(q_\perp^2)\alpha_s(q_\perp'^2)}{\alpha_s(Q^2)}, \tag{6.8}$$

so that the answer is still order α_s. This structure is sometimes referred to as a *triumvirate* of couplings. It was first postulated for the running-coupling corrections to the BFKL evolution by Braun (1995) and Levin (1995). It was explicitly derived for the BFKL, BK, and JIMWLK evolution equations by Balitsky (2007) and by Kovchegov and Weigert (2007a).

The detailed calculation of the scale Q with explicit demonstrations that $q_\perp$ and $q_\perp'$ set the scales for the other two couplings in (6.7), along with the Fourier transform of the answer into transverse coordinate space, are too technically involved to be presented here in any detail. We refer the interested reader to the papers Balitsky (2007), Kovchegov and Weigert (2007a, b), and Gardi *et al.* (2007). We simply quote here the final answer for the running-coupling BK equation.

Writing the BK evolution equation (4.137) as

$$\partial_Y S(\vec{x}_{1\perp}, \vec{x}_{0\perp}, Y) = \int d^2 x_2 \, K(\vec{x}_{1\perp}, \vec{x}_{0\perp}, \vec{x}_{2\perp})$$

$$\times \left[S(\vec{x}_{1\perp}, \vec{x}_{2\perp}, Y) S(\vec{x}_{2\perp}, \vec{x}_{0\perp}, Y) - S(\vec{x}_{1\perp}, \vec{x}_{0\perp}, Y) \right], \tag{6.9}$$

we note that the LO dipole kernel is

$$K_{LO}(\vec{x}_{1\perp}, \vec{x}_{0\perp}, \vec{x}_{2\perp}) = \frac{\alpha_s N_c}{2\pi^2} \frac{x_{10}^2}{x_{20}^2 x_{21}^2}. \tag{6.10}$$

The form of the running-coupling kernel depends on how one extracts the scale Q shown in Eqs. (6.6)–(6.8); while $\ln \mu^2$ in $\ln(Q^2/\mu^2)$ is identified unambiguously, it is less clear how to define uniquely the scale Q^2. The problem originates in the fact that the contribution to the evolution kernel coming from diagram E in Fig. 6.1 cannot even be cast into the form (6.9). In the large-N_c limit this diagram has two dipoles interacting with the target: the dipole 13, consisting of the original quark and antiquark of the $q\bar{q}$ pair fluctuation of the gluon, and the dipole 40, consisting of the quark in the pair and the antiquark in the parent dipole (the coordinates are defined in Fig. 6.1E). The two dipoles do not have a common transverse coordinate, therefore their contribution is not of the form (6.9) and actually includes integrals over both $\vec{x}_{3\perp}$ and $\vec{x}_{4\perp}$, with the kernel dependent on four points in the transverse plane, $\vec{x}_{1\perp}, \vec{x}_{0\perp}, \vec{x}_{3\perp}, \vec{x}_{4\perp}$. The UV divergence that we need stems from the region between $\vec{x}_{3\perp}$ and $\vec{x}_{4\perp}$, and can be extracted either by integrating over $\vec{x}_{3\perp}$ while keeping $\vec{x}_{4\perp}$ fixed (the Balitsky (2007) prescription) or by integrating over $\vec{x}_{3\perp}$ and $\vec{x}_{4\perp}$ keeping the gluon position $\vec{x}_{2\perp}$ (see Fig. 6.1E) fixed (the Kovchegov and Weigert (2007a) prescription). The gluon position is related to $\vec{x}_{3\perp}$ and $\vec{x}_{4\perp}$ via the following expression (cf. Eq. (1.87) along with the discussion after it, as well as Fig. 1.4):

$$\vec{x}_{2\perp} = z_3 \vec{x}_{3\perp} + (1 - z_3) \vec{x}_{4\perp} \tag{6.11}$$

with z_3 the fraction of the gluon's light cone momentum carried by quark 3. (Indeed, other extractions of the UV divergence are also possible but calculations have been done only for the two cases mentioned.)

The kernel of the running-coupling BK evolution (rcBK) in the Balitsky prescription is (Balitsky 2007)

$$K_{rc}^{Bal}(\vec{x}_{1\perp}, \vec{x}_{0\perp}, \vec{x}_{2\perp}) = \frac{N_c \alpha_s(x_{10}^2)}{2\pi^2} \left[\frac{x_{10}^2}{x_{20}^2 x_{21}^2} + \frac{1}{x_{20}^2} \left(\frac{\alpha_s(x_{20}^2)}{\alpha_s(x_{21}^2)} - 1 \right) \right.$$

$$\left. + \frac{1}{x_{21}^2} \left(\frac{\alpha_s(x_{21}^2)}{\alpha_s(x_{20}^2)} - 1 \right) \right], \tag{6.12}$$

where we have used the abbreviated notation

$$\alpha_s(x_\perp^2) = \alpha_s \left(\frac{4 e^{-5/3 - 2\gamma_E}}{x_\perp^2} \right) \tag{6.13}$$

and the coupling on the right is defined by Eq. (1.88) in the $\overline{MS}$ renormalization scheme.

In the Kovchegov–Weigert prescription the rcBK kernel is (Kovchegov and Weigert 2007a)

$$K_{rc}^{KW}(\vec{x}_{1\perp}, \vec{x}_{0\perp}, \vec{x}_{2\perp})$$

$$= \frac{N_c}{2\pi^2} \left[\alpha_s(x_{20}^2) \frac{1}{x_{20}^2} - 2 \frac{\alpha_s(x_{20}^2) \alpha_s(x_{21}^2)}{\alpha_s(R^2)} \frac{\vec{x}_{20} \cdot \vec{x}_{21}}{x_{20}^2 x_{21}^2} + \alpha_s(x_{21}^2) \frac{1}{x_{21}^2} \right], \tag{6.14}$$

with the scale R^2 given by

$$R^2 = x_{20} x_{21} \left(\frac{x_{21}}{x_{20}} \right)^{\Xi}, \tag{6.15}$$

where

$$\Xi = \frac{x_{20}^2 + x_{21}^2}{x_{20}^2 - x_{21}^2} - 2 \frac{x_{20}^2 x_{21}^2}{\vec{x}_{20} \cdot \vec{x}_{21}} \frac{1}{x_{20}^2 - x_{21}^2}.$$

The two prescriptions (6.12) and (6.15) neglect different contributions of the diagram in Fig. 6.1E; as was shown by Albacete and Kovchegov (2007b), when the neglected terms are put back in, the two calculations agree with each other. It was also shown by an explicit numerical evaluation that the Balitsky prescription, when used in the BK evolution, gives a result that is closer to the full answer obtained by using the full diagram in Fig. 6.1E in the kernel of the small-x evolution (Albacete and Kovchegov 2007b). This is probably related to the fact that in the Balitsky prescription one obtains the linear (BFKL) part of the equation exactly: it gives the contribution correctly when only one dipole in Fig. 6.1E (either 13 or 40) interacts with the target. In Sec. 4.5.1 we saw that a good approximation to the solution of the fixed-coupling BK equation can be constructed by solving the linear BFKL equation with a saturation boundary in the IR (Gribov, Levin, and Ryskin 1983, Mueller and Triantafyllopoulos 2002). Most probably the same is true in the running-coupling case (see Gribov, Levin, and Ryskin (1983), Section 2.3.2), justifying the fact that the Balitsky prescription gives the full answer more accurately.

The evolution kernel of the running-coupling JIMWLK (rcJIMWLK) equation has been calculated only using the Kovchegov–Weigert prescription and can be found in Kovchegov and Weigert (2007a).

Once one has the running-coupling corrections to the nonlinear evolution equations, it is possible to obtain the running-coupling version of the BFKL equation. We first define the unintegrated gluon distribution $\phi(k_\perp, Y)$, using the dipole amplitude N, by (cf. Eqs. (3.92), (4.98))

$$\int d^2 b \, N(\vec{x}_\perp, \vec{b}_\perp, Y) = \frac{2\pi}{N_c} \int d^2 k_\perp \left(1 - e^{i\vec{k}_\perp \cdot \vec{x}_\perp} \right) \frac{\alpha_s(k_\perp^2)}{k_\perp^2} \phi(k_\perp, Y) \tag{6.16}$$

(Levin and Ryskin 1987). This connection between ϕ and N follows from the two-gluon exchange depicted in Fig. 6.2 (in the notation of Fig. 4.3), and, while its validity in the nonlinear saturation regime may be questioned, it is valid in the linear regime in which we want to apply it.

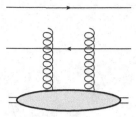

Fig. 6.2. The lowest-order diagram contributing to the relation between the dipole amplitude N and the unintegrated gluon distribution ϕ.

Using Eq. (6.16) in Eq. (6.9) with the kernel given by Balitsky prescription Eq. (6.12) (for which, in momentum space, $Q^2 = k_\perp^2$), linearizing the result, and Fourier-transforming it into momentum space, one obtains the running-coupling BFKL (rcBFKL) equation for the unintegrated gluon distribution (Kovchegov and Weigert 2007b):

$$\partial_Y \phi(k_\perp, Y) = \frac{N_c}{\pi^2} \int \frac{d^2q}{(\vec{k}_\perp - \vec{q}_\perp)^2}$$

$$\times \left[\alpha_s \left((\vec{k}_\perp - \vec{q}_\perp)^2 \right) \phi(q_\perp, Y) - \frac{k_\perp^2}{2q_\perp^2} \frac{\alpha_s \left(q_\perp^2 \right) \alpha_s \left((\vec{k}_\perp - \vec{q}_\perp)^2 \right)}{\alpha_s \left(k_\perp^2 \right)} \phi(k_\perp, Y) \right].$$

$$(6.17)$$

This equation was originally conjectured by Braun (1995) and Levin (1995) by requiring that the bootstrap property of BFKL is preserved after running-coupling corrections are included. Equation (6.17) can be compared with the fixed-coupling BFKL evolution of Eq. (3.94). One sees that, for the real term (the first term on the right-hand side of Eq. (6.17)), the coupling constant runs with the momentum in the rung (the s-channel gluon) of the BFKL ladder while in the virtual term (the second term on the right) a triumvirate structure arises for the three momenta involved in the color-octet gluon reggeization diagrams (see e.g. Fig. 3.11).

6.2.2 Impact of running coupling on small-x evolution

The effects of the running-coupling corrections on the small-x evolution can be summarized as follows.

(i) They slow down the evolution, by reducing the growth rates of the amplitude $N(x_\perp, Y)$ and of the saturation scale $Q_s(Y)$ with energy or rapidity.

(ii) They preserve geometric scaling in the vicinity of the saturation scale ($x_\perp \sim 1/Q_s(Y)$) while changing the profile of the dipole amplitude $N(x_\perp, Y)$ as a function of $x_\perp Q_s(Y)$.

(iii) They make the saturation scale Q_s independent of the atomic number A at very small x, thus eliminating the nuclear enhancement that we saw in the GGM model (Eq. (4.52)) and in the fixed-coupling small-x evolution (Eq. (4.156)).

These properties can be derived from a numerical solution of the rcBK equation or by analytical methods. The numerical solution of the rcBK equation with the kernel from

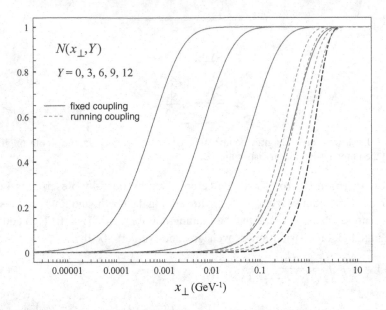

Fig. 6.3. The dipole amplitude $N(x_\perp, Y)$ as a function of the dipole size $x_\perp$, plotted for several rapidities Y generated by fixed-coupling BK evolution with $\alpha_s = 0.4$ (solid lines) and by running-coupling evolution (dashed lines) for the same initial condition (thick dashed line). (We thank Javier Albacete for providing us with this figure.) A color version of this figure is available online at www.cambridge.org/9780521112574.)

Eq. (6.12) (and with the strong coupling "frozen" in the IR at $\alpha_s = 1$) is shown in Fig. 6.3 by the dashed lines. It may be compared with the fixed-coupling BK evolution with $\alpha_s = 0.4$ (the solid lines) for the same initial condition from Eq. (4.205). The figure depicts the dipole scattering amplitude N plotted as a function of the dipole size $x_\perp$ for several different rapidities. It is clear that the fixed-coupling evolution, shown by the solid lines, is faster than the running-coupling evolution, shown by the dashed lines: the fixed-coupling curves grow faster with rapidity and the saturation scale corresponding to the fixed-coupling curves is clearly larger than that for the running coupling. Thus Fig. 6.3 illustrates property (i) in the above list.

Property (ii) is partially demonstrated in Fig. 4.32, where one can see that the two geometric scaling functions, for running and for fixed coupling, are in fact different in shape. Property (iii) is derived analytically below along with properties (i) and (ii).

Many qualitative and some quantitative features of the solution for rcBK evolution can be obtained analytically using an approximation in which a simple running of the coupling with the parent-dipole size, $\alpha_s(x_{10}^2)$, is used in the kernel (6.10) instead of the more complicated exact results seen in Eqs. (6.12) and (6.14). In an impact-parameter-independent approximation we can write the BK equation, by analogy with Eq. (4.175), as

$$\partial_Y \tilde{N}(\rho, Y) = \bar{\alpha}_s(\rho)\chi(-\partial_\rho)\tilde{N}(\rho, Y) - \bar{\alpha}_s \tilde{N}^2(\rho, Y), \qquad (6.18)$$

where now

$$\rho = \ln \frac{k_\perp^2}{\Lambda_{QCD}^2} \qquad (6.19)$$

and

$$\bar{\alpha}_s(\rho) = \frac{\alpha_s(\rho)N_c}{\pi} = \frac{N_c}{\pi\beta_2\rho} = \frac{N_c}{\pi\beta_2 \ln(k_\perp^2/\Lambda_{QCD}^2)}. \qquad (6.20)$$

Let us analyze Eq. (6.18) using the semiclassical approximation of Sec. 4.5.3. Writing $\tilde{N} = e^\Omega$ and neglecting the derivatives of Ω of second order and higher, we rewrite Eq. (6.18) as (cf. Eq. (4.178))

$$\partial_Y\Omega = \bar{\alpha}_s(\rho)\chi(-\partial_\rho\Omega) - \bar{\alpha}_s e^\Omega. \qquad (6.21)$$

Defining $-\gamma \equiv \Omega_\rho$ and $\omega \equiv \Omega_Y$ we get

$$F \equiv \omega - \bar{\alpha}_s(\rho)\chi(\gamma) + \bar{\alpha}_s(\rho)e^\Omega = 0. \qquad (6.22)$$

The equations for the characteristics are:

$$\frac{d\rho}{dt} = F_{-\gamma} = \bar{\alpha}_s(\rho)\frac{d\chi(\gamma)}{d\gamma}, \qquad (6.23a)$$

$$\frac{dY}{dt} = F_\omega = 1, \qquad (6.23b)$$

$$\frac{d\gamma}{dt} = -F_\rho - (-\gamma)F_\Omega = \frac{d\bar{\alpha}_s(\rho)}{d\rho}\chi(\gamma) + \left[\bar{\alpha}_s(\rho)\gamma - \frac{d\bar{\alpha}_s(\rho)}{d\rho}\right]e^\Omega, \qquad (6.23c)$$

$$\frac{d\omega}{dt} = -F_Y - \omega F_\Omega = -\bar{\alpha}_s(\rho)\,\omega\,e^\Omega, \qquad (6.23d)$$

$$\frac{d\Omega}{dt} = (-\gamma)F_{-\gamma} + \omega F_\omega = -\bar{\alpha}_s(\rho)\gamma\frac{d\chi(\gamma)}{d\gamma} + \omega. \qquad (6.23e)$$

Again, $Y = t$ results from Eq. (6.23b). Eliminating ω using Eq. (6.22) and noticing that $d\bar{\alpha}_s(\rho)/d\rho = -\bar{\alpha}_s(\rho)/\rho$ yields

$$\frac{d\rho}{dY} = \bar{\alpha}_s(\rho)\frac{d\chi(\gamma)}{d\gamma}, \qquad (6.24a)$$

$$\frac{d\gamma}{dY} = -\frac{\bar{\alpha}_s(\rho)}{\rho}\chi(\gamma) + \bar{\alpha}_s(\rho)\left(\gamma + \frac{1}{\rho}\right)e^\Omega, \qquad (6.24b)$$

$$\frac{d\Omega}{dY} = \bar{\alpha}_s(\rho)\left[\chi(\gamma) - \gamma\frac{d\chi(\gamma)}{d\gamma} - e^\Omega\right]. \qquad (6.24c)$$

Working in the linearized regime, we can neglect e^Ω in Eq. (6.24c) and obtain the critical trajectory along which $\Omega \approx$ const, with the critical value of the anomalous dimension γ

specified by

$$\chi(\gamma_{cr}) = \gamma_{cr} \frac{d\chi(\gamma_{cr})}{d\gamma_{cr}}, \tag{6.25}$$

so that $\gamma_{cr} \approx 0.6275$, just as in the fixed-coupling case.

Solving Eq. (6.24a) along the critical (saturation) trajectory yields

$$\rho_s^2(Y) = \rho_0^2 + \frac{2N_c}{\pi\beta_2} \frac{\chi(\gamma_{cr})}{\gamma_{cr}} Y, \tag{6.26}$$

where we have imposed the initial condition

$$\rho_s(Y = 0) = \ln \frac{Q_{s0}^2}{\Lambda_{QCD}^2} \equiv \rho_0. \tag{6.27}$$

Since $\rho_s(Y) = \ln(Q_s^2(Y)/\Lambda_{QCD}^2)$, we obtain the saturation scale in the running-coupling case (Gribov, Levin, and Ryskin 1983):

$$Q_s^2(Y) = \Lambda_{QCD}^2 \exp\left\{ \sqrt{\frac{2N_c}{\pi\beta_2} \frac{\chi(\gamma_{cr})}{\gamma_{cr}} Y + \ln^2 \frac{Q_{s0}^2}{\Lambda_{QCD}^2}} \right\}. \tag{6.28}$$

Comparing this result with the fixed-coupling saturation scale in Eq. (4.156) we see that the saturation scale in the running-coupling case grows *more slowly* with rapidity Y, confirming property (i) above stating that the running-coupling corrections slow down small-x evolution. This property of the running-coupling solution is very important: as the reader may remember, the fixed-coupling BFKL intercept (6.1) is too large to describe any data. The slower growth of the running-coupling solution makes phenomenological applications of rcBK and rcJIMWLK much more successful.

Equation (6.28) has another important property: at very large rapidity we can neglect the rapidity-independent logarithm squared under the square root, since it eventually becomes small compared with the term linear in Y. This gives

$$Q_s^2(Y) \approx \Lambda_{QCD}^2 \exp\left\{ \sqrt{\frac{2N_c}{\pi\beta_2} \frac{\chi(\gamma_{cr})}{\gamma_{cr}} Y} \right\}. \tag{6.29}$$

We see that all the Q_{s0}-dependence has disappeared. Since the dependence of the saturation scale on the atomic number A comes in only through $Q_{s0}^2 \sim A^{1/3}$, we conclude that at very large rapidity the running-coupling saturation scale becomes independent of A (Levin and Ryskin 1987, Mueller 2003). This demonstrates property (iii) above. Therefore, at extremely high energies the parton densities in the proton and in the nucleus will be the same. While this conclusion may be somewhat disappointing, note that our analysis applies to asymptotic energies: for the energies available in modern experiments the nuclei still provide a strong enhancement of the saturation scale.

A more careful evaluation of the high energy asymptotics of the saturation scale in the running-coupling case yields

$$
Q_s^2(Y) = \Lambda_{QCD}^2 \exp\left\{ \sqrt{\frac{2 N_c}{\pi \beta_2} \frac{\chi(\gamma_{cr})}{\gamma_{cr}} Y} + \frac{3}{4}\xi_1 \left[\frac{N_c}{2\pi \beta_2} \frac{\chi''(\gamma_{cr})}{\gamma_{cr}\chi(\gamma_{cr})} Y \right]^{1/6} \right.
$$

$$
\left. + \text{const} + O\left(Y^{-1/6}\right) \right\}, \tag{6.30}
$$

where $\xi_1 \approx -2.338$ is the first zero of the Airy function $\text{Ai}(\xi)$. The first term in the exponent of Eq. (6.30) was calculated by Gribov, Levin, and Ryskin (1983) (see also Iancu, Itakura, and McLerran (2002) and Mueller and Triantafyllopoulos (2002)), while the second term was found by Mueller and Triantafyllopoulos (2002) and by Munier and Peschanski (2004a). All the terms shown explicitly in Eq. (6.30) are universal (except for the constant): they do not depend on the initial conditions for the evolution. Several new higher-order universal terms in the expansion (6.30) were found recently by Beuf (2010).

For the constant $\gamma = \gamma_{cr}$ to be a solution of Eq. (6.24b) we need to require that the right-hand side of this equation is zero, which gives

$$
e^{\Omega_{cr}} = \frac{\chi(\gamma_{cr})}{\rho_s(Y)\gamma_{cr} + 1} \approx \frac{\chi(\gamma_{cr})}{\rho_s(Y)\gamma_{cr}}. \tag{6.31}
$$

This is indeed a small quantity at high energy, when $\rho_s(Y)$ is large, justifying the linearized approximation used in deriving the above results. Since $\rho_s(Y) \sim \sqrt{Y}$ we see that $e^{\Omega_{cr}}$ is a slowly varying function of Y, validating our treatment of it as a constant.

Finally, just as we did to obtain Eq. (4.187), we can expand Ω near the saturation trajectory keeping Y fixed, to get

$$
\Omega \approx \Omega_{cr} + \Omega_{\rho_s}(\rho - \rho_s) = \Omega_{cr} - \gamma_{cr}(\rho - \rho_s) \tag{6.32}
$$

so that

$$
\tilde{N}(\rho, Y) = e^{\Omega} \propto e^{-\gamma_{cr}(\rho-\rho_s)} = \left(\frac{Q_s^2(Y)}{k_\perp^2} \right)^{\gamma_{cr}}, \tag{6.33}
$$

where now $Q_s^2(Y)$ is given by Eq. (6.28). We see that the geometric scaling property of the solution persists when running-coupling corrections are included. This affirms part of the claim in property (ii) above. The anomalous dimension γ_{cr} obtained in the semiclassical approximation for the running-coupling coupling case is the same as for the fixed-coupling evolution: hence the dependence of $\tilde{N}$ on $k_\perp$ in Eq. (6.33) is the same as in Eq. (4.188). This appears to contradict the difference in $k_\perp$-dependence of the running- and fixed coupling BK evolution observed in the numerical simulation in Fig. 4.32. (This discrepancy was first observed by Albacete *et al.* (2005).) We believe that the accuracy of the semiclassical approximation is insufficient to detect this difference. Presumably more precise analytical solution techniques are needed to explain the difference.

6.2.3 Nonperturbative effects and renormalons*

Nonperturbative effects in the framework of perturbative QCD stem from the asymptotic nature of the perturbation series

$$\sum_n C_n \alpha_s^n \tag{6.34}$$

and from the fact that the coefficients C_n of this series increase as $n!$ for large n. To date there are three known sources of this $n!$ behavior of the perturbation-series coefficients in QCD: infrared (IR) and ultraviolet (UV) renormalons and instantons (see the review 't Hooft (1979) and the paper Mueller (1992)). A running QCD coupling generates renormalons. Since we now know how to include a running coupling in the BFKL, BK, and JIMWLK equations, we should be able to find the renormalon contribution to small-x evolution and estimate the contribution of nonperturbative QCD to small-x physics. Clearly the nonperturbative contribution stems from the IR renormalons, since the long distances (low momenta) corresponding to this case determine the nonperturbative corrections. Therefore we will concentrate on the IR renormalons in this section.

Since it is in line with the goal of this chapter to keep the calculations simple, let us illustrate the role of IR renormalons in saturation physics by the following toy-model example. We start with the relation between the dipole amplitude N and the unintegrated gluon distribution ϕ in Eq. (6.16). Assume that it is valid in the saturation region, where one can show that $\phi \propto k_\perp^2 / Q_s^2$. Then the contribution of dipole amplitude in the saturation region with $k_\perp < Q_s$ to the dipole amplitude outside the saturation region, i.e., for $x_\perp \ll 1/Q_s$, is proportional to

$$\int^{Q_s^2} \frac{d^2 k_\perp}{k_\perp^2} \left(1 - e^{i\vec{k}_\perp \cdot \vec{x}_\perp} \right) \alpha_s(k_\perp^2) \frac{k_\perp^2}{Q_s^2}. \tag{6.35}$$

In the $k_\perp < Q_s, x_\perp \ll 1/Q_s$ regime we have $k_\perp x_\perp \ll 1$, and the exponential in Eq. (6.35) can be expanded to yield after angular integration

$$\frac{x_\perp^2}{Q_s^2} \int_0^{Q_s^2} dk_\perp^2 \, k_\perp^2 \alpha_s(k_\perp^2), \tag{6.36}$$

where, for simplicity, we ignore overall constants.

Writing

$$\alpha_s(k_\perp^2) = \frac{\alpha_\mu}{1 + \alpha_\mu \beta_2 \ln(k_\perp^2 / \mu^2)}, \tag{6.37}$$

we substitute this into Eq. (6.36) and expand in powers of α_μ, obtaining

$$\frac{x_\perp^2}{Q_s^2} \alpha_\mu \sum_{n=0}^{\infty} (-\alpha_\mu \beta_2)^n \int_0^{Q_s^2} dk_\perp^2 k_\perp^2 \ln^n \frac{k_\perp^2}{\mu^2}. \tag{6.38}$$

Defining $\zeta = \ln(\mu^2/k_\perp^2)$ we rewrite Eq. (6.38) as

$$\frac{x_\perp^2}{Q_s^2}\mu^4\alpha_\mu \sum_{n=0}^\infty (\alpha_\mu\beta_2)^n \int_{\ln(\mu^2/Q_s^2)}^\infty d\zeta\,\zeta^n e^{-2\zeta}. \tag{6.39}$$

For large enough n, the integral in Eq. (6.39) is dominated by $\zeta \approx n/2$, so that its lower limit becomes irrelevant and can be set equal to zero. After that the ζ-integration can be easily performed, yielding

$$\frac{x_\perp^2}{2Q_s^2}\mu^4\alpha_\mu \sum_{n\gg 1}^\infty \left(\frac{\alpha_\mu\beta_2}{2}\right)^n n!, \tag{6.40}$$

which is a divergent perturbation series with coefficients proportional to $n!$. This is the effect of the IR QCD renormalons. If we define the applicability of the perturbation theory by the order n at which the $(n+1)$th term in the series is comparable with the nth term, we can conclude that perturbation theory breaks down for $n \approx n_0 = 2/(\alpha_\mu\beta_2)$. Thus the series (6.40) but terminating at $n = n_0$ would be perturbation theory's best guess at the exact answer.

We can also try to evaluate the series in Eq. (6.40) using the Borel resummation procedure. Namely, we rewrite the series as

$$\frac{x_\perp^2}{2Q_s^2}\mu^4 \int_0^\infty db\,e^{-b/\alpha_\mu} \sum_{n\gg 1}^\infty \left(\frac{\beta_2 b}{2}\right)^n, \tag{6.41}$$

where b is a dummy integration variable. Assuming that the series starts at $n = 0$, we resum it to obtain

$$-\frac{2}{\beta_2}\frac{x_\perp^2}{2Q_s^2}\mu^4 \int_0^\infty db\,e^{-b/\alpha_\mu}\frac{1}{b - 2/\beta_2}. \tag{6.42}$$

The pole at $b = 2/\beta_2$ is known as the IR renormalon pole in the complex-b Borel plane. The b-integral in Eq. (6.42) is divergent because of this renormalon pole along the integration contour: the series is not Borel-summable. While different $\pm i\epsilon$ regularizations of the pole can be proposed, it is not clear which such regularization would be correct. Instead the consensus in the community is that the difference between the various regularization prescriptions gives us an estimate of the uncertainty due to the renormalon singularity. Therefore, to evaluate the size of this uncertainty we simply need to take the residue of the renormalon pole, which gives

$$\frac{x_\perp^2}{Q_s^2}\mu^4 e^{-2/(\alpha_\mu\beta_2)} = \frac{x_\perp^2}{Q_s^2}\Lambda_{QCD}^4. \tag{6.43}$$

The fact that the result is proportional to Λ_{QCD}^4 indicates the nonperturbative origin of the uncertainty. The physical meaning of this phenomenon is well known (see Mueller (1985) and Zakharov (1992)). Indeed, the typical value of the momentum in the integral in

Eq. (6.38) is

$$k_\perp^2 \sim \mu^2 e^{-n/2} \tag{6.44}$$

and, regardless of the value of the renormalization scale μ, at sufficiently large n this momentum will become very small, so small that we would not be able to use perturbative QCD in our calculations ($k_\perp \approx \Lambda_{QCD}$). Of course we cannot trust our calculation for the low momenta of Eq. (6.44): instead we consider this equation as an indication that perturbation theory is breaking down and we should examine nonperturbative contributions to the observable.

The uncertainty (6.43) of our perturbative estimate should be compared with the perturbative estimate itself. If we forget the contribution of the Landau pole at $k_\perp = \Lambda_{QCD}$ in Eq. (6.36) (since, effectively, we have estimated the size of that contribution in Eq. (6.43)), the rest of the integral is clearly dominated by the upper limit of integration, giving an answer proportional to

$$\sim x_\perp^2 Q_s^2 \alpha_s(Q_s^2). \tag{6.45}$$

Comparing Eq. (6.43) with Eq. (6.45) we see that the relative contribution of the nonperturbative IR renormalon corrections is of order

$$\frac{\Lambda_{QCD}^4}{Q_s^4} \ll 1. \tag{6.46}$$

We conclude that saturation effects tend to suppress the renormalon contribution. Equation (6.46) is analogous to the conclusion by Mueller (1985) and Zakharov (1992) that the renormalon contribution in e^+e^- annihilation is of order Λ_{QCD}^4/Q^4, i.e., it is a higher-twist effect. In our case IR renormalons are also higher twist and, importantly, they are not enhanced by powers of $A^{1/3}$ or powers of $1/x$ and are therefore subleading compared with the perturbative saturation effects.

The qualitative conclusions of our toy model presented above are substantiated by more detailed calculations. The interested reader is referred to the papers by Levin (1995) and by Gardi *et al.* (2007) for much more detailed analytical and numerical investigations on the subject.

6.3 The next-to-leading order BFKL and BK equations

The NLO (order-α_s^2) corrections to the kernels of the BFKL and BK equations are now known. The NLO BFKL kernel was found by Fadin and Lipatov (1998) and Ciafaloni and Camici (1998), while the NLO BK equation was constructed by Balitsky and Chirilli (2008). Here we will briefly outline the calculational strategy and the main physical conclusions stemming from these calculations.

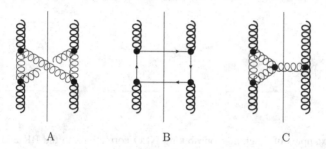

Fig. 6.4. Examples of diagrams contributing next-to-leading order corrections to the BFKL kernel. The bold lines denote reggeized gluons, and the circles denote regular QCD vertices. The vertical solid straight lines represent cuts.

6.3.1 Short summary of NLO calculations

It took almost a decade from the first papers on the subject (Fadin and Lipatov 1989, Ciafaloni 1988) to the last (Fadin and Lipatov 1998, Ciafaloni and Camici 1998) to solve the problem of finding the NLO BFKL kernel. As discussed above, the LO BFKL is a sum of ladder diagrams with Lipatov effective vertices and with reggeized gluons in the t-channel. The NLO corrections to this ladder include the running-coupling corrections to the vertices in the ladder (see for example Fig. 6.4C) and also processes involving the emission of two gluons with comparable rapidities in a single rung (see Fig. 6.4A). One also needs to calculate the quark–antiquark pair production (see Fig. 6.4B) and to find the reggeized gluon trajectory in the NLO approximation. (Of course one also has to prove that we can still use the reggeized gluon in the NLO approximation: this turns out to be the case.) The relative simplicity of the LO BFKL equation originates in part from the fact that in the LLA one can easily separate the longitudinal and transverse degrees of freedom. In the NLO approximation one has to take into account the fact that the limits of integration over longitudinal momenta depend also on the transverse momenta. The large number of extra diagrams, a tiny subset of which is shown in Fig. 6.4, along with the more sophisticated diagram evaluation required in a beyond-LLA approximation, are the two main reasons why it took so long to find the NLO BFKL kernel.

The exact NLO BFKL kernel is too cumbersome to be presented here. It can be found in the papers of Fadin and Lipatov (1998) and Ciafaloni and Camici (1998). The results of these calculations yielded some new features and new questions. The NLO corrections to the LO BFKL intercept turned out to be negative. Such negative corrections had been expected, since the LO BFKL intercept given in Eq. (6.1) is too large: this is why the LO BFKL evolution overestimates the rise of the DIS structure functions with energy as well as the size of the Bjorken scaling violation $dF_2/d\ln Q^2$ in comparison with the HERA experimental data. However, the size of the negative NLO corrections turned out to be too large. The BFKL pomeron intercept in the saddle point approximation for the LO and NLO orders is equal to (cf. the LO BFKL intercept in Eq. (3.86))

$$\alpha_P - 1 \approx 4\,\bar{\alpha}_s \ln 2\,(1 - 6.7\bar{\alpha}_s). \tag{6.47}$$

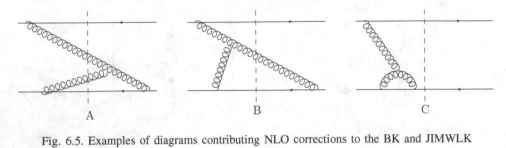

Fig. 6.5. Examples of diagrams contributing NLO corrections to the BK and JIMWLK kernels.

Equation (6.47) leads to a negative intercept for $\bar{\alpha}_s > 1/6.7 \approx 0.15$, which means that the NLO BFKL pomeron leads to structure functions that fall off with energy at all virtualities up to $Q^2 \approx 10^3\,\mathrm{GeV}^2$. This result points toward an instability of the NLO BFKL pomeron, and this is confirmed by the oscillatory behavior of the resulting NLO gluon density. (The reader is also referred to Ross (1998), where it is shown that, in the cross section mediated by NLO BFKL pomeron exchange in the diffusion approximation, the logarithm squared term in the exponent (see Eq. (3.85)) enters with a plus sign, indicating an instability of the solution due to the enhancement of transverse momentum fluctuations.)

The NLO BFKL kernel appears to lead to a number of serious questions, which may be resolved by the calculation of the higher-order corrections. Unfortunately, owing to the apparent complexity of the NNLO calculations, it looks unreasonable to expect the result soon. Fortunately, it turns out that the most numerically essential contribution in the NLO BFKL kernel comes from collinear singularities, which are resummed by the DGLAP evolution. The resummation of collinear corrections to the BFKL kernel appears to cure the instability of the NLO BFKL pomeron, as we will discuss in the next section.

It is also possible that some problems of NLO BFKL would be cured by a consideration of saturation effects. To check this, one has to study the nonlinear BK and JIMWLK evolution equations at NLO. The NLO BK equation was calculated by Balitsky and Chirilli (2008) (see also Balitsky and Belitsky (2002) for a calculation of part of the kernel). A small subset of the diagrams that one has to calculate in order to find the NLO BK kernel is shown in Fig. 6.5. In one step of NLO evolution one has to emit two s-channel gluons with comparable rapidities. The diagrams can be classified by the number of gluons interacting with the target: there may be zero, one, or two gluons, as shown in Figs. 6.5C, B, A respectively. Indeed, one should also include one-loop quark corrections, illustrated by the lowest-order (one-loop) case of the diagrams in Fig. 6.1. Note that, in the large-N_c limit, diagrams of the type shown in Fig. 6.5A imply that the parent dipole will split into *three* daughter dipoles, resulting in an evolution equation which is cubic in N (or S) (Balitsky and Belitsky 2002). Hence the quadratic structure of the LO BK evolution does not survive at higher orders.

As in the BFKL case, the calculation of the NLO BK evolution is too technically involved to be presented here; we refer the reader to the paper by Balitsky and Chirilli (2008) for details. In the linearized limit, the NLO BK evolution indeed reduces to the NLO BFKL evolution. Unfortunately, at this time the physical implications of NLO BK evolution are not completely understood.

6.3.2 Renormalization-group-improved NLO approach[*]

Let us now present a strategy to cure the problems of the NLO BFKL kernel by performing the resummation of collinear singularities to all orders. This procedure was suggested in the works of Salam (1998, 1999) and Ciafaloni, Colferai, and Salam (1999a). The key idea is based on the observation that the large NLO corrections to the BFKL kernel stem mostly from collinearly enhanced physical contributions. At the same time we know that the collinear singularities can be resummed with the help of the renormalization group (RG) and have been taken into account in the DGLAP evolution. The idea of finding a combined description that includes the BFKL anomalous dimension with the anomalous dimension of the DGLAP evolution equation has a history ranging from the first attempt by Gribov, Levin, and Ryskin, where the DGLAP anomalous dimension was simply added to the BFKL anomalous dimension, to the Catani, Ciafaloni, Fiorani and Marchesini (CCFM) evolution equation, in which a correct treatment of the coherence effect in the collinear kinematics was introduced. In the renormalization-group-improved NLO approach this problem was solved, and we will follow the paper of Ciafaloni, Colferai, and Salam (1999a) in our discussion of the theoretical approach. We will also present the next-to-leading-order resummed BFKL kernel in the simple form given in the paper of Khoze *et al.* (2004) to illustrate the numerical importance of the corrections that were introduced.

The starting point is the expression for the azimuthally symmetric Green function of the BFKL pomeron in the double Mellin representation, namely (cf. Eqs. (3.78) and (3.80) with $n = 0$)

$$G(k, k_0, Y) = \frac{1}{k^2} \int_{a-i\infty}^{a+i\infty} \frac{d\omega}{2\pi i} \int_{1/2-i\infty}^{1/2+i\infty} \frac{d\gamma}{2\pi^2 i} \left(\frac{s}{kk_0}\right)^\omega e^{\gamma\xi} \frac{1}{\omega - \kappa(\gamma, \omega)}, \tag{6.48}$$

where k and k_0 with $k > k_0$ are the transverse momentum scales at the ends of the ladder, with $\xi = \ln(k^2/k_0^2)$, and the integration contour over ω in Eq. (6.48) lies to the right of all the singularities of the integrand. We have also replaced the rapidity Y by $\ln(s/(kk_0))$. For the LO BFKL the function $\kappa(\gamma, \omega)$ reduces to $\kappa_{LO}(\gamma, \omega) = \bar{\alpha}_s \chi(\gamma)$, with the latter given by Eq. (4.174). In the LO case, integrating over ω in Eq. (6.48) yields Eq. (3.80) with $n = 0$.

It is clear from Eq. (6.48) that the traditional DGLAP moment-space representation (see Sec. 2.4.5) can be achieved after the integration over γ by closing the contour to pick up the singularity at $\gamma = \gamma(\bar{\alpha}_s, \omega)$, which is the solution of the equation

$$\omega = \kappa(\gamma, \omega). \tag{6.49}$$

The result of the calculation of the NLO BFKL equation allows one to write

$$\kappa(\gamma, \omega) = \bar{\alpha}_s \chi(\gamma) + \bar{\alpha}_s^2 \chi_1(\gamma, \omega) + \cdots, \tag{6.50}$$

where the exact form of the NLO correction $\chi_1(\gamma, \omega)$ is unimportant for our purposes. We only need to know that an inspection of the explicit form of $\chi_1(\gamma, \omega)$ shows that numerically large contributions stem from the regions $\gamma \to 0$ and $\gamma \to 1$, where $\chi_1(\gamma, \omega) \propto 1/\gamma^2$ and

$1/(1 - \gamma)^2$ respectively. These are the regions of γ where the DGLAP equation governs the energy and transverse momentum evolution of the parton densities, which will be utilized shortly.

Another large contribution originates from the dependence of the QCD coupling on the transverse momenta. This contribution can be incorporated in the framework of the RG approach if we first integrate over γ in Eq. (6.48); then, solving Eq. (6.49) to find the pole at $\gamma(\bar{\alpha}_s, \omega)$ and, for $k > k_0$, making the replacement

$$\gamma(\bar{\alpha}_s(\xi), \omega)\, \xi \longrightarrow \int_0^\xi d\xi'\, \gamma\left(\bar{\alpha}_s(\xi'), \omega\right) \tag{6.51}$$

in the exponent of Eq. (6.48) would give us the answer as a single integral over ω with the running-coupling corrections included.

Our aim is to find a function $\kappa(\gamma, \omega)$ that describes both the LO and NLO BFKL kernels and DGLAP evolution for $k^2 > k_0^2$ (or $k_0^2 > k^2$). A comparison of Eq. (6.48) with the DGLAP equation shows that since the latter is written in terms of the distribution functions, the definition of Bjorken x (and consequently of the rapidity $Y = \ln 1/x$) in the DGLAP picture is given by $x = k^2/s$ for $k^2 \gg k_0^2$ and by $x = k_0^2/s$ for $k^2 \ll k_0^2$; these expressions are different from those for an up–down symmetric (for a vertically drawn ladder) variable $x = e^{-Y} = (k_0 k)/s$ used in Eq. (6.48). Indeed, in the LLA small-x evolution such differences were outside the approximation's control and were not important: now they are crucial for matching NLO BFKL onto DGLAP evolution. Changing the momentum scale in the definition of rapidity and Bjorken x leads to the shift $\gamma \to \gamma \pm \omega/2$:

$$G(k, k_0, Y) = \frac{1}{k^2} \int_{a-i\infty}^{a+i\infty} \frac{d\omega}{2\pi i} \int_{1/2-i\infty}^{1/2+i\infty} \frac{d\gamma}{2\pi^2 i} \left(\frac{s}{k^2}\right)^\omega e^{(\gamma + \omega/2)\xi} \frac{1}{\omega - \kappa(\gamma, \omega)}$$

$$= \frac{1}{k_0^2} \int_{a-i\infty}^{a+i\infty} \frac{d\omega}{2\pi i} \int_{1/2-i\infty}^{1/2+i\infty} \frac{d\gamma}{2\pi^2 i} \left(\frac{s}{k_0^2}\right)^\omega e^{(1-\gamma + \omega/2)\xi_0} \frac{1}{\omega - \kappa(\gamma, \omega)}, \tag{6.52}$$

where we have defined $\xi_0 = \ln(k_0^2/k^2)$ to replace ξ in the case $k_0 > k$. We see that one effect of DGLAP evolution would be to replace γ by $\gamma + \omega/2$ near the singularity $\gamma = 0$ and γ by $\gamma - \omega/2$ near the singularity $\gamma = 1$.

The LO BFKL kernel has singularities for all integer values of γ, corresponding to different powers of k^2/k_0^2 in the Green function G, that is, to different twists. Singling out the leading-twist singularities at $\gamma = 0$ and $\gamma = 1$ we write the LO BFKL kernel as a sum of the leading-twist contribution and the higher-twist terms:

$$\chi(\gamma) = \frac{1}{\gamma} + \frac{1}{1 - \gamma} + \chi^{HT}(\gamma) \tag{6.53}$$

where the higher-twist part is given by

$$\chi^{HT}(\gamma) = 2\psi(1) - \psi(1 + \gamma) - \psi(2 - \gamma). \tag{6.54}$$

The two terms of the leading-twist part of $\chi(\gamma)$ describe two different branches of the leading-twist evolution: the $1/\gamma$ term corresponds to a DGLAP evolution from low k_0 to high k with ordering in the transverse momenta of emitted partons $k_0 \ll k_{1\perp} \ll \cdots \ll k_{i\perp} \ll \cdots \ll k$, while the $1/(1-\gamma)$ term leads to an evolution from low k to high k_0 with the opposite ordering, $k \ll \cdots \ll k_{i\perp} \ll \cdots \ll k_{1\perp} \ll k_0$. The higher-twist contributions play a significant role in the LO BFKL evolution: for example, they change the leading-twist value of the pomeron intercept from $[1/\gamma + 1/(1-\gamma)]_{\gamma=1/2} = 4$ to $\chi(\gamma = 1/2) = 4 \ln 2 \approx 2.8$. However, the DGLAP evolution or, in other words, the anomalous dimensions of the operators giving rise to the higher-twist contributions are entirely unknown. Fortunately, on scrutinizing the NLO BFKL kernel one can see that the large problematic contribution does not come from these higher-twist terms and so we can concentrate on the leading-twist terms alone.

The next step is to replace the residue 1 in the first term in Eq. (6.53) by the full DGLAP gluon–gluon anomalous dimension $\gamma_{GG}(\omega)$ from Eq. (2.121d), in order to incorporate the DGLAP effects at finite ω:

$$\frac{\bar{\alpha}_s}{\gamma} \longrightarrow \frac{\bar{\alpha}_s}{2N_c} \frac{\omega \gamma_{GG}(\omega)}{\gamma}. \tag{6.55}$$

Performing the same replacement for the term $1/(1-\gamma)$ in Eq. (6.53) and using the shifts in γ incorporated into Eq. (6.52), we obtain the RG-improved BFKL kernel (Ciafaloni, Colferai, and Salam 1999a):

$$\kappa_{RG}(\gamma, \omega) = \frac{\bar{\alpha}_s}{2N_c} \left[\frac{\omega \gamma_{GG}(\omega)}{\gamma + \omega/2} + \frac{\omega \gamma_{GG}(\omega)}{1 - \gamma + \omega/2} \right] + \bar{\alpha}_s \chi^{HT}(\gamma) + \cdots, \tag{6.56}$$

where the ellipsis stand for order-α_s^2 terms that are nonsingular at $\gamma = 0$ and $\gamma = 1$. The expansion of Eq. (6.56) in powers of ω would give us the correct collinear singularities ($1/\gamma^2$ and $1/(1-\gamma)^2$ terms) at the NLO at order ω, while higher orders in ω capture the collinear singularities of the higher-order BFKL kernels.

The kernel in Eq. (6.56) contains the full LO and NLO BFKL kernels, with the leading-twist parts of the LO BFKL kernel enhanced by DGLAP evolution, which resums all the leading (transverse) logarithmic collinear singularities to all orders.

To impose energy conservation one has to make sure that the kernel (6.56) vanishes at $\omega = 1$ (see Exercise 2.3). This can be achieved in a crude way by simply multiplying $\chi^{HT}(\gamma)$ and other terms in Eq. (6.56) that are a priori nonvanishing at $\omega = 1$ by $1 - \omega$ (Ellis, Kunszt, and Levin 1994). The expression that results from this procedure, after the terms denoted by the ellipses in Eq. (6.56) have been discarded,

$$\kappa_{RG}(\gamma, \omega) = \frac{\bar{\alpha}_s}{2N_c} \left[\frac{\omega \gamma_{GG}(\omega)}{\gamma + \omega/2} + \frac{\omega \gamma_{GG}(\omega)}{1 - \gamma + \omega/2} \right] + (1 - \omega) \bar{\alpha}_s \chi^{HT}(\gamma), \tag{6.57}$$

was used by Khoze et al. (2004), who showed that Eq. (6.57) describes the full RG-resummed NLO BFKL kernel (6.56) within 7% accuracy. The most interesting aspect of the result is that even such a simple modification of the NLO BFKL kernel leads to a stable result for the Green function, and for the resulting amplitudes and cross sections, and considerably reduces the value of the NLO corrections. The kernel (6.56) has a minimum at $\gamma = 1/2$, at which the value of the BFKL pomeron intercept is about 0.25 for $\bar{\alpha}_s = 0.15$. The diffusion logarithm squared term (similar to Eq. (3.85)) again has a negative coefficient, so that the instability of transverse momentum fluctuations is now avoided. In the region $\gamma < 1/2$ the NLO kernel (6.57) is very close to the DGLAP kernel. For a detailed comparison of the modified BFKL kernel with the experimental data as well as with other approaches we recommend the paper Ciafaloni (2005).

We have demonstrated that our knowledge of DGLAP evolution allows us to understand the sources of the large NLO contributions to the BFKL equation and allows us to formulate a more stable approach to higher-order corrections for small-x evolution.

The effect of the NLO corrections on the value of the saturation scale was considered by Gotsman *et al.* (2005); not unexpectedly, their conclusion was that the NLO corrections, while lowering the value of the BFKL intercept, also slow down the growth in the saturation scale with energy, leading to lower values of the saturation scale than those given by the fixed-coupling evolution.

Further reading

The first attempts to include the running QCD coupling in the BFKL equation by resumming powers of $\alpha_s N_f$ were made by Braun (1995) and Levin (1995) and were based on the bootstrap equation (see Sec. 3.3.5) at next-to-leading order, obtaining the triumvirate of running couplings for the first time. Their original conjecture was proved at NLO by Fadin and Fiore (1998). The direct calculations in the BK/JIMWLK formalism discussed above were performed by Balitsky (2007) and by Kovchegov and Weigert (2007a).

For more information on the nonperturbative corrections to BFKL, BK, and JIMWLK evolution coming from the IR renormalons, we refer the reader to Gardi *et al.* (2007) and Levin (1995). Some aspects of the nonperturbative effects due to instantons in the CGC were studied by Kharzeev, Kovchegov, and Levin (2002). The possibility that the BFKL pomeron could reach the nonperturbative region of small momenta through "tunneling" was suggested by Ciafaloni *et al.* (2003b). Whether the nonlinear evolution can withstand this type of nonperturbative correction and remain perturbative is still an open question.

The NLO BFKL kernel has been calculated by two groups: Fadin, Lipatov, and their collaborators and Camici, Ciafaloni, and their collaborators. All references for these works can be found in the papers with the final results: Fadin and Lipatov (1998) and Ciafaloni and Camici (1998). The NLO BK equation was found by Balitsky and Chirilli (2008).

In our presentation of the RG-improved BFKL kernel we described the key ideas proposed by Salam (1998, 1999) and by Ciafaloni, Colferai, and Salam (1999a). We discussed the simplest possible example of a resummed kernel, that from the paper of Khoze *et al.* (2004). More recent developments in this area can be found in the papers by Ciafaloni *et al.* (2003a) and by Altarelli, Ball, and Forte (2006). We need to remember that we have no information on the anomalous dimensions of the higher-twist contributions and, therefore, the NLO corrections to the part of the BFKL kernel that is responsible for the higher-twist corrections cannot be improved on the basis of the existing renormalization group. This problem is not for further reading but rather for further research. Alternatives to the

RG-improved BFKL kernel can be found in the papers by Brodsky *et al.* (1999) and by Schmidt (1999). For a description of the experimental DIS data in the NLO BFKL approximation we refer the reader to the papers White and Thorne (2007), Ciafaloni (2005), and Peschanski, Royon, and Schoeffel (2005). The effect of the NLO BFKL kernel on the saturation scale was studied by Khoze *et al.* (2004). The impact of NLO corrections on the solution of the BK equation was studied by Gotsman *et al.* (2004). The large contribution of the NLO correction to the saturation scale possibly calls for a generalization of the nonlinear equation and could be a good subject for further investigations.

Exercises

6.1 Using Eq. (6.56) (dropping the ellipses) with the gluon–gluon splitting function γ_{GG} given by Eq. (2.121d), calculate the correction to the intercept of the BFKL pomeron, as follows.

   (a) Solve

$$\omega = \kappa_{RG}(\gamma, \omega) \tag{6.58}$$

for ω by assuming that $\omega = O(\bar{\alpha}_s)$ and expanding the right-hand side to the quadratic order in ω.

   (b) Find the saddle point of the resulting expression for $\omega(\gamma)$; the value of $\omega(\gamma)$ at the saddle point yields the new BFKL pomeron intercept.

   (c) Find the Green function (6.48) in the diffusion approximation using the results of parts (a) and (b). Show that the diffusion term is negative and, therefore, the solution is stable.

6.2 Consider the analogue of the series from Eq. (6.40) in QED:

$$\sum_{n=0}^{\infty} \left(-\frac{\alpha_{EM}}{6\pi} \right)^n n!. \tag{6.59}$$

Resum the series using the Borel resummation procedure outlined in Sec. 6.2.3. Comment on the analyticity of the function of α_{EM} that is obtained.

7

Diffraction at high energy

The observables discussed in this book so far have been limited to total cross sections and the related structure functions. To calculate these quantities one does not need to impose any constraints on the final state. We now present a small-x calculation of a more exclusive quantity, the cross section for diffractive dissociation, where one requires that the final state has at least one rapidity gap, i.e., a region of rapidity where no particles are produced. We again tackle the problem using the two-step formalism of Chapters 4 and 5: we first calculate the cross sections for quasi-elastic processes using the classical MV/GGM approximation and then include small-x evolution corrections in the resulting expression. For diffractive dissociation where the produced hadrons have large invariant mass, we develop a nonlinear evolution equation that describes the process.

7.1 General concepts

7.1.1 Diffraction in optics

Diffraction is a typical process in which we can see the wave nature of particles. When thinking of diffraction one usually pictures the diffraction of light, when a plane wave is incident on an aperture or an obstacle (see Fig. 7.1) and forms a diffraction pattern on the screen behind. The diffraction pattern consists of a bright spot in the middle and a series of maxima and minima of light intensity around it, as shown schematically in Fig. 7.1. The positions of these maxima and minima depend on the size R of the obstacle or aperture (the target), the distance d between the target and the screen (the detector), and the light wavelength λ. Depending on the values of these three parameters one usually distinguishes three types of diffraction, as follows.

(i) Fraunhofer diffraction, when $R^2/(\lambda d) \ll 1$, which corresponds to the scattering at very small angles;
(ii) Fresnel diffraction, when $R^2/(\lambda d) \approx 1$, which corresponds to the scattering at small (but not very small) angles;
(iii) geometrical optics, when $R^2/(\lambda d) \gg 1$ and we recover the light-ray picture.

Keeping λ and R fixed, one can see that when the screen is close to the obstacle or aperture (i.e., at small d) we have geometrical optics. As we move the screen further away from the

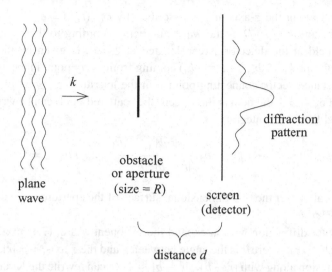

Fig. 7.1. The diffraction pattern for the scattering of light.

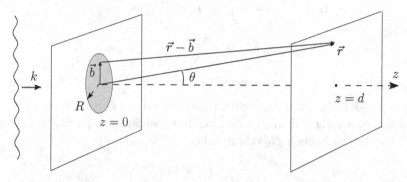

Fig. 7.2. The diffraction of a plane wave with momentum (wave vector) k by a circular aperture of radius R in a plane screen at $z = 0$. We observe the diffraction pattern on another plane screen at $z = d$.

target we go through a region of Fresnel diffraction, and, far away from the target, we enter the region of Fraunhofer diffraction. As our goal here is to build an optical analogy to the high energy scattering of particles, when the detector is far from the target (compared with the particle size and wavelength), we see that we need to study Fraunhofer diffraction.

To be more specific, consider diffraction on a circular aperture in an infinite plane screen, with the detector also an infinite plane screen, as shown in Fig. 7.2. We are interested in the shape of the diffraction pattern on the screen. Our calculation is going to be rather brief, as we assume that the reader is familiar with the theory of diffraction in optics. For a more detailed theoretical discussion of the diffraction of light we refer the reader to the book of Jackson (1998) (Sections 10.5–10.11). For simplicity we also imagine that instead of electromagnetic light we have some massless scalar field $\phi(t, \vec{r})$. The plane wave is incident along the z-axis and is described by $\phi_{inc}(t, \vec{r}) = e^{-i\omega t + ikz}\phi_0$, where ω is the frequency and $k = 2\pi/\lambda$ is the wave number. According to the Huygens–Fresnel principle, the field at the detector screen located at $z = d$, is given by the sum of the spherical waves $\exp(ik|\vec{r} - \vec{b}|)/(4\pi|\vec{r} - \vec{b}|)$ coming from every point in the aperture. Here $\vec{r} - \vec{b}$ is the distance vector connecting point $\vec{b}$ in the aperture and point $\vec{r}$ on the screen (see Fig. 7.2). For $d \gg \lambda$ the field at the screen (the scattered field) can be written in terms of a generalized Kirchhoff integral as

$$\phi_{sc}(\vec{r}) = \frac{k}{2\pi i} \int_{\text{aperture}} d^2 b_\perp \frac{e^{ik|\vec{r} - \vec{b}|}}{|\vec{r} - \vec{b}|} \frac{(\vec{r} - \vec{b})_z}{|\vec{r} - \vec{b}|} \phi_{inc}\left(\vec{b}_\perp, z = 0\right), \qquad (7.1)$$

where the integral is over the two-dimensional surface of the aperture located at $z = 0$ and $\phi_{sc}(t, \vec{r}) = e^{-i\omega t}\phi_{sc}(\vec{r})$.

For Fraunhofer diffraction we can expand the exponent of Eq. (7.1) to give $|\vec{r} - \vec{b}| = r - b\sin\theta + O(b^2/r)$, where θ is the angle between $\vec{r}$ and the z-axis, $r = |\vec{r}|$, and $b = |\vec{b}|$. Using this expansion along with $(\vec{r} - \vec{b})_z/|\vec{r} - \vec{b}| \approx 1$ we can rewrite the Kirchhoff integral of Eq. (7.1) in the form

$$\phi_{sc}(\vec{r}) = \frac{k}{2\pi i} \frac{e^{ikr}}{r} \int_{\text{aperture}} d^2 b_\perp e^{-ikb\sin\theta}\phi_0$$

$$= \frac{k}{2\pi i} \frac{e^{ikr}}{r} \int_{\text{aperture}} d^2 b_\perp e^{-i\vec{q}\cdot\vec{b}}\phi_0 = \frac{e^{ikr}}{r} f(\vec{q}), \qquad (7.2)$$

where $\vec{q} = \vec{k}' - \vec{k}$ is the recoil momentum (wave vector) and $\vec{k}'$ is the wave vector of the scattered spherical wave with magnitude equal to k and direction parallel to $\vec{r}$. Note that $\vec{b} = (\vec{b}_\perp, z = 0)$. The function $f(\vec{q})$ is defined as

$$f(\vec{q}) = \frac{k}{2\pi i} \int_{\text{aperture}} d^2 b_\perp e^{-i\vec{q}\cdot\vec{b}}\phi_0 \qquad (7.3)$$

and corresponds to the scattering amplitude.

The (time-averaged) energy densities in the incident and scattered waves are given by

$$I_{inc} = \frac{\omega^2|\phi_{inc}|^2}{2} = \frac{\omega^2\phi_0^2}{2} \quad \text{and} \quad I_{sc} = \frac{\omega^2|\phi_{sc}|^2}{2} = \frac{\omega^2\phi_0^2}{2} \frac{|f(\vec{q})|^2}{r^2}. \qquad (7.4)$$

Defining the differential scattering cross section as the ratio of the outgoing energy in an infinitesimal solid angle $d\Omega$ and the flux of energy in the incoming wave we obtain

$$d\sigma = \frac{I_{sc}r^2 d\Omega}{I_{inc}} = |f(\vec{q})|^2 d\Omega. \qquad (7.5)$$

For the circular aperture of radius R in Fig. 7.2 we have

$$f(\vec{q}) = \frac{k}{2\pi i} \int d^2 b_\perp e^{-i\vec{q}\cdot\vec{b}} \theta(R - b) = -ik \int_0^R db\, b J_0(bq) = -i\frac{kR}{q} J_1(qR), \qquad (7.6)$$

where $q = |\vec{q}|$ and $b = |\vec{b}_\perp|$. Using this result in Eq. (7.5) and writing $d\Omega = 2\pi d \cos\theta$ with $\sin\theta \approx \theta \approx q/k$ leads to the cross section

$$\frac{d\sigma}{dt} = \pi R^2 \frac{J_1^2\left(\sqrt{|t|}R\right)}{|t|}, \tag{7.7}$$

where we have introduced the Mandelstam variable $t = -\vec{q}^{\,2}$ in direct analogy with the scattering of particles.

From Eq. (7.7) one can see that the differential cross section has a series of minima (zeroes) at $R\sqrt{|t|} = Rq = Rk\sin\theta = x_{1n}$, where x_{1n} are the zeroes of $J_1(z)$, and a series of maxima between the minima. These minima and maxima give the diffraction pattern shown in Fig. 7.1. In t-space the positions of the minima and maxima are determined solely by the target size R.

Diffraction in the context of optics, as discussed here, demonstrates that the diffraction is a direct consequence of the wave nature of light and has the characteristic structure of a differential cross section $d\sigma/dt$ as a function of t, with minima and maxima whose positions depend only on the inverse size of the target.

Certainly this introduction to the subject of light diffraction is rather short, and the analogy between optics and particle scattering is much richer and more instructive than has been demonstrated here. For the interested reader we recommend the book of Barone and Predazzi (2002), which presents the fascinating history of diffraction in optics along with discussions of optical diffraction in a framework that corresponds to particle scattering.

7.1.2 Elastic scattering and inelastic diffraction

Now we return to high energy scattering in QCD. Let us begin by considering the scattering of a projectile such as a color dipole on a nuclear target. The elastic, inelastic, and total scattering cross sections for the process can be found using Eqs. (3.119) if one knows the S-matrix for the process in impact parameter space. Equation (3.119b) can be rewritten as

$$\frac{d\sigma_{el}}{d^2b} = \left|1 - S(s, \vec{b}_\perp)\right|^2 = \left|T(s, \vec{b}_\perp)\right|^2, \tag{7.8}$$

where T is the T-matrix. Going into momentum space we obtain

$$T(s, \vec{b}_\perp) = \int \frac{d^2q}{(2\pi)^2} e^{i\vec{q}_\perp \cdot \vec{b}_\perp} \tilde{T}(s, \vec{q}_\perp) \tag{7.9}$$

and, noticing that $t = -q_\perp^2$, one can readily derive that

$$\frac{d\sigma_{el}}{dt} = \frac{1}{4\pi}\left|\tilde{T}(s, \vec{q}_\perp)\right|^2 = \frac{1}{4\pi}\left|\int d^2b\, e^{-i\vec{q}_\perp \cdot \vec{b}_\perp} T(s, \vec{b}_\perp)\right|^2. \tag{7.10}$$

Now consider the scattering on a target that is circular with radius R in the impact parameter plane. Moreover, assume that the black-disk limit has been reached for all impact parameters inside the target, i.e., for all $b_\perp < R$. By analogy with the dipole scattering studied earlier, we see that in this limit $T(s, \vec{b}_\perp) = iN(s, \vec{b}_\perp) = i\theta(R - b_\perp)$, where N is

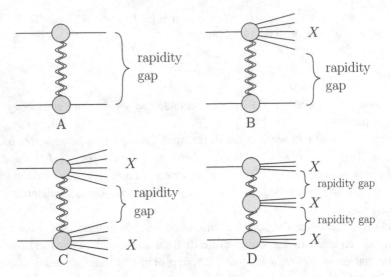

Fig. 7.3. The main types of diffraction. The double wavy lines denote pomeron exchange.

the generalization of the imaginary part of the dipole forward scattering amplitude to the case of an arbitrary projectile. Using this in Eq. (7.10) we obtain

$$\frac{d\sigma_{el}}{dt} = \pi R^2 \frac{J_1^2 \left(\sqrt{|t|} R\right)}{|t|},\tag{7.11}$$

in complete agreement with Eq. (7.7). We see that elastic scattering on a black disk in high energy physics is mathematically identical to the diffraction of light.

Indeed, in arriving at Eq. (7.7) we considered scattering on an aperture while here we have analyzed scattering on a disk, which is a different object but of a complementary shape (that is, the disk and the plane with the aperture together form a complete plane). However, Babinet's principle in optics tells us that diffraction patterns for recoil momentum $q \neq 0$ are identical for the obstacle and its complement: hence the optical diffraction pattern (7.7) is the same for a black disk of radius R, as can be verified by an explicit calculation.

The term "diffraction" in high energy scattering and the first theoretical ideas on the subject were introduced in the early 1950s by Landau, Pomeranchuk, Feinberg, Ahiezer, Ter-Mikaelyan, and Sitenko (see the review by Feinberg and Pomeranchuk (1956)). These ideas were crystallized and put into an elegant theoretical framework by Good and Walker (1960).

At high energy the term diffraction covers a much broader range than the elastic processes considered so far in this chapter. An event is considered diffractive if it contains a rapidity gap. A rapidity gap is an interval in rapidity (usually at least a few units wide) over which no particles are produced. Clearly, in elastic collisions no new particles are produced in the rapidity interval between the target and the projectile; in this case the rapidity gap covers the whole interval in rapidity.

Scattering amplitudes for the main types of diffractive event are shown in Fig. 7.3 for hadron–hadron scattering. The elastic process we have discussed above is presented

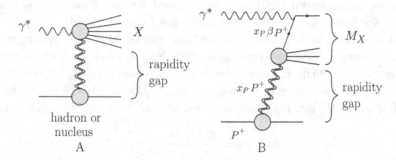

Fig. 7.4. Single diffractive dissociation in DIS: (A) a general representation of the process and (B) an explanation of the kinematic variables. The double wavy lines represent pomeron exchange, while the single wavy line denotes a virtual photon.

in panel A. The double wavy line denotes the exchange of a color-singlet object (the pomeron). Panels B through D represent various cases of inelastic diffraction, where one or both hadrons break up in the collision, that still lead to a rapidity gap or gaps. Panel B shows the process of single diffractive dissociation, where one hadron breaks up into several other hadrons denoted by X while the other stays intact. A single rapidity gap is generated in this process. One also distinguishes low- and high-mass diffraction, depending on whether the invariant mass M_X of the produced particles X is small or large. Processes with a single rapidity gap in which both hadrons dissociate, as shown in panel C of Fig. 7.3, are referred to as double diffractive dissociation. Finally, one may have more than one rapidity gap in the event: an example of a process with two rapidity gaps is shown in panel D, where hadrons are produced at mid-rapidity and are flanked by a rapidity gap on either side. Such processes are called central diffraction. If a single particle is produced at mid-rapidity then the process is referred to as central exclusive diffraction.

While diffraction in hadronic scattering represents an interesting and often challenging problem in itself, here we will concentrate on diffractive dissociation in DIS.

7.2 Diffractive dissociation in DIS

Consider single diffractive dissociation in DIS. This is a process in which a virtual photon interacts with the target, producing a number of hadrons and jets in the final state (denoted by X) but leaving the target intact and generating a rapidity gap. The process is illustrated in Fig. 7.4A. The particles X with net invariant mass M_X produced as a result of the target's breakup do not fill the whole rapidity interval; they leave a rapidity gap between the target and the "slowest" produced particle. This rapidity gap is of order $\Delta Y_{gap} = \ln[(\hat{s} + Q^2)/(M_X^2 + Q^2)]$, where $\hat{s}$ is the center-of-mass energy squared of the virtual photon–target collision (see Eq. (2.5)). (The net rapidity interval for $\hat{s} \gg Q^2$ is $Y = \ln(\hat{s}/Q^2)$, while the produced hadrons fill in the rapidity range $\Delta Y_{filled} = Y - \Delta Y_{gap} = \ln[(M_X^2 + Q^2)/Q^2]$.) No particles are produced in the rapidity gap; the existence of such a rapidity gap is indeed the characteristic signature of diffractive processes.

Diffractive dissociation in DIS is usually described in terms of the kinematic variables x_P and β, which originate in pomeron phenomenology and are explained by Fig. 7.4B. Treating the pomeron as an effective "parton" one may describe it as carrying a fraction x_P of the incoming proton's light cone momentum. Neglecting the mass of the proton and working in the IMF with the proton moving in the light cone plus direction, we see that the pomeron will carry momentum $x_P P^\mu$, with P^μ given by Eq. (2.26) with $m = 0$. Requiring that $(x_P P + q)^2 = M_X^2$ (see Fig. 7.4B) and remembering that $\hat{s} = (P + q)^2$, we obtain

$$x_P = \frac{Q^2 + M_X^2}{Q^2 + \hat{s}}. \tag{7.12}$$

The variable β is defined as the fraction of the pomeron's light cone momentum carried by the quark that is struck by the virtual photon; thus (see Fig. 7.4B again)

$$\beta = \frac{x_{Bj}}{x_P} = \frac{Q^2}{Q^2 + M_X^2}. \tag{7.13}$$

Below we will distinguish low- and high-mass diffraction in DIS. When the invariant mass of the produced hadrons, M_X, is low, $M_X \ll Q$, which is usually the case when few hadrons are produced, we see that $\beta \approx 1$ and $x_P \approx x_{Bj}$, so that $\Delta Y_{gap} = \ln 1/x_P \approx \ln 1/x_{Bj} = Y$; the rapidity gap covers much of the net rapidity interval, as expected. When the mass M_X is large, $M_X \gg Q$, we have $\beta \ll 1$ and $x_P \approx M_X^2/\hat{s} \approx e^{-\Delta Y_{gap}}$; the rapidity gap may still be large but the rapidity interval filled by the produced hadrons is large as well.

A particularly interesting aspect of diffraction is that the typical momentum transfer appears to be of order $|t| \sim 1/R^2$, as follows from Eq. (7.11). The momentum is of order the inverse size of the target, which is in the nonperturbative QCD region (This is why t was neglected in the derivation of Eqs. (7.12) and (7.13).). It would seem, on the one hand, that diffraction is a non-perturbative process and cannot be studied within the perturbative QCD framework. On the other hand, a main postulate of the saturation or CGC approach is that saturation effects generate the saturation scale $Q_s(Y)$; this screens the IR physics, making the cross sections and other observables perturbative. On top of that, in DIS one has a hard scale Q^2, which may also be perturbatively large. We see that diffraction becomes a cross-check of the saturation approach, the main question being whether saturation physics makes diffraction a perturbative process.

7.2.1 Low-mass diffraction

To describe low-mass diffraction in DIS at high energies it is natural to start with the dipole picture of DIS presented in Sec. 4.1. Again we have a separation of scales: a virtual photon will decay into a $q\bar{q}$ pair long before hitting the target and the $q\bar{q}$ dipole interacts with the target in due course. The dipole–target interaction can be either inelastic or elastic. In Chapter 4 we showed that the total DIS cross section can be written as (see Eqs. (4.6) and (4.24))

$$\sigma_{tot}^{\gamma^* A} = \int \frac{d^2 x_\perp}{2\pi} d^2 b_\perp \int_0^1 \frac{dz}{z(1-z)} \, |\Psi^{\gamma^* \to q\bar{q}}(\vec{x}_\perp, z)|^2 \, N(\vec{x}_\perp, \vec{b}_\perp, Y). \tag{7.14}$$

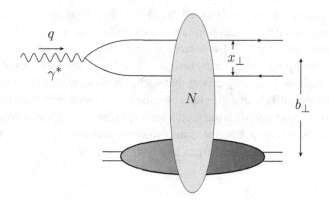

Fig. 7.5. The quasi-elastic DIS amplitude that leads to Eq. (7.17).

Omitting the part of the expression related to the virtual photon wave function, we obtain for the dipole–nucleus total cross section

$$\sigma_{tot}^{q\bar{q}A}(\vec{x}_\perp, Y) = 2 \int d^2 b \, N(\vec{x}_\perp, \vec{b}_\perp, Y) = 2 \int d^2 b \left[1 - S(\vec{x}_\perp, \vec{b}_\perp, Y) \right]. \quad (7.15)$$

We see that this is exactly Eq. (3.119a) (as S is real). By analogy with Eq. (3.119b) the elastic dipole–nucleus cross section is then

$$\sigma_{el}^{q\bar{q}A}(\vec{x}_\perp, Y) = \int d^2 b \left[1 - S(\vec{x}_\perp, \vec{b}_\perp, Y) \right]^2 = \int d^2 b \, N^2(\vec{x}_\perp, \vec{b}_\perp, Y). \quad (7.16)$$

The corresponding quasi-elastic DIS cross section is obtained by convoluting (7.16) with the square of the virtual photon's wave function:[1]

$$\sigma_{el}^{\gamma^* A} = \int \frac{d^2 x_\perp}{4\pi} d^2 b_\perp \int_0^1 \frac{dz}{z(1-z)} |\Psi^{\gamma^* \to q\bar{q}}(\vec{x}_\perp, z)|^2 N^2(\vec{x}_\perp, \vec{b}_\perp, Y). \quad (7.17)$$

Equation (7.17) was derived in the quasi-classical GGM/MV approximation, that is, with N given by Eq. (4.51), by Buchmuller, Gehrmann, and Hebecker (1997) (see also Buchmuller, McDermott, and Hebecker (1999)) and by Kovchegov and McLerran (1999). However, one can see that our derivation assumes only the decomposition of the interaction into the virtual photon's wave function and the amplitude N, with the latter independent of the light cone momentum fraction z. This assumption is also true in the LLA: hence the elastic DIS cross section (7.17) is also valid in the case when the LLA quantum evolution is included. Therefore Eq. (7.17) is true whether N is found from the BK equation (4.138) in the large-N_c limit or from the JIMWLK equation for the dipole S-matrix (5.98) when the large-N_c limit is relaxed.

The quasi-elastic DIS process corresponding to Eq. (7.17) is illustrated in Fig. 7.5. Here the virtual photon splits into a $q\bar{q}$ pair, after which the pair interacts with the target nucleus

---

[1] Note that the high energy $\gamma^* A$ cross section at order α_{EM} cannot be elastic, since we do not have a photon in the final state (see Fig. 7.5): we will refer to this process as quasi-elastic, to distinguish it from high-mass diffraction.

elastically, as denoted by the oval labeled N. In the LLA approximation this means that before the interaction the dipole develops a gluon cascade, as in, say, Fig. 4.23; this cascade interacts with the target by GGM coulomb–gluon exchange and after the interaction is reabsorbed into the dipole, so that in the final state one finds only the original $q\bar{q}$ pair along with the intact nucleus. One may expect that such a process would be very rare, since it would appear highly unlikely that the gluon cascade would recombine back into the original dipole. An amazing property of elastic scattering and diffraction is that it demonstrates that such an intuition is incorrect when it comes to cross sections. Indeed, in the saturation regime, when the black-disk limit is reached and $N = 1$, we see from Eqs. (7.15) and (7.16) that

$$\frac{\sigma_{el}^{q\bar{q}A}}{\sigma_{tot}^{q\bar{q}A}} = \frac{\int d^2 b N^2}{2 \int d^2 b N} \longrightarrow \frac{1}{2}. \tag{7.18}$$

Elastic dipole–nucleus scattering constitutes half the total cross section in very high energy collisions!

Note that in Fig. 7.5 we have two quarks in the final state: they are likely to fragment into one or several hadrons, leading to a low-M_X diffractive final state.

Using Eqs. (7.14) and Eq. (7.17) we can compare the main properties of the total and diffractive (quasi-elastic) DIS cross sections, looking for their common and different features. The large-Q^2 behavior is particularly instructive. Analyzing the virtual-photon wave functions squared in Eqs. (4.18) and (4.21) we see that, since the modified Bessel functions K_1 and K_0 fall off exponentially at large values of the argument, at large Q^2 the main contribution to the z-integral in both Eqs. (7.14) and (7.17) comes from the region $a_f x_\perp \leq 1$ with a_f given by Eq. (4.17). Neglecting for simplicity the quark masses m_f we see that this implies $\sqrt{z(1-z)} \leq 1/(Qx_\perp)$, and, since $z(1-z) < 1/4$, we either have $z \ll 1$ or $1 - z \ll 1$ if $Qx_\perp \gg 2$. Either the quark or the antiquark carries most of the virtual photon's light cone momentum. This configuration is known as the aligned-jet configuration and is the basis for the aligned-jet model (Bjorken and Kogut 1973, Nikolaev and Zakharov 1975, Frankfurt and Strikman 1988) since, in the case when the produced quark and antiquark in, say, Fig. 7.5, fragment into jets, one jet is aligned with the momentum of the virtual photon.

Concentrating on the $z \ll 1$ region (and multiplying the expression by 2 to account for the $1 - z \ll 1$ region), we can integrate over z explicitly with the help of Eqs. (4.18) and (4.21) to obtain (note that N is z-independent)

$$\int\limits_0^1 \frac{dz}{z(1-z)} |\Psi_T^{\gamma^* \to q\bar{q}}(\vec{x}_\perp, z)|^2 \approx 4N_c \sum_f \frac{\alpha_{EM} Z_f^2}{\pi} Q^2 \int\limits_0^\infty dz\, z \left[K_1(x_\perp Q \sqrt{z}) \right]^2$$

$$= \frac{16 N_c}{3} \sum_f \frac{\alpha_{EM} Z_f^2}{\pi} \frac{1}{Q^2 x_\perp^4} \tag{7.19}$$

and

$$\int_0^1 \frac{dz}{z(1-z)} |\Psi_L^{\gamma^* \to q\bar{q}}(\vec{x}_\perp, z)|^2 \approx 16 N_c \sum_f \frac{\alpha_{EM} Z_f^2}{\pi} Q^2 \int_0^\infty dz \, z^2 \left[K_0(x_\perp Q \sqrt{z}) \right]^2$$

$$= \frac{512}{15} N_c \sum_f \frac{\alpha_{EM} Z_f^2}{\pi} \frac{1}{Q^4 x_\perp^6}. \qquad (7.20)$$

One can show that for $N \sim x_\perp^2$ at small $x_\perp$, which is true in the GGM (4.51) and DLA (4.150) approximations, the longitudinal contribution (7.20) is more suppressed at high Q^2 than the transverse contribution, both for $\sigma_{tot}^{\gamma^* A}$ and $\sigma_{el}^{\gamma^* A}$; therefore, we can neglect it in the large-Q^2 limit.

Noting that $Q x_\perp \gg 2$ and using Eq. (7.19) in Eqs. (7.14) and (7.17) yields, after integration over the angles of $\vec{x}_\perp$,

$$\sigma_{tot}^{\gamma^* A} \approx \frac{4 N_c \alpha_{EM}}{3\pi Q^2} \sum_f Z_f^2 \int_{4/Q^2}^\infty \frac{dx_\perp^2}{x_\perp^4} \int d^2 b_\perp \, 2N(x_\perp, \vec{b}_\perp, Y), \qquad (7.21a)$$

$$\sigma_{el}^{\gamma^* A} \approx \frac{4 N_c \alpha_{EM}}{3\pi Q^2} \sum_f Z_f^2 \int_{4/Q^2}^\infty \frac{dx_\perp^2}{x_\perp^4} \int d^2 b_\perp \, N^2(x_\perp, \vec{b}_\perp, Y), \qquad (7.21b)$$

where we have assumed that N is independent of the direction of $\vec{x}_\perp$.

To evaluate the integrals in Eqs. (7.21) we use the explicit leading-twist expression for the dipole amplitude N from Eq. (4.32), which is valid for $x_\perp \ll 1/Q_s$, obtaining

$$\sigma_{tot}^{\gamma^* A} \approx \frac{4 \alpha_s \alpha_{EM} \pi}{3 N_c Q^2} \sum_f Z_f^2 \int_{4/Q^2}^{1/Q_s^2} \frac{dx_\perp^2}{x_\perp^2} x G_A \left(x, \frac{1}{x_\perp^2} \right), \qquad (7.22a)$$

$$\sigma_{el}^{\gamma^* A} \approx \frac{\alpha_s^2 \alpha_{EM} \pi^3}{3 N_c^3 Q^2} \sum_f Z_f^2 \int_{4/Q^2}^{1/Q_s^2} dx_\perp^2 \left[x G_A \left(x, \frac{1}{x_\perp^2} \right) \right]^2, \qquad (7.22b)$$

where we have used the fact that $\int d^2 b T(\vec{b}_\perp) = A$ and, in the spirit of the GGM approximation, replaced $A x G_N$ with the nuclear gluon distribution $x G_A$.

Assuming that $x G_A$ is a slowly varying function of $x_\perp$, we can see from Eqs. (7.22) that the total cross section depends on the upper limit of the $x_\perp$-integral logarithmically, while the quasi-elastic cross section depends on it quadratically. In the absence of saturation effects the integrals would have to be cut off by the nonperturbative physics in the IR, that is, we should replace Q_s by Λ_{QCD}: in such a case both cross sections would be nonperturbative, the elastic one being more so than the total. We see that saturation effects make both cross sections perturbative, even for $t = 0$ in the diffractive case, yet again justifying the perturbative QCD approximation.

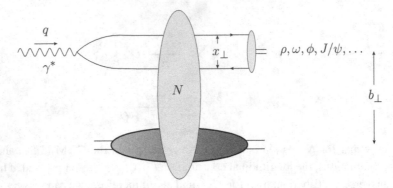

Fig. 7.6. Exclusive vector meson production amplitude in DIS.

We can make another crude approximation leading to a physical insight if we notice that in the GGM model the saturation scale squared Q_s^2 is proportional to the gluon distribution function, as follows from Eq. (4.47). Assuming that xG_A is a slowly varying function of $x_\perp$, we integrate over $x_\perp$ in Eqs. (7.22) to obtain[2]

$$\sigma_{tot}^{\gamma^* A} \propto xG_A(x, Q_s^2) \ln \frac{Q^2}{4Q_s^2}, \tag{7.23a}$$

$$\sigma_{el}^{\gamma^* A} \propto \frac{1}{Q_s^2} \left[xG_A(x, Q_s^2) \right]^2 \propto xG_A(x, Q_s^2), \tag{7.23b}$$

so that, neglecting logarithms, which are outside the precision of our approximation, the ratio $\sigma_{el}^{\gamma^* A}/\sigma_{tot}^{\gamma^* A}$ is approximately independent of Bjorken x (Kovchegov and McLerran 1999). This appears to be in approximate agreement with the data collected at HERA (see Fig. 9.2 along with Abramowicz and Dainton (1996), H1 collaboration (1997), and ZEUS collaboration (1999)).

Another important low-mass diffractive DIS process to consider is exclusive vector meson production. It is related to the quasi-elastic scattering of Eq. (7.17): in exclusive vector meson production the outgoing quark and antiquark recombine into a single vector meson. One still has a rapidity gap with the target intact in the final state. This is illustrated in Fig. 7.6. More precisely the reaction is

$$\gamma^* + p \to V(\rho, \omega, \phi, J/\Psi, \ldots) + p \tag{7.24}$$

for DIS on a proton; V denotes a vector meson.

The cross section of this reaction can be easily calculated using LCPT; one simply needs to find the overlap of the $q\bar{q}$ state with the (complex conjugate) light cone wave function of the vector meson Ψ^V. To see this, let us calculate the diagram in Fig. 7.6 in LCPT. The additional contributions, not present in Eq. (7.17), come from the $q\bar{q} \to V$ process: this is exactly the reverse of what is described by the vector meson's light cone wave function.

---

[2] Note that, while in Eq. (4.45) xG_N does not depend on x, Eq. (4.32) is still valid in the DLA region even after small-x evolution is included (cf. Eq. (4.150)), with xG depending on x.

The cross section for exclusive vector meson production is equal to (see Ryskin (1993) for the J/Ψ production case and Brodsky *et al.* (1994) for the general case of vector meson production)

$$\sigma^{\gamma^*+A\to V+A}$$

$$= \int d^2b_\perp \left| \int \frac{d^2x_\perp}{4\pi} \int\limits_0^1 \frac{dz}{z(1-z)} \, \Psi^{\gamma^*\to q\bar{q}}(\vec{x}_\perp, z) \, N(\vec{x}_\perp, \vec{b}_\perp, Y)\Psi^V(\vec{x}_\perp, z)^* \right|^2 . \quad (7.25)$$

Note that now the dipole transverse sizes in the amplitude and in the complex conjugate amplitude are different, and are integrated over separately.

In general the wave function Ψ^V is nonperturbative. In practice the main contribution of the $x_\perp$-integral in Eq. (7.25) comes from short distances, owing to the large values of Q^2 and Q_s^2 there; thus we can approximate $\Psi^V(\vec{x}_\perp, z)$ by $\Psi^V(0, z)$ (the wave function at the origin). For vector mesons $\Psi^V(0, z)$ is known from their decay into an electron–positron pair, $V \to e^+e^-$.

One can also show that the typical transverse distances in this process are of order $1/Q_s$ and so the process is, therefore, perturbative. However, the diffractive minima and maxima in the corresponding t-distribution are determined by the size of the target. To see this, we write (analogously to the transition from Eq. (7.8) to Eq. (7.10))

$$\frac{d\sigma^{\gamma^*+A\to V+A}}{dt} = \frac{1}{4\pi} \left| \int d^2b \, e^{-i\vec{q}_\perp\cdot\vec{b}_\perp} T^{q\bar{q}A}(\hat{s}, \vec{b}_\perp) \right|^2 , \quad (7.26)$$

where now the T-matrix element is given by

$$T^{q\bar{q}A}(\hat{s}, \vec{b}_\perp) = i \int \frac{d^2x_\perp}{4\pi} \int\limits_0^1 \frac{dz}{z(1-z)} \, \Psi^{\gamma^*\to q\bar{q}}(\vec{x}_\perp, z) \, N(\vec{x}_\perp, \vec{b}_\perp, Y)\Psi^V(\vec{x}_\perp, z)^*, \quad (7.27)$$

with $Y = \ln \hat{s}x_\perp^2$. We see that the $\vec{b}_\perp$-dependence of the T-matrix in Eq. (7.27) is given by that of N, which in turn is largely determined by the geometry of the target. Hence the positions of the diffractive maxima and minima of vector meson production are proportional to $1/R$, with R the nuclear radius. In addition, the size of the target (or, more precisely, the size of the interaction region) increases with energy; however, this increase depends mostly on nonperturbative corrections as we will discuss later.

As suggested by Munier, Stasto, and Mueller (2001), we can use Eq. (7.26) to extract the $\vec{b}_\perp$-dependence of the T-matrix (and, consequently, the S-matrix) from the experimental data, since we can invert it to write

$$T^{q\bar{q}A}(\hat{s}, \vec{b}_\perp) = \frac{i}{2\pi^{3/2}} \int d^2q \, e^{i\vec{q}_\perp\cdot\vec{b}_\perp} \sqrt{\frac{d\sigma^{\gamma^*+A\to V+A}}{dt}}. \quad (7.28)$$

This relation relies on the assumption that the T-matrix is purely imaginary at high energies (and hence the S-matrix is real). This assumption is certainly correct for small t, in the LLA, but at large values of t the real part of the T-matrix may not be small. In spite of this uncertainty, Eq. (7.28) shows that we can observe the T-matrix at fixed b, which, in turn,

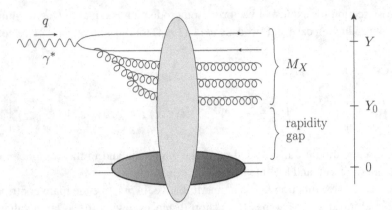

Fig. 7.7. An example of a diagram contributing to high-mass diffraction in DIS. The vertical axis on the right measures the rapidities of the particles.

can be calculated in the saturation approach. Note that the nonperturbative light cone wave function Ψ^V enters Eq. (7.27). Above we suggested a rough approximation in which the vector meson wave function was replaced by its value at the origin. For mesons consisting of heavy quarks, such as J/Ψ, or even for the Υ-meson, consisting of a $b\bar{b}$-pair, one may hope to improve on that approximation by using perturbative QCD to calculate the wave functions.

7.2.2 Nonlinear evolution equation for high-mass diffraction

So far we have considered only low-mass diffractive processes, when the gluon cascade developed by the dipole before the interaction with the nucleus is reabsorbed back into the dipole after the interaction, so that we only have the original $q\bar{q}$ dipole in the final state (along with the target), as shown in Fig. 7.5. The quark and the antiquark in the pair cannot be far from each other in rapidity, since the quark emission at small x, unlike that for gluons, is not enhanced by a logarithm of x: this is why we neglected the quark contribution to small-x evolution in the LLA. Therefore, since the invariant mass of the produced particles is related to the rapidity interval that they fill by $M_X^2 = Q^2(e^{\Delta Y_{filled}} - 1)$, we see that elastic $q\bar{q}$ pair production mainly leads to low-mass diffraction. If one wants to produce a high-mass state with a rapidity gap, one has to augment the picture of Fig. 7.5 by allowing some "fast" gluons to survive in the final state, as shown in Fig. 7.7. The process is now more complicated than that in Fig. 7.5. While the incoming dipole still develops a dipole cascade, not all the gluons in the cascade recombine back by the time the system reaches the final state: only gluons with rapidities between 0 and Y_0 recombine back, so that a rapidity gap $\Delta Y_{gap} = Y_0 - 0 = Y_0$ is formed and the target nucleus remains intact, as illustrated in Fig. 7.7. Gluons with rapidities $y > Y_0$ do not need to recombine back and thus can become "produced" gluons, as shown in Fig. 7.7. In fact some gluons with

$y > Y_0$ may even be emitted after the GGM interaction with the target, as we will see shortly.

First let us define the observable we would like to calculate. Considering dipole–nucleus scattering, we denote by $N^D(\vec{x}_\perp, \vec{b}_\perp, Y, Y_0)$ the cross section per unit impact parameter for diffractive dissociation with a single rapidity gap stretching from 0 (the target) to some rapidity greater than or equal to Y_0. The corresponding single diffractive cross section with rapidity gap greater than or equal to Y_0 in the dipole–nucleus scattering is

$$\sigma_{diff}^{q\bar{q}A} = \int d^2b \, N^D(\vec{x}_\perp, \vec{b}_\perp, Y, Y_0). \tag{7.29}$$

If we want to find the diffraction cross section for a given fixed rapidity gap Y_0 we simply have to differentiate N^D with respect to Y_0, obtaining

$$M_X^2 \frac{d\sigma_{diff}^{q\bar{q}A}}{dM_X^2} = -\int d^2b \frac{\partial N^D(\vec{x}_\perp, \vec{b}_\perp, Y, Y_0)}{\partial Y_0}, \tag{7.30}$$

where the minus sign is due to the fact that $Y_0 = \Delta Y_{gap} \approx \ln \hat{s}/M_X^2$ for $\hat{s}, M_X^2 \gg Q^2$, so that $dY_0 = -dM_X^2/M_X^2$ for fixed $\hat{s}$.

By analogy with Eq. (7.17) we can use Eq. (7.29) to write down the following expression for the single diffractive cross section in DIS:

$$\sigma_{diff}^{\gamma^* A} = \int \frac{d^2x_\perp}{4\pi} d^2b_\perp \int_0^1 \frac{dz}{z(1-z)} \, |\Psi^{\gamma^* \to q\bar{q}}(\vec{x}_\perp, z)|^2 N^D(\vec{x}_\perp, \vec{b}_\perp, Y, Y_0). \tag{7.31}$$

The differential cross section is

$$M_X^2 \frac{d\sigma_{diff}^{\gamma^* A}}{dM_X^2} = -\int \frac{d^2x_\perp}{4\pi} d^2b_\perp \int_0^1 \frac{dz}{z(1-z)} \, |\Psi^{\gamma^* \to q\bar{q}}(\vec{x}_\perp, z)|^2 \frac{\partial N^D(\vec{x}_\perp, \vec{b}_\perp, Y, Y_0)}{\partial Y_0}. \tag{7.32}$$

Sometimes we will use the notation $N^D(\vec{x}_{1\perp}, \vec{x}_{0\perp}, Y, Y_0)$ for the diffractive cross section N_D of a dipole with the quark at $\vec{x}_{1\perp}$ and the antiquark at $\vec{x}_{0\perp}$; this is similar to the notation used in the forward dipole amplitude (see e.g. Eq. (4.141)).

We will assume that Y, Y_0, and $Y - Y_0$ are all large, so that $\alpha_s Y \sim 1$, $\alpha_s Y_0 \sim 1$ and $\alpha_s(Y - Y_0) \sim 1$ are all important and one has to devise a small-x resummation procedure to calculate N^D in the LLA by resumming all these parameters. We want to derive a (nonlinear) evolution equation for N^D (Kovchegov and Levin 2000). We will work in the frame where the nucleus is moving in the light cone plus direction, while the incoming dipole is moving along the minus axis. We will also employ the $A^- = 0$ light cone gauge of the (dipole) projectile.

We first observe that when $Y_0 = Y$ the rapidity gap becomes equal to the whole rapidity interval, and we have returned to the case of elastic dipole–nucleus scattering considered in Sec. 7.2.1. The elastic dipole–nucleus cross section is given in Eq. (7.16), which we can

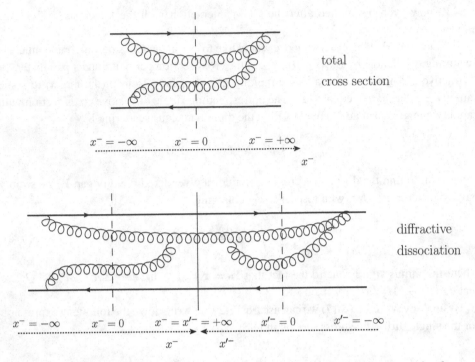

total

cross section

$x^- = -\infty \qquad x^- = 0 \qquad x^- = +\infty$

x^-

diffractive

dissociation

$x^- = -\infty \qquad x^- = 0 \qquad x^- = x'^- = +\infty \qquad x'^- = 0 \qquad x'^- = -\infty$

$x^- \qquad x'^-$

Fig. 7.8. Small-x evolution of the total cross section (upper panel) and the cross section for diffractive dissociation (lower panel).

use to write

$$N^D(\vec{x}_\perp, \vec{b}_\perp, Y = Y_0, Y_0) = N^2(\vec{x}_\perp, \vec{b}_\perp, Y_0). \qquad (7.33)$$

Again in the LLA, N is given by the solution of the BK/JIMWLK equations. We will use Eq. (7.33) as the initial condition for the evolution equation that we will construct below.

From the point of view of the space–time picture, the process of diffractive production is quite different from the total cross section considered above. In calculating the total cross section we needed to find the forward scattering amplitude. For a dipole moving along the x^--axis, this means that we had to follow its evolution from $x^- = -\infty$ to the time of the GGM-type interactions at $x^- = 0$ and then again from $x^- = 0$ to the final state at $x^- = +\infty$, which, for the forward amplitude, is identical to the initial state. There are two main differences in the diffractive case (and, as we will see later, for the inclusive production cross sections as well). First, the final state at $x^- = +\infty$ is no longer identical to the initial state. Second, we cannot use the optical theorem to find the diffractive cross section: instead we have to square the scattering amplitude. This means that we have to follow the evolution of the dipole and its gluon cascade from $x^- = -\infty$ to $x^- = +\infty$ both in the amplitude and in the complex conjugate amplitude.

We illustrate the differences between the calculations of the total and diffractive cross sections in Fig. 7.8. Using the notation of Fig. 5.8 we denote by a vertical dashed line the

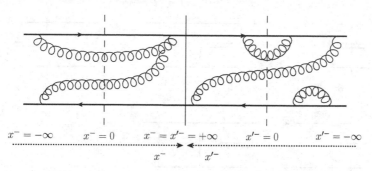

$$x^- = -\infty \qquad x^- = 0 \qquad x^- = x'^- = +\infty \qquad x'^- = 0 \qquad x'^- = -\infty$$

$$x^- \qquad x'^-$$

Fig. 7.9. Diagrammatic representation of the elastic scattering in Eq. (7.33).

GGM interactions with the target (along with the subsequent small-x evolution) for all the particles that cross the dashed line. The interactions with the target (and the subsequent evolution) are instantaneous compared with the lifetimes of the s-channel gluons and quarks. The forward scattering amplitude in the upper panel of Fig. 7.8 is used to obtain the total dipole–nucleus cross section. There is a single light cone time, which varies from $x^- = -\infty$ to $x^- = +\infty$, with the GGM interaction at $x^- = 0$. The lower panel shows the scattering amplitude squared: the time varies from $x^- = -\infty$ to $x^- = +\infty$ in the amplitude and in the complex conjugate amplitude (the light cone time in the latter is denoted by x'^- to distinguish it from the time in the amplitude). The vertical solid line in the lower panel of Fig. 7.8 denotes a cut; gluons may be emitted in the amplitude and then, crossing the cut, be absorbed into the complex conjugate amplitude. Such gluons exist in the final state and so are "produced". (Indeed, they fragment into hadrons in the final state: here we are using perturbative slang, in which the term "produced" implies eventual convolution with the fragmentation functions, according to the standard perturbative QCD prescription.) The gluons could be also emitted and then reabsorbed in the (complex conjugate) amplitude: such gluons are not "produced" and may contribute to the formation of the rapidity gap in which we are interested. Readers familiar with finite-temperature field theory may draw an analogy between the two light cone times x^- and x'^- and the time contour in the Schwinger–Keldysh formalism.

To illustrate further this two-time formalism we show a diagram contributing to the elastic dipole–target scattering given by Eq. (7.33) in Fig. 7.9. No gluon in Fig. 7.9 is "produced" since none crosses the cut. Instead, the evolution and the interaction with the target happen separately and independently in the amplitude and in the complex conjugate amplitude, each giving a factor N generated by the BK/JIMWLK evolution. (The interaction with the target has been assumed to be elastic in all the diagrams in this subsection; we have to use the forward, i.e., elastic dipole amplitude in the GGM approximation from Eq. (4.139) as the initial condition for the evolution for N on each side of the cut.)

An interesting question arises about the connection between the two-time amplitude-squared approach and the forward amplitude in the upper panel of Fig. 7.8. After all, we do not have to use the optical theorem to find the total cross section: we can simply square the sum of all the possible scattering amplitudes. This type of calculation would follow

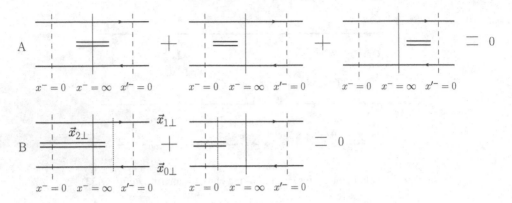

Fig. 7.10. Classes of diagrams with final-state interactions that cancel. The dotted vertical lines denote the intermediate states discussed in the text.

the logic of the lower panel in Fig. 7.8 and would include many more diagrams than the forward scattering amplitude in the upper panel of the same figure. Yet the answer for the total cross section should be the same regardless of the calculational strategy. One would like to see explicitly how a summation over all possible emissions and absorptions in the lower panel of Fig. 7.8 (without any constraints on the final state such as rapidity gaps) would lead to the diagram in the upper panel of Fig. 7.8. Understanding this question would give us a new insight into how the optical theorem works.

The answer to this puzzle is that all the emissions and absorptions for times after the interaction with the target, $x^- > 0$ and $x'^- > 0$, simply cancel. This cancellation of the final-state interactions was first proven by Chen and Mueller (1995) using a diagrammatic approach. We present the canceling diagrams in the large-N_c language of the dipole model in Fig. 7.10 using the notation introduced in Figs. 4.18 and 4.20. The cancellations are of two types. Type-A cancellations, shown in the first line of Fig. 7.10, involve a gluon that is emitted and absorbed in the final state, i.e., at $x^- > 0$, $x'^- > 0$. Type-B cancellations, shown in the second line of Fig. 7.10, involve a gluon that is emitted at $x^- < 0$ but is absorbed either at $x^- > 0$ or at $x'^- > 0$. (Cancellations in diagrams that are the complex conjugates of those of type B take place as well but are not shown, for brevity.)

The proof of the cancellations in Fig. 7.10 can be performed diagrammatically, following the original derivation of Chen and Mueller (1995). This is based on the fact that, for instance, the only difference between the two diagrams of type B is the sign of the LCPT energy denominators of the intermediate states denoted by the dotted vertical lines. When constructing an energy denominator the rules of LCPT require us to subtract the energy of the incoming state. However, since the energy of the whole process is conserved, the energy of the incoming state is equal to the energy of the outgoing state and we can equally well subtract the latter from the energy denominators. Since for $x^- > 0$ the energy of the target does not change any more, as the interactions with the target are over, we only need to consider the s-channel gluons. The intermediate state in the diagram on the left in row B in Fig. 7.10 then brings in a denominator with the energies of the quark and antiquark minus the energy of the same $q\bar{q}$ pair and a gluon in the final state. Since in the LLA the energies of the quark and antiquark change little, we get

$$\frac{1}{p_1^- + p_0^- - p_1^- + p_0^- - k_2^-} = -\frac{1}{k_2^-}, \tag{7.34}$$

where p_1^-, p_0^- are the quark energies while k_2^- is the gluon light cone energy. Similarly, the intermediate state denoted by the dotted line on the right in row B gives

$$\frac{1}{p_1^- + p_0^- + k_2^- - p_1^- + p_0^-} = \frac{1}{k_2^-}, \tag{7.35}$$

exactly equal to Eq. (7.34) in magnitude but opposite in sign. Since, as we have already mentioned, the remainders of the type-B diagrams are identical, we obtain the cancellation illustrated in Fig. 7.10B.

Similar sign changes in the energy denominators apply in the type-A case along with symmetry factors; the result is that the second and the third graphs are equal to minus one-half the first graph. (Note that a complete analysis of the diagrams of type A should include instantaneous terms like those shown in Fig. 4.14.)

One may also use the language of Chapter 5 to argue for the cancellation. Define the fundamental Wilson line over an arbitrary interval $[x_1^-, x_2^-]$ along the x^--axis (see Eq. (5.78)) by

$$V_{\vec{y}_\perp}[x_2^-, x_1^-] = \text{P} \exp \left\{ \frac{ig}{2} \int_{x_1^-}^{x_2^-} dx^- t^a A^{a+}(x^+ = 0, x^-, \vec{y}_\perp) \right\} \tag{7.36}$$

where the gluon field A^+ is either the classical field of the target or an effective field taking into account the small-x evolution corrections. The contribution of the final-state part ($x^- \in [0, +\infty]$) of a dipole–nucleus scattering diagram is then given by

$$V_{\vec{x}_{1\perp}}[+\infty, 0] \otimes V_{\vec{x}_{0\perp}}^\dagger[+\infty, 0], \tag{7.37}$$

where, as usual, the quark is at $\vec{x}_{1\perp}$ and the antiquark is at $\vec{x}_{0\perp}$, as shown in Fig. 7.10B. Squaring the amplitude we get

$$V_{\vec{x}_{1\perp}}[+\infty, 0]V_{\vec{x}_{1\perp}}^\dagger[+\infty, 0] \otimes V_{\vec{x}_{0\perp}}^\dagger[+\infty, 0]V_{\vec{x}_{0\perp}}[+\infty, 0] = \mathbf{1} \otimes \mathbf{1}. \tag{7.38}$$

All the interactions cancel and we end up with a noninteraction unit contribution. This cancellation is akin to the unitarity argument presented in Sec. 2.4.2, which was also used for construction of the dipole wave function in Sec. 4.3.

The cancellation of interactions at $x^- > 0$, $x'^- > 0$ and the identification of the $x'^- < 0$ part of the diagram in the lower panel of Fig. 7.8 with the $x^- > 0$ part of the upper panel reduces the amplitude squared to twice the imaginary part of the forward amplitude (according to the Cutkosky rules), in agreement with the optical theorem.

We are now ready to construct an evolution equation for the diffractive cross section $N^D(\vec{x}_{1\perp}, \vec{x}_{0\perp}, Y, Y_0)$. As we have seen before, with regard to the derivation of BK evolution, it is a little easier to construct the equation for the S-matrix than for the T-matrix. The

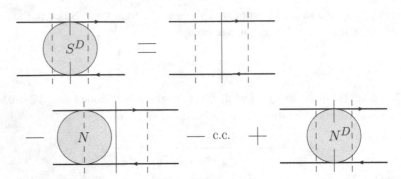

Fig. 7.11. Diagrammatic representation of the definition of S^D in Eq. (7.39). Here the dashed lines simply denote the times $x^- = 0$ or $x'^- = 0$. Interactions with the target are included only when these lines cross the shaded circles.

cross section N^D by definition contains the interaction with the target both in the amplitude and in the complex conjugate amplitude. We can complete it to an S-matrix-like object S^D by adding terms where there is no interaction either in the amplitude or in the complex conjugate amplitude (which gives $-N(\vec{x}_{1\perp}, \vec{x}_{0\perp}, Y)$ for each) and also terms where there is no interaction on either side of the cut (which gives 1). We thus define

$$S^D(\vec{x}_{1\perp}, \vec{x}_{0\perp}, Y, Y_0) = 1 - 2N(\vec{x}_{1\perp}, \vec{x}_{0\perp}, Y) + N^D(\vec{x}_{1\perp}, \vec{x}_{0\perp}, Y, Y_0). \qquad (7.39)$$

The quantity S^D includes both the interacting and the noninteracting contributions to the left and to the right of the cut with the constraint that in the final state there is always a rapidity gap greater than or equal to Y_0. The definition (7.39) of S^D is illustrated in Fig. 7.11.

When $Y = Y_0$, using Eq. (7.33) we have

$$S^D(\vec{x}_{1\perp}, \vec{x}_{0\perp}, Y = Y_0, Y_0) = S^2(\vec{x}_{1\perp}, \vec{x}_{0\perp}, Y_0) \qquad (7.40)$$

with S given by the BK/JIMWLK equations. We now want to derive the Y evolution for S^D for $Y > Y_0$, with the initial condition at $Y = Y_0$ given by Eq. (7.40). Suppose that in one step of the evolution we emit a gluon with rapidity $y > Y_0$. Since there are no final-state restrictions on gluons with $y > Y_0$ the gluon may or may not cross the cut and be present in the final state. (The rapidity gap is greater than or equal to Y_0; if the $y > Y_0$ gluon is not present in the final state this would simply extend the gap to rapidity y, which is still included in the definition of S^D.) A gluon with $y > Y_0$ may be emitted and absorbed at $x^- \lessgtr 0$ and $x'^- \lessgtr 0$. However, owing to the cancellations in Fig. 7.10, all emissions or absorptions at $x^- > 0$, $x'^- > 0$ cancel out, and we are left with the normal dipole evolution of the forward amplitude that we used in deriving the BK evolution equation, with all the emissions and absorptions taking place at $x^- < 0$, $x'^- < 0$. We conclude that the evolution for S^D is equivalent to that of the S-matrix, which in the large-N_c limit is given by Eq. (4.137) and

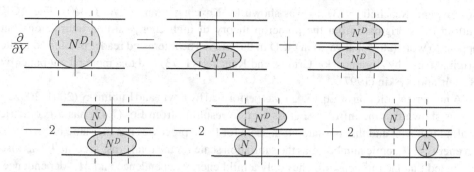

Fig. 7.12. A graphic form of the equation for the cross section of diffractive production. The dashed-line notation is the same as in Fig. 7.11.

illustrated in Fig. 4.26. Thus we can write

$$\partial_Y S^D(\vec{x}_{1\perp}, \vec{x}_{0\perp}, Y, Y_0)$$

$$= \frac{\bar{\alpha}_s}{2\pi} \int d^2 x_2 \frac{x_{10}^2}{x_{20}^2 x_{21}^2} \left[S^D(\vec{x}_{1\perp}, \vec{x}_{2\perp}, Y, Y_0) S^D(\vec{x}_{2\perp}, \vec{x}_{0\perp}, Y, Y_0) - S^D(\vec{x}_{1\perp}, \vec{x}_{0\perp}, Y, Y_0) \right].$$

$$(7.41)$$

Substituting Eq. (7.39) into Eq. (7.41) we readily obtain (Kovchegov and Levin 2000)

$$\partial_Y N^D(\vec{x}_{1\perp}, \vec{x}_{0\perp}, Y, Y_0)$$

$$= \frac{\bar{\alpha}_s}{2\pi} \int d^2 x_2 \frac{x_{10}^2}{x_{20}^2 x_{21}^2}$$

$$\times \left[N^D(\vec{x}_{1\perp}, \vec{x}_{2\perp}, Y, Y_0) + N^D(\vec{x}_{2\perp}, \vec{x}_{0\perp}, Y, Y_0) - N^D(\vec{x}_{1\perp}, \vec{x}_{0\perp}, Y, Y_0) \right.$$

$$+ N^D(\vec{x}_{1\perp}, \vec{x}_{2\perp}, Y, Y_0) N^D(\vec{x}_{2\perp}, \vec{x}_{0\perp}, Y, Y_0)$$

$$- 2N(\vec{x}_{1\perp}, \vec{x}_{2\perp}, Y) N^D(\vec{x}_{2\perp}, \vec{x}_{0\perp}, Y, Y_0)$$

$$\left. - 2N^D(\vec{x}_{1\perp}, \vec{x}_{2\perp}, Y, Y_0) N(\vec{x}_{2\perp}, \vec{x}_{0\perp}, Y) + 2N(\vec{x}_{1\perp}, \vec{x}_{2\perp}, Y) N(\vec{x}_{2\perp}, \vec{x}_{0\perp}, Y) \right].$$

$$(7.42)$$

This is a nonlinear evolution equation with the initial condition specified at $Y = Y_0$ in Eq. (7.33). To solve it one has first to solve the BK equation (4.138) to find the dipole amplitude N, which is then used in Eq. (7.42) to find N_D. Owing to the apparent complexity of both Eq. (7.42) and the BK equation, no analytic solution of Eq. (7.42) exists.

Equation (7.42) is illustrated diagrammatically (and, perhaps, somewhat schematically) in Fig. 7.12. We can see that in the nonlinear terms the factors 2 arise from adding diagrams that are mirror-reflected with respect to the cut diagrams shown. The coefficients in front of the nonlinear terms on the right-hand side of Eq. (7.42) turn out to be in agreement with the Abramovsky–Gribov–Kancheli (AGK) cutting rules (Abramovsky,

Gribov, and Kancheli 1973), as was shown by Kovchegov and Levin (2000). The AGK cutting rules originate from the pomeron theory of high energy strong interactions but appear to work (almost always) in QCD in the LLA. The interested reader is referred to the original paper by Abramovsky, Gribov, and Kancheli (1973) and to a more recent paper by Bartels and Ryskin (1997).

A numerical solution of Eq. (7.42) was performed by Levin and Lublinsky (2001, 2002a). It was shown that the diffractive cross section resulting from Eq. (7.42) has a geometric scaling behavior and that the saturation scale for these processes has the same dependence on energy and atomic number A as the saturation scale for the total cross section. It was also predicted that the ratio $\sigma_{diff}/\sigma_{tot}$ has only a mild energy dependence and M_X^2-dependence, in agreement with an estimate from Eqs. (7.23) and, more importantly, with the HERA data (see Abramowicz and Dainton (1996), H1 collaboration (1997), and ZEUS collaboration (1999)).

Further reading

For further reading on diffraction we recommend the books by Barone and Predazzi (2002), Donnachie, Dosch, and Landshoff (2005), and Forshaw and Ross (1997). In these books a wide spectrum of different problems is discussed, from the wave nature of diffractive scattering to the practical phenomenology based on the reggeon approach. Several reviews of diffraction in DIS cover the topics that have been discussed here in more detail: Wusthoff and Martin (1999) (perturbative QCD), Hebecker (2000) (perturbative QCD and beyond, including the semiclassical approach to diffraction), and Weigert (2005) (the CGC approach to diffractive processes). We also recommend the review by Boreskov, Kaidalov, and Kancheli (2006), which gives an outline of diffractive processes using pomeron phenomenology, as well as the paper by Bartels and Kowalski (2001), where the space–time picture of diffractive processes is explored.

Low-mass diffraction with the production of a $q\bar{q}$-pair and a $q\bar{q}G$ state was discussed originally by Buchmuller, Gehrmann, and Hebecker (1997), Buchmuller, McDermott, and Hebecker (1999), and Kovchegov and McLerran (1999). We also recommend the papers by Munier and Shoshi (2004), Marquet (2005), and Golec-Biernat and Marquet (2005), Kopeliovich, Potashnikova, and Schmidt (2007), and Golec-Biernat and Luszczak (2009). In them one can find comparisons of the theory with the experimental data on diffraction.

For exclusive vector meson production in DIS we recommend the papers by Ryskin (1993) and Brodsky *et al.* (1994), along with the more recent paper of Marquet, Peschanski, and Soyez (2007).

As we have shown, high-mass diffraction is intimately related to the BFKL pomeron interaction, and the calculation of this process has been a main subject of interest in the community over several decades, starting from the Gribov, Levin, and Ryskin (1983) paper (see also Levin and Wusthoff (1994)).

High-mass diffraction in perturbative QCD was studied in the following papers: Levin and Wusthoff (1994), Bartels and Wusthoff (1995), Braun and Vacca (1997), Bartels, Braun, and Vacca (2005), and Bartels and Kutak (2008). Using a dipole approach, high-mass diffraction has been considered in papers by Bialas, Navelet, and Peschanski (1998, 1999), Korchemsky (1999), and Navelet and Peschanski (2001). Equation (7.42) has been rederived using methods different from those of Kovchegov and Levin (2000) in the following papers: Kovner and Wiedemann (2001) (using the eikonal approach), Hentschinski, Weigert, and Schafer (2006) (in the JIMWLK approach, generalizing Eq. (7.42) beyond the large-N_c limit), Kovner, Lublinsky, and Weigert (2006), and Hatta *et al.* (2006) (using general CGC concepts).

Exercises

7.1 Pick any diagram with a specific quark–gluon couplings in the type-B class from Fig. 7.10 and show explicitly that the cancellation does happen, as shown in the figure.

7.2 **(a)** Solve the following zero-transverse-dimensional equation for $S^D(Y, Y_0)$:

$$\partial_Y S^D(Y, Y_0) = \alpha_s [S^D(Y, Y_0)]^2 - \alpha_s S^D(Y, Y_0) \tag{7.43}$$

with the initial condition $S^D(Y = Y_0, Y_0) = [1 - N(Y_0)]^2$ (see Eq. (7.40)) with $N(Y)$ as found in Exercise 4.5(b).

(b) Using the result of part (a) find the diffractive cross section

$$M_X^2 \frac{d\sigma_{diff}}{dM_X^2} = -\frac{\partial N^D(Y, Y_0)}{\partial Y_0} = -\frac{\partial S^D(Y, Y_0)}{\partial Y_0}. \tag{7.44}$$

Plot it as a function of Y_0. You should get a function of Y_0 that peaks at a large Y_0 value that is close to Y: this maximum in the diffractive cross section is a direct manifestation of the saturation/CGC dynamics.

8

Particle production in high energy QCD

We now turn to the question of particle production in the saturation framework. We will work out explicitly the case of inclusive gluon production in high energy collisions. Again the calculation follows the two-step formalism we have seen before: we first find the gluon production in the quasi-classical approximation and then include quantum small-x evolution effects in the result. The methods that we present can be applied to the calculation of other particle production observables as well.

8.1 Gluon production at the lowest order

Consider gluon production in onium–onium collisions, studied in Chapter 3. The lowest-order gluon production takes place at order α_s^3 in the amplitude squared (order g^3 in the amplitude). The corresponding Feynman diagrams for high energy gluon production in quark–quark scattering are shown on the right-hand side of the equation in Fig. 3.6: in the case of onium–onium scattering one should also include a contribution where the antiquark line replaces either one or both quark lines. For quark–quark scattering the gluon production cross section can be obtained from Eq. (3.39): noticing that $dy = dk^+/k^+$ we write for the differential cross section

$$\frac{d\sigma}{d^2k_T dy} = \frac{2\alpha_s^3 C_F}{\pi^2} \frac{1}{k_T^2} \int d^2q_\perp \frac{1}{q_\perp^2 (\vec{k}_\perp - \vec{q}_\perp)^2}. \tag{8.1}$$

Using the lowest-order unintegrated gluon distribution of a quark (cf. Eqs. (4.26) and (5.55)),

$$\phi_{LO}(k_T^2) = \frac{\alpha_s C_F}{\pi} \frac{1}{k_T^2}, \tag{8.2}$$

one can rewrite Eq. (8.1) in the following form:

$$\frac{d\sigma}{d^2k_T dy} = \frac{2\alpha_s}{C_F} \frac{1}{k_T^2} \int d^2q_\perp \phi_{LO}(q_T^2)\, \phi_{LO}\left((\vec{k}_\perp - \vec{q}_\perp)^2\right). \tag{8.3}$$

We see that the gluon production cross section consists of factorized contributions of the unintegrated gluon distributions ϕ_{LO} for each quark, convoluted with a q_T-integral (Gribov, Levin, and Ryskin 1983, Catani, Ciafaloni, and Hautmann 1991, Collins and Ellis 1991).

272

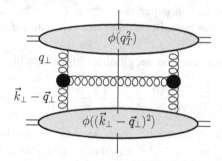

Fig. 8.1. Diagrammatic representation of k_T-factorization in Eq. (8.3). The vertical solid straight line denotes the final-state cut.

Such a factorization containing an integral over transverse momenta is known as a k_T-factorization as opposed to the more standard collinear factorization in perturbative QCD, in which there is no transverse momentum integral (Collins, Soper, and Sterman 1985a, b, 1988a, b).[1]

To obtain the gluon production cross section at order α_s^3 in the case of onium–onium scattering one has to sum over all possible diagrams of the type shown in Fig. 3.6, including the antiquark contributions both in the amplitude and in the complex conjugate amplitude (cf. Fig. 3.7). Eventually we see that Eq. (8.3) remains the answer for gluon production in the case of onium–onium scattering also, but with the unintegrated gluon distribution of an onium given by Eq. (3.92) instead of Eq. (8.2) and with the Green function G given by Eq. (3.59), so that

$$\phi_{LO}^{onium}(k_T^2) = \frac{\alpha_s C_F}{\pi} \frac{1}{k_T^2} \int d^2 x_\perp \int_0^1 dz \, |\Psi(\vec{x}_\perp, z)|^2 \left(2 - e^{-i\vec{k}_\perp \cdot \vec{x}_\perp} - e^{i\vec{k}_\perp \cdot \vec{x}_\perp} \right), \qquad (8.4)$$

where $\Psi(\vec{x}_\perp, z)$ is the bare onium wave function.

Equation (8.3) is illustrated in Fig. 8.1, where the shaded ovals denote the unintegrated distributions of the two onia and the solid circles denote Lipatov vertices. We see that, at this lowest order, the gluon production is given by factorized diagrams like that shown in Fig. 8.1, leading to the k_T-factorization expression (8.3).

Equation (8.1) can be integrated over $\vec{q}_\perp$ explicitly, with the help of the integral performed in appendix section A.3. This yields (cf. Gunion and Bertsch 1982)[2]

$$\frac{d\sigma}{d^2 k_T dy} = \frac{8\alpha_s^3 C_F}{\pi} \frac{1}{k_T^4} \ln \frac{k_T}{\Lambda}, \qquad (8.5)$$

with Λ an IR cutoff as usual. There is a problem with Eq. (8.5): if we integrate both sides over $\vec{k}_\perp$ to obtain the integrated cross section $d\sigma/dy$ we would get $d\sigma/dy \sim (1/\Lambda^2) \ln \Lambda$.

---

[1] A proper discussion of particle production in the collinear factorization framework is beyond the scope of this book. Instead we refer the reader to the textbook by Sterman (1993) and the monograph by Collins (2011) and the references therein.

[2] In the framework of the MV model, Eq. (8.5) was derived by Kovner, McLerran, and Weigert (1995a, b) and by Kovchegov and Rischke (1997).

Thus the integrated cross section depends on the nonperturbative IR cutoff in a power-law way, indicating that one probably should not calculate $d\sigma/dy$ using a perturbative approach.

A similar problem remains for the net gluon multiplicity, which is defined by

$$\frac{dN}{d^2k_T dy} = \frac{1}{\sigma_{inel}} \frac{d\sigma}{d^2k_T dy} \tag{8.6}$$

with σ_{inel} the total inelastic scattering cross section. At the lowest-order, two-gluon-exchange, level, we have $\sigma_{inel} \sim 1/\Lambda^2$. Combining this with $d\sigma/dy \sim (1/\Lambda^2) \ln \Lambda$ we see that the gluon–gluon multiplicity scales as $dN/dy \sim \ln \Lambda$, i.e., it is also IR-cutoff-dependent, though the dependence is logarithmic and so is much softer than that for $d\sigma/dy$.

These IR problems in $d\sigma/dy$ and dN/dy are remedied by saturation physics, which again enables the use of the perturbative approximation. Above we have seen several examples of how saturation effects screen the IR region, making observables much less dependent on nonperturbative physics; we will now see how this happens for the inclusive gluon production.

8.2 Gluon production in DIS and pA collisions

8.2.1 Quasi-classical gluon production

Let us now try to generalize the leading-order gluon production cross section we have just obtained to the case of gluon production in DIS on a nucleus. Our observable is the single-particle inclusive gluon production: we want to measure a gluon in the final state without any constraints on what else can be present in the final state. (Because of the lack of final-state constraints this observable is referred to as "inclusive", as opposed to exclusive processes such as diffraction.) We will be working in the quasi-classical MV/GGM approximation. Just as for the total DIS cross section we can factor out the light cone wave function for a virtual photon splitting into a $q\bar{q}$ pair and consider only the dipole–nucleus scattering. According to the MV model, the gluon production in DIS should be obtained by finding the classical gluon field solution of the Yang–Mills equations (5.4) with the source current now given by the nucleus and the $q\bar{q}$ dipole. One may also consider gluon production in proton–nucleus (pA) collisions: in this case the source current is given by the nucleus and the proton. The difference between a nucleus and a proton in the saturation picture is that the proton's saturation scale Q_{s1} is much smaller than the nuclear saturation scale Q_{s2}, i.e., $Q_{s1} \ll Q_{s2}$. Therefore, when one is considering gluon production with transverse momentum $k_T \gg Q_{s1}$ and making no assumption about the relation between k_T and Q_{s2}, one can neglect multiple interactions with the proton and other saturation effects in the proton wave function while keeping multiple rescatterings to all orders in the nucleus. As we saw in Chapter 5 the classical MV treatment resums the parameter $\alpha_s^2 A^{1/3}$ resulting from two-gluon exchanges with each nucleon and in this way is equivalent to the GGM approach. We conclude that, working either in the light cone gauge of the dipole (proton),

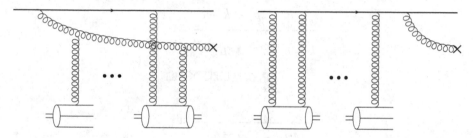

Fig. 8.2. Diagrams contributing to the gluon production amplitude in quark–nucleus collisions. The cross denotes the measured gluon.

or in the covariant gauge, when resumming the diagrams for gluon production we have to include all the multiple two-gluon rescatterings on the nucleons in the nucleus.

For simplicity we begin by calculating gluon production in quark–nucleus scattering. We work in a frame in which the nucleus is moving along the x^+-direction while the quark is moving along the x^--direction. The calculation is simpler in the $A^- = 0$ gauge, which we will be using. This gauge is equivalent to the $\partial_\mu A^\mu = 0$ covariant gauge for the nucleus. The LCPT diagrams contributing to gluon production in quark–nucleus collisions in the $A^- = 0$ gauge are shown in Fig. 8.2. Analogously to the case of small-x evolution considered above, the gluon's minus component of momentum is much smaller than that of the incoming quark. Gluon emission may take place either before or after the interaction with the target (cf. Fig. 4.24 and the accompanying explanation of why emissions during the interaction are suppressed by a power of s). Interactions with the nucleons in the target may be both elastic and inelastic.

Just as before, calculation of the diagrams is easier to carry out in transverse coordinate space. However, in the end we need to find the differential production cross section in momentum space. To connect the momentum-space cross section to transverse coordinate space let us go back to the lowest-order gluon production from the previous section. The lowest-order term in Eq. (8.1) is given by (see Sec. 3.3.1)

$$\frac{d\sigma}{d^2k_T dy} = \frac{1}{2(2\pi)^3} \int \frac{d^2q}{(2\pi)^2} \frac{1}{4s^2} \langle |M_{qq \to qqG}(\vec{k}_\perp, \vec{q}_\perp)|^2 \rangle \qquad (8.7)$$

where the scattering amplitude $M_{qq \to qqG}$ is given in Eq. (3.35) and s is the center-of-mass energy squared of the collision. Defining a rescaled amplitude by (cf. Eq. (B.22))

$$A(\vec{k}_\perp, \vec{q}_\perp) = \frac{M_{qq \to qqG}(\vec{k}_\perp, \vec{q}_\perp)}{2s}, \qquad (8.8)$$

we can rewrite Eq. (8.7) as

$$\frac{d\sigma}{d^2k_T dy} = \frac{1}{2(2\pi)^3} \int \frac{d^2q}{(2\pi)^2} \langle |A(\vec{k}_\perp, \vec{q}_\perp)|^2 \rangle. \qquad (8.9)$$

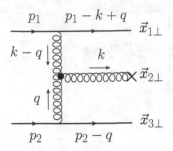

Fig. 8.3. Gluon production at the lowest order with the transverse coordinates shown explicitly.

To Fourier-transform the amplitude $A(\vec{k}_\perp, \vec{q}_\perp)$ into transverse coordinate space we use

$$A(\vec{x}_{23}, \vec{x}_{13}) = \int \frac{d^2k_\perp}{(2\pi)^2} \frac{d^2q_\perp}{(2\pi)^2} e^{i\vec{k}_\perp \cdot \vec{x}_{21} + i\vec{q}_\perp \cdot \vec{x}_{13}} A(\vec{k}_\perp, \vec{q}_\perp), \tag{8.10}$$

where as usual $\vec{x}_{ij} = \vec{x}_{i\perp} - \vec{x}_{j\perp}$. Equation (8.10) is illustrated in Fig. 8.3 for one of the lowest-order diagrams from Fig. 3.6. Remember that in eikonal interactions the transverse coordinates of the particles remain unchanged: hence the colliding quarks have transverse coordinates $\vec{x}_{1\perp}$ and $\vec{x}_{3\perp}$ both before and after the interaction. Also note that the amplitude $M_{qq \to qqG}$ and with it $A(\vec{k}_\perp, \vec{q}_\perp)$ already have momentum conservation imposed on them: this leads to translational invariance in coordinate space, making the coordinate-space amplitude $A(\vec{x}_{23}, \vec{x}_{13})$ in Eq. (8.10) a function of the differences between the transverse vectors only.

Inverting Eq. (8.10) we get

$$A(\vec{k}_\perp, \vec{q}_\perp) = \int d^2x_2 d^2x_0 e^{-i\vec{x}_{2\perp} \cdot \vec{k}_\perp - i\vec{x}_{1\perp} \cdot (\vec{q}_\perp - \vec{k}_\perp)} A(\vec{x}_{23}, \vec{x}_{13}). \tag{8.11}$$

Using this in Eq. (8.9) and integrating over $\vec{q}_\perp$ we obtain

$$\frac{d\sigma}{d^2k_T dy} = \frac{1}{2(2\pi)^3} \int d^2x_2 d^2x_{2'} d^2x_1 e^{-i\vec{k}_\perp \cdot \vec{x}_{22'}} \langle A(\vec{x}_{23}, \vec{x}_{13}) A^*(\vec{x}_{2'3}, \vec{x}_{13}) \rangle, \tag{8.12}$$

where $\vec{x}_{2\perp}$ and $\vec{x}_{2'\perp}$ are the transverse coordinates of the gluon in the amplitude and in the complex conjugate amplitude respectively. We see that the reason that the two coordinates are different is that we are keeping the transverse momentum of the gluon $\vec{k}_\perp$ fixed: if we integrated Eq. (8.12) over $\vec{k}_\perp$ this would make $\vec{x}_{2\perp}$ and $\vec{x}_{2'\perp}$ equal. One can think of the difference between $\vec{x}_{2\perp}$ and $\vec{x}_{2'\perp}$ as due to the uncertainty principle: if we know the momentum $\vec{k}_\perp$ exactly, we cannot have precise knowledge of the transverse position of the gluon. Note that since we do not keep the transverse momenta of the two quarks in the final state fixed and, rather, allow them to be arbitrary (that is, we integrate over all their values), the transverse coordinates of the quarks are the same in the amplitude and in the complex conjugate amplitude.

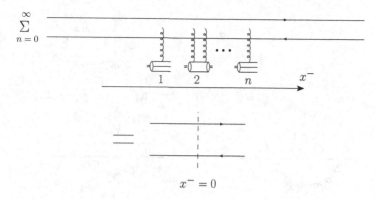

Fig. 8.4. An extension of the notation of Fig. 5.8 to include inelastic interactions between the projectile and the nucleons in the target nucleus.

Note that for a nuclear target with the quark at $\vec{x}_{3\perp}$ part of a nucleon in the nucleus, the angle brackets in Eq. (8.12) would imply an average over all positions of the nucleon,

$$\int d^2x_3 T(\vec{x}_{3\perp}),\qquad(8.13)$$

where $T(\vec{x}_{3\perp})$ is the nuclear profile function defined in Eq. (4.31).

Equation (8.12), while derived for the lowest-order gluon production, is applicable for gluon production in any process involving the eikonal scattering of a quark (or any other projectile, such as an onium or a proton) on a target nucleus, including multiple rescatterings and small-x evolution. In deriving Eq. (8.12) we had to rescale the scattering amplitude in Eq. (8.8): however, it is possible to see that in the GGM multiple rescattering case each additional nucleon enters with the same normalization factor, so that the net scattering amplitude (4.51) is energy independent. In addition, in eikonal scattering and LLA small-x evolution the transverse coordinates of the gluons and quarks do not change, which implies that the transverse vectors $\vec{x}_{2\perp}$, $\vec{x}_{2'\perp}$, and $\vec{x}_{1\perp}$ in Eq. (8.12) will not change if we allow the quark and the gluon to multiply rescatter and branch out into more gluons in the LLA approximation. We conclude that Eq. (8.12) is the required relation between the momentum-space cross section and the coordinate-space scattering amplitude.

Squares of the diagrams in Fig. 8.2 giving the production cross section are shown in Fig. 8.5 using the slight extension defined in Fig. 8.4 of the notation from Fig. 5.8 for the interaction of the projectile with the target: now the vertical dashed line includes both elastic and inelastic interactions with the target. Beyond the diagrams, the notation in Fig. 5.8 includes the averaging of Wilson lines over the target fields while the notation of Fig. 8.4 does not. The graphs shown in Fig. 8.5 represent the amplitude squared, with two time axes, one in the amplitude and one in the complex conjugate amplitude, just as in the case of diffraction considered above.

Figure 8.5 demonstrates all the main cases one has to consider. If x_{em}^- is the light cone time of the gluon emission in the amplitude while $x_{em}'^-$ is that in the complex conjugate amplitude (see diagram C in Fig. 8.5), we have four cases represented in Fig. 8.5, as follows.

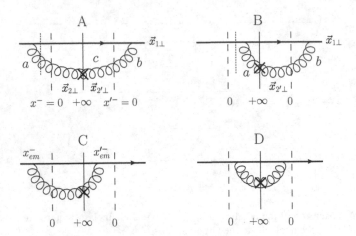

Fig. 8.5. Diagrams contributing to gluon production in quark–nucleus collisions. The cross denotes the measured gluon.

(A) $x_{em}^- < 0$, $x_{em}'^- < 0$: the gluon is emitted before the interaction with the target both in the amplitude and in the complex conjugate amplitude.

(B) $x_{em}^- > 0$, $x_{em}'^- < 0$: the gluon is emitted after the interaction with the target in the amplitude but before the interaction with the target in the complex conjugate amplitude.

(C) $x_{em}^- < 0$, $x_{em}'^- > 0$: the gluon is emitted before the interaction with the target in the amplitude but after the interaction with the target in the complex conjugate amplitude.

(D) $x_{em}^- > 0$, $x_{em}'^- > 0$: the gluon is emitted after the interaction with the target both in the amplitude and in the complex conjugate amplitude.

We need to find the amplitude squared in transverse coordinate space to use in Eq. (8.12). To do this we have to calculate the diagrams in Fig. 8.5. The transverse coordinates of the quark ($\vec{x}_{1\perp}$) together with the coordinates $\vec{x}_{2\perp}$ and $\vec{x}_{2'\perp}$ of the gluon to the left and to the right of the final state cut are shown in panel A of Fig. 8.5. The calculation is easier to carry out using LCPT, in which the process factorizes into the light cone wave function for a quark splitting into a quark and a gluon and the amplitude for the interaction with the target.

The soft gluon emission brings in a factor (cf. Eq. (4.60))

$$i\frac{g t^a}{\pi} \frac{\vec{\epsilon}_\perp^{\lambda *} \cdot \vec{x}_{21}}{x_{21}^2} \tag{8.14}$$

in the amplitude and the same (but conjugate) factor with the index 2 replaced by 2' in the complex conjugate amplitude, along with $a \rightarrow b$ (the gluon colors allocation is given in Fig. 8.5A). In Eq. (8.14), λ is the polarization of the gluon, which remains unchanged in the eikonal interaction with the target and is therefore the same on both sides of the cut. Emissions with either $x_{em}^- > 0$ or $x_{em}'^- > 0$ lead to an extra minus sign compared with the early-time emissions: this can be deduced from panels A and B of Fig. 8.5. There the vertical dashed lines denote the intermediate states in the amplitudes A and B used in calculating the quark splitting corresponding to Eq. (8.14). In panel B the intermediate state contains only a quark (with the rest of the target, not shown): in building up the energy denominator for this state we have to take the light cone energy of the quark (and the outgoing target) and subtract the light cone energy of the incoming state, or, equivalently, the energy of the outgoing state. We thus obtain the light cone energy of the intermediate-state quark minus the energies of the outgoing quark and gluon. This is the negative of the energy denominator for the intermediate state highlighted by the dotted line in panel A of Fig. 8.5. We see that the late-time emissions bring in an extra minus sign. This conclusion is similar to that regarding the sign difference used in deriving the cancellations in panel B of Fig. 7.10 (see Eqs. (7.34) and (7.35)). To summarize, diagrams B and C in Fig. 8.5 have an extra minus sign compared with diagrams A and D in the same figure.

We now need to calculate the interactions with the target in each panel of Fig. 8.5. This is easiest to do using the language of Wilson lines defined in Eqs. (5.43) and (5.44). The interaction with the target in Fig. 8.5A (along with the color factors resulting from the gluon emission) gives

$$\left\langle \frac{1}{N_c} \text{tr} \left[t^b V^\dagger_{\vec{x}_{1\perp}} V_{\vec{x}_{1\perp}} t^a \right] U^{bc}_{\vec{x}_{2'\perp}} U^{ca}_{\vec{x}_{2\perp}} \right\rangle = C_F S_G(\vec{x}_{2\perp}, \vec{x}_{2'\perp}, y), \tag{8.15}$$

where we have used Eq. (5.33) and definition (5.46) to simplify the expression. As in the case of the calculation in Sec. 3.3.1 we assume that the target nucleus has rapidity 0 while the produced gluon along with the quark that emits it have rapidity y. Note that the quark's interaction with the target cancels out in diagram A: the quark becomes a "spectator" and the interaction is given by the gluon dipole S-matrix.

The contribution of Fig. 8.5B is obtained similarly, yielding, with the help of Eq. (5.31),

$$\left\langle \frac{1}{N_c} \text{tr} \left[t^a V^\dagger_{\vec{x}_{1\perp}} t^b V_{\vec{x}_{1\perp}} \right] U^{ba}_{\vec{x}_{2'\perp}} \right\rangle = C_F S_G(\vec{x}_{1\perp}, \vec{x}_{2'\perp}, y). \tag{8.16}$$

By analogy diagram C brings in

$$C_F S_G(\vec{x}_{2\perp}, \vec{x}_{1\perp}, y) \tag{8.17}$$

and diagram D contributes only a factor C_F.

Combining the emission contribution (8.14) with the interaction terms we have just found, and using it all in Eq. (8.12) while keeping in mind that diagrams B and C have an extra minus sign, we obtain, after summation over the gluon polarizations (Kovchegov and Mueller 1998a):

$$\frac{d\sigma^{qA}}{d^2k_1\, dy} = \frac{1}{(2\pi)^3} \int d^2x_2\, d^2x_{2'}\, d^2x_1 e^{-i\vec{k}_\perp \cdot \vec{x}_{22'}} \frac{\alpha_s C_F}{\pi^2} \frac{\vec{x}_{21} \cdot \vec{x}_{2'1}}{x^2_{21} x^2_{2'1}}$$

$$\times \left[S_G(\vec{x}_{2\perp}, \vec{x}_{2'\perp}, y) - S_G(\vec{x}_{1\perp}, \vec{x}_{2'\perp}, y) - S_G(\vec{x}_{2\perp}, \vec{x}_{1\perp}, y) + 1 \right]. \tag{8.18}$$

This result was independently confirmed by Kopeliovich, Tarasov, and Schafer (1999) and by Dumitru and McLerran (2002). Equation (8.18) can also be rewritten in terms of the imaginary part of the gluon dipole–nucleus forward scattering amplitude,

$$N_G(\vec{x}_{1\perp}, \vec{x}_{0\perp}, Y) = 1 - S_G(\vec{x}_{1\perp}, \vec{x}_{0\perp}, Y), \tag{8.19}$$

as follows:

$$\frac{d\sigma^{qA}}{d^2k_T dy} = \frac{1}{(2\pi)^2} \int d^2x_2 \, d^2x_{2'} \, d^2x_1 e^{-i\vec{k}_\perp \cdot \vec{x}_{2'1}} \frac{\alpha_s C_F}{\pi^2} \frac{\vec{x}_{21} \cdot \vec{x}_{2'1}}{x_{21}^2 x_{2'1}^2}$$

$$\times \left[N_G(\vec{x}_{1\perp}, \vec{x}_{2'\perp}, y) + N_G(\vec{x}_{2\perp}, \vec{x}_{1\perp}, y) - N_G(\vec{x}_{2\perp}, \vec{x}_{2'\perp}, y) \right]. \tag{8.20}$$

While our derivation of Eq. (8.20) was quite general, in the MV/GGM quasi-classical approximation one should use the amplitude N_G given by Eq. (5.51) in Eq. (8.20). In that approximation for N_G, Eq. (8.20) would give us the gluon production in quark–nucleus scattering in the quasi-classical MV/GGM approximation.

It is interesting to note that Eq. (8.20) can be rewritten in the k_T-factorized form seen in Eq. (8.3). To see this we integrate over $\vec{x}_{2\perp}$ in the first term in the square brackets of Eq. (8.20), over $\vec{x}_{2'\perp}$ in the second term in the same brackets, and over $\vec{x}_{1\perp}$ in the third term, with the help of Eqs. (A.10) and (A.12). This yields, after some coordinate relabeling,

$$\frac{d\sigma^{qA}}{d^2k_T dy} = \frac{\alpha_s C_F}{2\pi^3} \int d^2x_2 \, d^2x_{2'} e^{-i\vec{k}_\perp \cdot \vec{x}_{22'}} \left(2i \frac{\vec{k}_\perp}{k_\perp^2} \cdot \frac{\vec{x}_{22'}}{x_{22'}^2} - \ln \frac{1}{x_{22'}\Lambda} \right)$$

$$\times N_G(\vec{x}_{2\perp}, \vec{x}_{2'\perp}, y) \tag{8.21}$$

with Λ again an IR cutoff. Using the fact that $N_G(\vec{x}_{2\perp} = \vec{x}_{2'\perp}, \vec{x}_{2'\perp}, Y) = 0$ (a zero-size dipole does not interact), we can rewrite Eq. (8.21) as

$$\frac{d\sigma^{qA}}{d^2k_T dy} = \frac{\alpha_s C_F}{2\pi^3} \frac{1}{k_T^2} \int d^2x_2 \, d^2x_{2'} N_G(\vec{x}_{2\perp}, \vec{x}_{2'\perp}, y) \nabla_{\vec{x}_{2\perp}}^2 \left(e^{-i\vec{k}_\perp \cdot \vec{x}_{22'}} \ln \frac{1}{x_{22'}\Lambda} \right), \tag{8.22}$$

where $\nabla_{\vec{x}_\perp}^2 = \partial_{x^1}^2 + \partial_{x^2}^2$ is the two-dimensional Laplace operator.

Let us write Eq. (8.22) in a projectile–target-symmetric way. The expression already contains the dipole–nucleus scattering amplitude N_G. Let us include the dipole–projectile scattering amplitude in it as well. Since our projectile is a single quark, we will construct the scattering amplitude of a gluon dipole on a quark in the quasi-classical approximation by expanding Eq. (5.51) to the lowest nontrivial order and using $Q_{sG}^2(\vec{b}_\perp) = 4\pi\alpha_s^2 T_q(\vec{b}_\perp)$ with the "nuclear profile function" of a single quark, normalized such that

$$\int d^2b_\perp T_q(\vec{b}_\perp) = 1. \tag{8.23}$$

(The factor 2 difference between the value of Q_{sG}^2 that we are using now and that in Eq. (5.41) arises because the saturation scale in Eq. (5.41) is due to scattering on an onium "nucleon" while now we are dealing with a single-quark "nucleon".) We obtain for the dipole–quark scattering amplitude n_G^q at the two-gluon exchange level, integrated over all

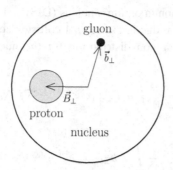

Fig. 8.6. Gluon production in proton(quark)–nucleus collisions as seen in the transverse plane. To make the figure clearer, the gluon has been placed far from the proton; this is, however, a highly improbable configuration.

impact parameters $\vec{b}_\perp$,

$$\int d^2 b_\perp \, n_G^q(\vec{x}_\perp, \vec{b}_\perp, Y = 0) = \pi \alpha_s^2 x_\perp^2 \ln \frac{1}{x_\perp \Lambda},\tag{8.24}$$

where $\vec{x}_\perp$ is the dipole's transverse size and $\vec{b}_\perp$ is the location of its center-of-mass.

Neglecting constants under the logarithms (which were neglected in arriving at Eq. (8.24) anyway), we write

$$\nabla_{\vec{x}_\perp}^2 \left(x_\perp^2 \ln \frac{1}{x_\perp \Lambda} \right) = 4 \ln \frac{1}{x_\perp \Lambda}.\tag{8.25}$$

Using this formula together with Eq. (8.24) in Eq. (8.22) yields, after integration by parts in the latter,

$$\frac{d\sigma^{qA}}{d^2 k_T dy} = \frac{C_F}{\alpha_s \pi (2\pi)^3} \frac{1}{k_T^2} \int d^2 B_\perp d^2 b_\perp d^2 x_\perp \left[\nabla_{\vec{x}_\perp}^2 n_G^q(\vec{x}_\perp, \vec{B}_\perp - \vec{b}_\perp, 0) \right]$$

$$\times e^{-i\vec{k}_\perp \cdot \vec{x}_\perp} \left[\nabla_{\vec{x}_\perp}^2 N_G(\vec{x}_\perp, \vec{b}_\perp, y) \right],\tag{8.26}$$

where $\vec{x}_\perp = \vec{x}_{22'}$ and we have modified the notation as follows:

$$N_G(\vec{x}_{2\perp}, \vec{x}_{2'\perp}, Y) \to N_G(\vec{x}_\perp, \vec{b}_\perp, Y).\tag{8.27}$$

Now $\vec{B}_\perp$ and $\vec{b}_\perp$ are the impact parameters of the proton and the produced gluon respectively, measured with respect to the center of the nucleus, as shown in Fig. 8.6.

Equation (8.26) describes gluon production in quark–nucleus scattering. However, it can be generalized to the onium–nucleus (DIS) and proton–nucleus (pA) cases if n_G^q in it is replaced by the dipole scattering amplitude on the onium (see Eq. (3.139)) or on a proton, labeled n_G. (The proton can be modeled as three valence quarks in a color-singlet state or as any other number of partons moving along the light cone.) Proving this statement for the onium–nucleus scattering is left as an exercise for the reader (see Exercise 8.2). We can conclude that Eq. (8.26) with N_G taken at $y = 0$ and n_G^q replaced by n_G describes quasi-classical gluon production in onium–nucleus (DIS) and pA collisions. We will refer to it as quasi-classical gluon production in the pA collision context.

Defining the unintegrated gluon distribution for the nucleus (in the quasi-classical MV/GGM approximation) as

$$\phi_A(k_T^2) = \frac{C_F}{\alpha_s (2\pi)^3} \int d^2 b_\perp d^2 x_\perp e^{-i\vec{k}_\perp \cdot \vec{x}_\perp} \nabla_{x_\perp}^2 N_G \left(\vec{x}_\perp, \vec{b}_\perp, 0 \right) \tag{8.28}$$

and that for the proton as

$$\phi_p(k_T^2) = \frac{C_F}{\alpha_s (2\pi)^3} \int d^2 b_\perp d^2 x_\perp e^{-i\vec{k}_\perp \cdot \vec{x}_\perp} \nabla_{x_\perp}^2 n_G(\vec{x}_\perp, \vec{b}_\perp, 0), \tag{8.29}$$

transforms Eq. (8.26) into a k_T-factorized form of Eq. (8.3) (Kovchegov and Tuchin 2002):

$$\frac{d\sigma^{pA}}{d^2 k_T\, dy} = \frac{2\alpha_s}{C_F} \frac{1}{k_T^2} \int d^2 q_\perp\, \phi_p(q_T^2) \phi_A \left((\vec{k}_\perp - \vec{q}_\perp)^2 \right), \tag{8.30}$$

where now we use the superscript pA to signify the broader validity range of the derived cross section. (Again we have put $y = 0$ in the argument of N_G in Eq. (8.28) to reduce Eq. (8.26) to the purely quasi-classical case.)

Equation (8.30) may come as a surprise: remember that k_T-factorization usually results from factorizing the diagrams into the form shown in Fig. 8.1, separating them into distribution functions with a Lipatov vertex (squared) in the middle. The diagrams in Fig. 8.5 which led to Eq. (8.30) have no such factorization, since they include direct interactions between the target nucleus and the projectile quark, violating the factorization picture expected from Fig. 8.1. One may hope that perhaps in a different gauge the relevant diagrams would factorize, yielding the picture of Fig. 8.1; to date, however, no such gauge has been found and the physical origin of the k_T-factorization in Eq. (8.30) remains a mystery.

Notice that to achieve the factorized expression (8.30) we had to define the unintegrated gluon distributions (8.28) and (8.29). These definitions are different from the WW distribution of Eq. (5.50) and are more in line with Eq. (6.16) with the quark dipole replaced by the gluon one. (All the distributions agree at the lowest order of Eq. (8.2) up to quark or gluon dipole color factors.) While the definitions (8.28) and (8.29) allow us to achieve the k_T-factorization formula (8.30), it is *a priori* not clear why one has to use these gluon distributions in the gluon production formula. The interplay between the two unintegrated gluon distributions has recently been explored by Dominguez *et al.* (2011).

To assess the impact of (8.30) on the k_T-distribution of the produced gluons (also known as the gluon transverse momentum spectrum) let us evaluate its large-k_T and small-k_T asymptotics. At large k_T we expand Eq. (5.51) to the lowest nontrivial order to get $N_G = x_\perp^2 Q_{sG}^2(\vec{b}_\perp)/4$, which we use in Eq. (8.28) to obtain the lowest-order ϕ_A. For simplicity, modeling the proton again as a single quark, we use Eq. (8.24) in Eq. (8.29) to obtain the same expression for ϕ_p as in (8.2). Finally, substituting all this into Eq. (8.30) and remembering that $Q_{sG}^2(\vec{b}_\perp) = 4\pi \alpha_s^2 T(\vec{b}_\perp)$, we obtain

$$\left. \frac{d\sigma^{pA}}{d^2 k_T\, dy} \right|_{k_T \gg Q_{sG}} \approx \frac{8\alpha_s^3 C_F A}{\pi} \frac{1}{k_T^4} \ln \frac{k_T}{\Lambda}, \tag{8.31}$$

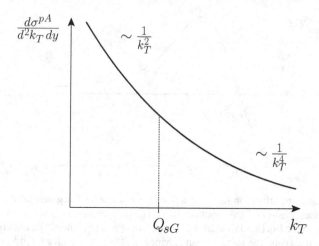

Fig. 8.7. A sketch of the gluon production spectrum generated by Eq. (8.30).

in agreement with Eq. (8.5) up to a factor A, signifying that now we have A nucleons in the nucleus that can interact with the incoming projectile quark to produce the gluon that we are measuring (again with each nucleon modeled by a single quark). Conversely, at $k_T \ll Q_{sG}$ we note that $N_G \approx 1$ for $x_\perp > 1/Q_{sG}$. Using this approximation in Eq. (8.21) and integrating over $\vec{x}_{2\perp}, \vec{x}_{2'\perp}$ yields

$$\left. \frac{d\sigma^{pA}}{d^2k_T\, dy} \right|_{k_T \ll Q_{sG}} \approx \frac{\alpha_s C_F S_\perp}{\pi^2} \frac{1}{k_T^2} \tag{8.32}$$

with $S_\perp$ the transverse area of the nucleus. We see that the produced gluon spectrum has been modified by saturation effects from the factor $1/k_T^4$ in Eq. (8.31) at $k_T \gg Q_{sG}$ to the factor $1/k_T^2$ in Eq. (8.32) at $k_T \ll Q_{sG}$. This is illustrated in Fig. 8.7 (cf. Fig. 5.7). We conclude that saturation effects in one scattering particle (the nucleus) tend to soften the IR divergence, making the integrated cross section dependent on the IR cutoff only logarithmically, $d\sigma^{pA}/dy \sim \ln(Q_{sG}/\Lambda)$.

To better understand the result (8.30) it is useful to construct the *nuclear modification factor*, which in pA collisions can be defined as

$$R^{pA}(k_T, y) = \frac{d\sigma^{pA}/d^2k_T\, dy}{A\, d\sigma^{pp}/d^2k_T\, dy}. \tag{8.33}$$

and has the meaning of the ratio of the number of particles produced in a pA collision per individual proton–nucleon collision and the number of particles produced in a proton–proton (pp) collision in the same kinematic region. Deviations of the nuclear modification factor from unity measure collective nuclear effects in the collision.

In the quasi-classical approximation one can evaluate R^{pA} by using N_G from Eq. (5.51) in Eq. (8.20) to find the gluon production in pA collisions and by using n_G^q from Eq. (8.24), instead of N_G, in the same formula to find the gluon production in pp. The resulting nuclear modification factor is plotted in Fig. 8.8 (Kopeliovich, Tarasov, and Schafer 1999, Baier,

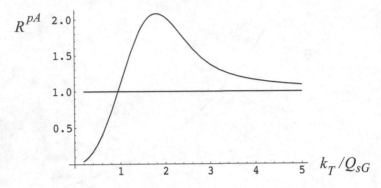

Fig. 8.8. Nuclear modification factor as a function of k_T/Q_{sG} for gluon production in pA collisions in the quasi-classical approximation. The horizontal line at $R^{pA} = 1$ is shown to guide the eye. Since Eq. (8.33) includes an integral over $b_\perp$, Q_{sG} should be understood as the typical saturation scale: the same Cronin enhancement is also seen for fixed $b_\perp$. (Reprinted with permission from Kharzeev, Kovchegov, and Tuchin (2003). Copyright 2003 by the American Physical Society.)

Kovner, and Wiedemann 2003, Kharzeev, Kovchegov, and Tuchin 2003): we see a depletion in the produced gluons at low k_T ($k_T \ll Q_{sG}$) and an enhancement at large k_T ($k_T \gtrsim Q_{sG}$). We conclude that multiple rescatterings tend to broaden the k_T-distribution of the produced gluons, effectively pushing the gluons out to large k_T. The enhancement shown in Fig. 8.8 was observed experimentally in hadron production in pA collisions and is known as the *Cronin effect* (Cronin *et al.* 1975).

The suppression of gluons at low k_T may be identified with the manifestation of nuclear shadowing in the saturation/CGC framework. In the case of gluons the shadowing is quantified by the ratio

$$R_A(x, Q^2) = \frac{xG_A(x, Q^2)}{AxG_N(x, Q^2)} \tag{8.34}$$

(with an analogous ratio for quarks). The shadowing ratio R_A measures the number of gluons per nucleon in the nucleus divided by the number of gluons in a proton (in the same kinematic region). At small x the ratio R_A is known to be below 1, which means that the number of gluons per nucleon in a nucleus is less than the number of gluons in a free proton: this effect is known as nuclear shadowing. (For more on nuclear shadowing see Frankfurt and Strikman (1988).) This depletion of gluons in the nuclear wave function leads to the suppression of produced gluons as measured by the nuclear modification factor. In fact, one can show that the suppression of produced gluons at low k_T in Fig. 8.8 results from the suppression of low-k_T gluons in the unintegrated gluon distribution of Eq. (8.28).

8.2.2 Including nonlinear evolution

We want to include the effects of nonlinear LLA small-x evolution in the inclusive gluon production cross section (8.20) and in the more general cross section (8.30). Again we

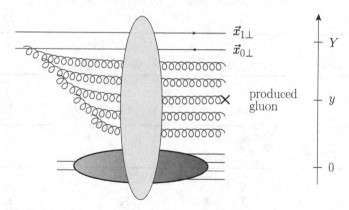

$\vec{x}_{1\perp}$

$\vec{x}_{0\perp}$

Y

produced gluon

y

0

Fig. 8.9. Gluon production in dipole–nucleus scattering with small-x evolution emissions both between the produced gluon (marked by a cross) and the target nucleus and between the produced gluon and the projectile dipole.

assume that the target nucleus has zero rapidity and that the produced gluon has rapidity y. However, now the incoming projectile has rapidity Y such that there are large rapidity intervals between the projectile and the produced gluon and between the gluon and the target. The process is illustrated in Fig. 8.9 for the case of dipole–nucleus scattering: gluon emissions (and absorptions) are allowed everywhere in the rapidity interval from 0 to Y. The target nucleus may indeed break up in this inclusive process.

To include BK/JIMWLK evolution effects in the gluon production formula, let us consider the dipole–nucleus scattering case rather than the quark–nucleus scattering considered in the previous subsection. We start by generalizing Eq. (8.20) to the case of a dipole projectile (Kovchegov 2001):

$$\frac{d\sigma^{q\bar{q}A}}{d^2k_T dy}(\vec{x}_{10}) = \int d^2x_2\, d^2x_{2'}\, d^2x_1 e^{-i\vec{k}_\perp \cdot \vec{x}_{22'}} \frac{\alpha_s C_F}{4\pi^4} \sum_{i,j=0}^{1} (-1)^{i+j} \frac{\vec{x}_{2i} \cdot \vec{x}_{2'j}}{x_{2i}^2 x_{2'j}^2}$$

$$\times \left[N_G(\vec{x}_{i\perp}, \vec{x}_{2'\perp}, y) + N_G(\vec{x}_{2\perp}, \vec{x}_{j\perp}, y) \right.$$

$$\left. - N_G(\vec{x}_{2\perp}, \vec{x}_{2'\perp}, y) - N_G(\vec{x}_{i\perp}, \vec{x}_{j\perp}, y) \right], \tag{8.35}$$

where the quark in the dipole is located at $\vec{x}_{1\perp}$ and the antiquark is at $\vec{x}_{0\perp}$, as shown in Fig. 8.9. (The reader is invited to derive Eq. (8.35) in Exercise 8.2.)

Including the evolution in the rapidity interval between the gluon and the target is straightforward: in fact this was already done in deriving Eq. (8.20). Indeed, as we have seen above, the Wilson-line formalism used in arriving at (8.20) applies equally well to GGM multiple rescatterings and to the LLA evolution. We therefore conclude that N_G evaluated at rapidity y in Eq. (8.35) (and in Eq. (8.20) as well) does not need to come from multiple rescatterings only but may also contain the nonlinear BK/JIMWLK LLA evolution. In the large-N_c limit of BK evolution one can readily show that

$$N_G(\vec{x}_\perp, \vec{b}_\perp, y) = 2N(\vec{x}_\perp, \vec{b}_\perp, y) - N^2(\vec{x}_\perp, \vec{b}_\perp, y), \tag{8.36}$$

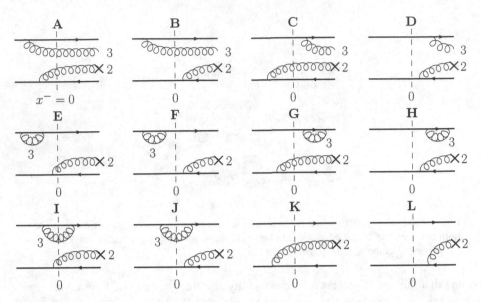

Fig. 8.10. One step of small-x evolution between the projectile onium and the produced gluon, marked by a cross and labeled 2.

where N is the quark dipole amplitude found from Eq. (4.138). The formula (8.36) is due to the fact that in the large-N_c a gluon dipole limit can be thought of as a pair of quark dipoles, with either one or both dipoles interacting directly with the target.

Including the evolution in the rapidity interval between the projectile and the produced gluon requires a bit more work. We will carry out the calculation in the large-N_c limit of Mueller's dipole model. We show one step of such an evolution in Fig. 8.10, where the gluon labeled 3 is harder than the produced gluon 2, that is $z_3 \gg z_2$ with z_2 and z_3 the fractions of the longitudinal momentum of the onium carried by the two gluons respectively. For simplicity we consider a case of particular couplings of the gluons to the onium: the gluon 3 is emitted (and absorbed) by the quark while gluon 2 is emitted by the antiquark. Diagrams A–D in Fig. 8.10 have two real emissions while in the diagrams E–J gluon 3 is virtual. For completeness, we show in diagrams K and L of Fig. 8.10 the emission of gluon 2 without the corrections due to gluon 3.

It is important to stress that we have only given the leading-logarithmic diagrams in Fig. 8.10. Indeed, a diagram similar to A but with the emission of gluon 2 before gluon 3 is possible but will not lead to leading logarithms of $1/x$ (see Fig. 4.17 and its evaluation for an analogous calculation). Similarly one can demonstrate that a diagram similar to D but with the emission of gluon 3 before gluon 2 also lies outside the LLA. This is an important observation: while for $x^- < 0$ the harder gluons are emitted before the softer ones, for $x^- > 0$ the harder gluons would need to be emitted *later* than the softer ones to give an LLA contribution. This rule is also valid for the virtual diagrams; this is illustrated by the fact that graph H in Fig. 8.10 gives an LLA contribution. (Indeed, we do not integrate over the rapidity y of gluon 2 since we are tagging this gluon (i.e., it is the gluon of interest): however, the leading contribution to the production cross section with all rapidity intervals large is given by the LLA approximation as if we were about to integrate over y.)

Using the cancellations of Fig. 7.10 (which are valid for the inelastic interactions with the target shown in Fig. 8.4 as well), one can write down the following relations between the squares and interference terms of the diagrams in Fig. 8.10:

$$|C|^2 + GK^* + G^*K = 0, \quad |D|^2 + HL^* + H^*L = 0, \tag{8.37a}$$

$$CD^* + GL^* + KH^* = 0, \qquad AC^* + IK^* = 0, \tag{8.37b}$$

$$BC^* + JK^* = 0, \quad AD^* + IL^* = 0, \quad BD^* + JL^* = 0. \tag{8.37c}$$

With the help of Eqs. (8.37) we see that

$$|A + B + \cdots + L|^2 \bigg|_{O(\alpha_s^2)} = |A + B|^2 + (E + F)(K + L)^*$$

$$+ (E + F)^*(K + L), \tag{8.38}$$

that is, only emissions or absorptions of gluon 3 with $x^- < 0$ both in the amplitude and in the complex conjugate amplitude remain. This conclusion can be generalized to the case of other gluon couplings and, more importantly, to the case of higher-order hard gluon emissions: the terms which survive the cancellations of Fig. 7.10 are those with all the emissions and absorptions at $x^- < 0$ and $x'^- < 0$; all late-time ($x^-, x'^- > 0$) emissions cancel. We conclude that the evolution in the rapidity interval between the projectile and the produced gluon is the evolution of Mueller's dipole model! In fact an analysis of the higher-order diagrams shows that not all the nonlinear dipole evolution contributes: rather, the nonlinearities cancel leaving only the linear part of the evolution describing the generation by the original incoming onium of the dipole in which gluon 2 was emitted.[3] The quantity n_1 describing such a distribution is defined in Eq. (4.81) and is found from the dipole BFKL equation (4.82) with the initial condition (4.83).

We have thus arrived at the following physical picture of the gluon production process in DIS and pA at large N_c: the evolution in the projectile wave function generates a distribution of single dipoles. A gluon that we will tag (measure) is then emitted by one such dipole, making the evolution between the projectile and the produced gluon linear. The gluon along with the dipole from which it was emitted then interact with the target nonlinearly.

Formally the LLA evolution between the projectile and the produced gluon is included in Eq. (8.35) by the following replacement:

$$\frac{d\sigma^{q\bar{q}A}}{d^2k_T dy d^2 B_\perp}(\vec{x}_{10}) \rightarrow \int d^2 b_\perp d^2 x_{1'0'} \, n_1(\vec{x}_{10}, \vec{x}_{1'0'}, \vec{B}_\perp - \vec{b}_\perp, Y - y)$$

$$\times \frac{d\sigma^{q\bar{q}A}}{d^2k_T dy d^2 b_\perp}(\vec{x}_{1'0'}), \tag{8.39}$$

---

[3] For instance, one can see that all subsequent evolution in the dipole formed by the (antiquark line of the) gluon 3 and the quark line in Fig. 8.10A cancels.

where the impact parameters are labeled as in Fig. 8.6; $\vec{B}_\perp = (\vec{x}_{1\perp} + \vec{x}_{0\perp})/2$ is the center of the projectile dipole 10 and $\vec{b}_\perp = (\vec{x}_{1'\perp} + \vec{x}_{0'\perp})/2$ is the center of the dipole $1'0'$. Note that the integration over $\vec{b}_\perp$ for fixed $\vec{x}_{1'0'}$ in Eq. (8.39) is equivalent to the integration over $\vec{x}_{11}$ in Eq. (8.35). Equation (8.35), which does not contain any evolution between the projectile and the produced gluon, is recovered from Eq. (8.39) by inserting Eq. (4.83) into it.

We can now write down the final expression for the gluon production in dipole–nucleus scattering. Combining Eq. (8.39) with Eq. (8.35) yields (Kovchegov and Tuchin 2002)

$$\frac{d\sigma^{q\bar{q}A}}{d^2k_T dy}(\vec{x}_{10})$$

$$= \int d^2B_\perp d^2b_\perp d^2x_{1'0'}\, n_1(\vec{x}_{10}, \vec{x}_{1'0'}, \vec{B}_\perp - \vec{b}_\perp, Y - y)$$

$$\times \int d^2x_2\, d^2x_{2'}\, e^{-i\vec{k}_\perp \cdot \vec{x}_{22'}}\, \frac{\alpha_s C_F}{4\pi^4} \sum_{i,j=0'}^{1'} (-1)^{i+j}\, \frac{\vec{x}_{2i} \cdot \vec{x}_{2'j}}{x_{2i}^2 x_{2'j}^2}$$

$$\times \left[N_G(\vec{x}_{i\perp}, \vec{x}_{2'\perp}, y) + N_G(\vec{x}_{2\perp}, \vec{x}_{j\perp}, y) - N_G(\vec{x}_{2\perp}, \vec{x}_{2'\perp}, y) - N_G(\vec{x}_{i\perp}, \vec{x}_{j\perp}, y) \right].$$

$$(8.40)$$

This is the gluon production cross section including the LLA small-x evolution both between the projectile and the produced gluon and between the gluon and the target. While it has been derived in the large-N_c limit, Eq. (8.40) is also valid for any N_c as was shown by Kovner and Lublinsky (2006).

Equation (8.40) can also be written in a k_T-factorized form. Defining the unintegrated gluon distribution of the nucleus (cf. Eq. (8.28)) as

$$\phi_A(y, k_T^2) = \frac{C_F}{\alpha_s(2\pi)^3} \int d^2b_\perp d^2x_\perp e^{-i\vec{k}_\perp \cdot \vec{x}_\perp}\, \nabla_{\vec{x}_\perp}^2 N_G\left(\vec{x}_\perp, \vec{b}_\perp, y\right) \qquad (8.41)$$

and that of the onium (cf. Eq. (8.4)) as

$$\phi_{q\bar{q}}(y, k_T^2) = \frac{\alpha_s C_F}{\pi} \frac{1}{k_T^2} \int d^2b_\perp d^2x_\perp \left(2 - e^{-i\vec{k}_\perp \cdot \vec{x}_\perp} - e^{i\vec{k}_\perp \cdot \vec{x}_\perp} \right)$$

$$\times n_1\left(\vec{x}_{10}, \vec{x}_\perp, \vec{b}_\perp, y\right) \qquad (8.42)$$

and evaluating Eq. (8.40) following steps similar to those in Sec. 8.2.1, we obtain (Braun 2000c, Kovchegov and Tuchin 2002)

$$\frac{d\sigma^{q\bar{q}A}}{d^2k_T\, dy} = \frac{2\alpha_s}{C_F} \frac{1}{k_T^2} \int d^2q_\perp\, \phi_{q\bar{q}}(Y - y, q_T^2)\, \phi_A\left(y, (\vec{k}_\perp - \vec{q}_\perp)^2\right). \qquad (8.43)$$

We see that the LLA small-x evolution preserves the k_T-factorization of Eq. (8.30)!

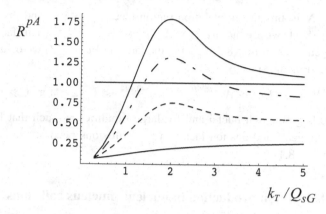

Fig. 8.11. A sketch of the nuclear modification factor as a function of k_T/Q_{sG}, where Q_{sG} is the typical saturation scale (see the caption to Fig. 8.8), for gluon production in pA collisions: the upper solid line corresponds to the quasi-classical approximation from Fig. 8.8, while the dashed-and-dotted, dashed, and lower solid lines demonstrate the change in R^{pA} with increasing rapidity, owing to small-x evolution. (Reprinted with permission from Kharzeev, Kovchegov, and Tuchin (2003). Copyright 2003 by the American Physical Society.)

Equation (8.43) can be generalized to the case of any projectile, in particular the proton, for which we define the unintegrated gluon distribution

$$\phi_p(y, k_T^2) = \frac{C_F}{\alpha_s (2\pi)^3} \int d^2 b_\perp d^2 x_\perp e^{-i \vec{k}_\perp \cdot \vec{x}_\perp} \, \nabla_{\vec{x}_\perp}^2 \, n_G(\vec{x}_\perp, \vec{b}_\perp, y), \qquad (8.44)$$

with n_G the gluon dipole–proton forward scattering amplitude evolved by the linear BFKL evolution equation. One then writes for the inclusive gluon production cross section (Braun 2000c, Kharzeev, Kovchegov, and Tuchin 2003)

$$\frac{d\sigma^{pA}}{d^2 k_T \, dy} = \frac{2\alpha_s}{C_F} \frac{1}{k_T^2} \int d^2 q_\perp \, \phi_p(Y - y, q_T^2) \, \phi_A \left(y, (\vec{k}_\perp - \vec{q}_\perp)^2 \right). \qquad (8.45)$$

The gluon spectrum generated by Eq. (8.45) is qualitatively similar to that obtained in the quasi-classical approximation in Fig. 8.7: the IR divergence is softened to $1/k_T^2$, though it is not removed completely. The nuclear modification factor R^{pA} resulting from using Eq. (8.45) is very different from the quasi-classical one in Fig. 8.8 and is shown in Fig. 8.11; the various curves correspond to different values of rapidity y, such that the lower the curve the higher is the rapidity y. The effect of small-x evolution and saturation is such that the quasi-classical Cronin enhancement at large k_T is replaced by suppression at all values of k_T (Kharzeev, Levin, and McLerran 2003, Albacete *et al.* 2004, Kharzeev, Kovchegov, and Tuchin 2003). The diagram in Fig. 8.11 was confirmed by the precise numerical evaluation of R^{pA} by Albacete *et al.* (2004).

To understand this result analytically we use the approximate solution of the fixed-coupling BK equation immediately outside the saturation region given by Eq. (4.161). Concentrating on its A-dependence we see that $N_G \sim A^{(1+2i\nu_0)/6}$, where we employ the facts that $N_G \sim N$ in this linear-evolution region (see Eq. (8.36)) and that $Q_{s0} \sim A^{1/6}$. Using this in Eq. (8.41) we see that $\phi_A \sim S_\perp A^{(1+2iv_0)/6} \sim A^{5/6+iv_0/3}$. This factor is the only source of A-dependence in Eq. (8.45). Using the latter in Eq. (8.33) we obtain (Kharzeev, Levin, and McLerran 2003)

$$R^{pA}(k_T \gtrsim Q_{sG}) \sim A^{-1/6+iv_0/3} = A^{-0.124} \ll 1 \qquad \text{for } A \gg 1. \qquad (8.46)$$

We see that for large enough nuclei and for large rapidities y (such that Eq. (4.161) is applicable) the nuclear modification factor in pA collisions becomes smaller than 1, in agreement with Fig. 8.11.

8.3 Gluon production in nucleus–nucleus collisions

An important problem, both from the standpoint of saturation physics and for the physics of ultrarelativistic heavy ion collisions, is to find the gluon production cross section for nucleus–nucleus (AA) collisions. Practically, the problem means that one has to find the gluon transverse momentum spectrum in the case when neither the saturation scale of the projectile Q_{s1} nor the saturation scale of the target Q_{s2} is negligibly small in the k_T-range of interest. The solution of the problem would involve first constructing a quasi-classical solution for the MV model with multiple rescatterings in both colliding nuclei. On top of that one would have to include quantum small-x evolution. Phenomenological applications would also require fixing the scales of all the coupling constants in the expression.

At the time of writing none of the above steps has been done analytically. For some theoretical developments see Balitsky (2004), Blaizot and Mehtar-Tani (2009), and Kovchegov (2001). One may perhaps expect that the persistence of the k_T-factorization formula (8.45) for various approximations of gluon production in pA and DIS would indicate that this formula could be valid for AA collisions as well. However, in AA collisions both nuclei come in with fully saturated wave functions, which are completely screened in the IR (see Eq. (5.56) and Fig. 5.7): one would therefore expect that owing to the lack of low-k_T partons the produced-gluon spectrum would have no power-law divergence at small k_T, making it unlikely that k_T-factorization gives the right answer for gluon production in AA collisions. To see this, note that, an inspection of Eq. (8.45) shows that it always leads to $1/k_T^2$ divergence at small k_T, thus contradicting the physical argument we have just presented. Furthermore, the k_T-factorization formula does not appear to agree with the results of numerical solutions of the quasi-classical AA problem.

The quasi-classical gluon field in AA collision was found numerically in the works of Krasnitz and Venugopalan (2000, 2001), Lappi (2003), and Krasnitz, Nara, and Venugopalan (2003a, b). In Fig. 8.12 we show the resulting gluon spectrum and, for comparison, the predictions of the k_T-factorization formula (8.30) (Blaizot, Lappi, and Mehtar-Tani 2010). The solid line in Fig. 8.12 gives the gluon spectrum multiplied by k_T^2 as a function of k_T/Q_{sG} for a numerical solution of the classical Yang–Mills equations with the two nuclei giving the source current ($Q_{sG} = 2$ GeV; the IR cutoff $\Lambda = 0.1$ GeV). The prediction of the k_T-factorization formula (8.30) is shown by the dotted line in Fig. 8.12: one can clearly see that while the two curves are close to each other for $k_T \gtrsim Q_{sG}$, Eq. (8.30) deviates from

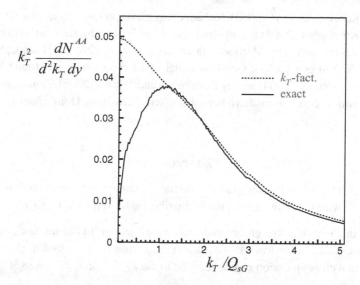

Fig. 8.12. The gluon spectrum multiplied by k_T^2 in AA collisions in the quasi-classical approximation as a function of k_T/Q_{sG}, given by a numerical solution (solid line). The dotted line represents the prediction of the k_T-factorization formula (8.30). (Reprinted from Blaizot, Lappi, and Mehtar-Tani (2010), with permission from Elsevier.) A color version of this figure is available online at www.cambridge.org/9780521112574.

the full solution for $k_T \lesssim Q_{sG}$. In fact $k_T^2 dN^{AA}/d^2k_T dy$ for the full numerical solution goes to zero as $k_T \to 0$, so that the total multiplicity dN^{AA}/dy is independent of the IR cutoff, in agreement with the physical argument presented above.

A promising strategy for including small-x evolution corrections in numerical simulations of the quasi-classical gluon production was proposed recently by Gelis, Lappi, and Venugopalan (2007, 2008a, b, 2009). It involves a new type of k_T-factorization in which the JIMWLK evolutions of both nuclei factorize, each providing sources for quasi-classical gluon production.

Further reading

In this chapter we presented theoretical developments addressing single inclusive gluon production in high energy collisions in the saturation/CGC framework. The techniques presented here have been used to calculate other inclusive observables in DIS and pA collisions. Single inclusive valence quark production was found by Dumitru and Jalilian-Marian (2002) for a hard quark, while the valence quark production at mid-rapidity was calculated by Albacete and Kovchegov (2007a). The prompt photon production cross section was derived by Gelis and Jalilian-Marian (2002a). Two-particle inclusive production (and hence particle correlations) can also be determined. Di-lepton pair production (the Drell–Yan process) was found by Gelis and Jalilian-Marian (2002b), Baier, Mueller, and Schiff

(2004), and Kopeliovich *et al.* (2003). Inclusive two-gluon production was found by Jalilian-Marian and Kovchegov (2004) and by Kovner and Lublinsky (2006). Gluon–valence-quark production was calculated by Marquet (2007) and, using a different technique, by Jalilian-Marian and Kovchegov (2004). Quark–antiquark pair production was found by Blaizot, Gelis, and Venugopalan (2004) and by Kovchegov and Tuchin (2006). A number of properties of two-particle correlations have been discussed recently by Dominguez *et al.* (2011).

Exercises

8.1 Show that the lowest-order gluon production in onium–onium scattering is given by Eq. (8.3) with the unintegrated gluon distributions given by Eq. (8.4).

8.2 Show that the inclusive gluon production cross section in onium–nucleus scattering and in the quasi-classical MV/GGM approximation is given by Eq. (8.35). Demonstrate that this expression can be reduced to the k_T-factorized form (8.30).

8.3 Using a simplified model for N_G,

$$N_G(\vec{x}_\perp, \vec{b}_\perp, 0) = 1 - \exp\left\{ -\frac{x_\perp^2 Q_{sG}^2}{4} \right\}, \tag{8.47}$$

and assuming that Q_{sG}^2 is $\vec{b}_\perp$-independent in a very large circle of radius R and is zero outside the circle, evaluate Eq. (8.20) exactly. Use the obtained expression to construct the nuclear modification factor R^{pA} defined in Eq. (8.33). Plot the resulting R^{pA} as a function of k_T/Q_{sG} and compare the plot with Fig. 8.8.

8.4 Prove the cancellations in Eqs. (8.37).

8.5* Reduce Eq. (8.40) to Eq. (8.43) using the unintegrated gluon distributions defined in Eqs. (8.41) and (8.42).

9
Instead of conclusions

The research on saturation/CGC physics is ongoing, with a number of open theoretical and phenomenological questions. Therefore, instead of conclusions, in this chapter we briefly review the phenomenology of saturation/CGC physics and list some important open theoretical problems.

9.1 Comparison with experimental data

In this section we will give a brief overview of how high energy QCD theory compares with the current experimental data. The reader may wonder whether such a comparison is possible to fit into one short section; indeed, a comparison of saturation/CGC physics with experiment could be a subject for a separate book. However, a serious quantitative comparison with experiment suffers from two major difficulties. The first is that a well-developed theoretical approach exists only for the scattering of a dilute parton system on a dense one; the key examples are DIS on nuclei (eA) and the proton–nucleus (pA) collisions considered earlier. At the same time, much of the data exist either for the scattering of a dense parton system on another dense parton system, as is the case in nucleus–nucleus (AA) collisions, or for the scattering of two dilute systems on each other, like DIS on a proton (ep) or proton–proton (pp) collisions. The theoretical progress in the description of these reactions in the saturation/CGC framework is rather limited, with many open questions and opportunities for further research (see Sec. 8.3 for a brief summary of the existing AA results). Hence, in describing the AA, ep, and pp data using existing theoretical knowledge one is often forced to make assumptions whose validity is hard to verify.

The second difficulty in comparing the saturation/CGC physics with experiment is in the fact that many experimental observations allow alternative descriptions, usually in the framework of DGLAP evolution within the standard collinear factorization framework. The key problem is that an *experimentum crucis* that would allow us to unambiguously differentiate the nonlinear saturation physics from the linear DGLAP evolution has not been found. This is also partly a theoretical problem.

On the positive side, high energy QCD leads to a unified description of various experimental observations providing not only a possible understanding of the underlying physics but also suggestions and directions for future experiments. The phenomenological picture

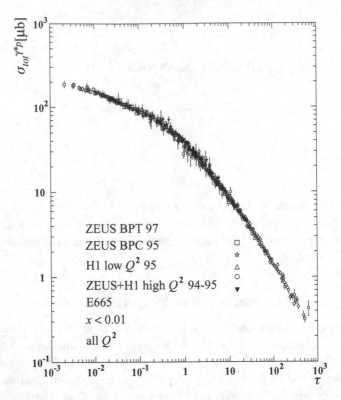

Fig. 9.1. Geometric scaling in the total $\gamma^* p$ cross section in DIS. (Reprinted with permission from Stasto, Golec-Biernat, and Kwiecinski (2001). Copyright 2001 by the American Physical Society.)

resulting from the theoretical developments discussed in this book is so beautiful, universal, and self-consistent that we cannot finish the book without sharing it with our reader.

9.1.1 Deep inelastic scattering

As discussed earlier (see Sec. 4.5), one of the most striking predictions of high energy QCD is that the DIS structure functions should depend on only one variable, $\tau = Q^2/Q_s^2$: this is the phenomenon of *geometric scaling*. This scaling behavior is a manifestation of the simple idea that the only relevant dimensionful scale at high energy is the saturation momentum. Geometric scaling was first observed by Stasto, Golec-Biernat, and Kwiecinski (2001) and is shown in Fig. 9.1 for a compilation of HERA data on the total $\gamma^* p$ cross section for $x < 0.01$.

Another important consequence of saturation physics is that the ratio of the diffractive and total cross sections should be independent of energy (see Eqs. (7.23) and the discussion around them). The experimental data from HERA shown in Fig. 9.2 appears to agree with this prediction.

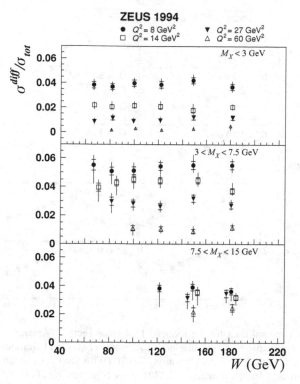

Fig. 9.2. The ratio of the cross section for diffractive production in DIS and the total cross section, as a function of energy W, for different intervals in the mass of the produced hadrons M_X. (Reprinted with permission from Abramowicz and Caldwell (1999). Copyright 1999 by the American Physical Society.)

On the more quantitative side, the first comprehensive fit of the HERA DIS data based on a variation of the GGM/MV formula (4.51) for the dipole amplitude was carried out by Golec-Biernat and Wusthoff (1999a, b) and is known as the GBW model (see Eqs. (4.12), (4.24), and (4.10) for the relation between F_2, F_1, and the dipole amplitude N). A more recent fit, based on rcBK evolution, of the combined data on the DIS cross section reported by the H1 and ZEUS experiments at HERA is shown in Fig. 9.3. It can be seen that the "reduced" DIS cross section, defined by

$$\sigma_r = F_2 - \frac{y^2}{1 + (1 - y)^2} F_L \qquad (9.1)$$

(with y from Eq. (2.2)) is well described by the rcBK evolution.

9.1.2 Proton(deuteron)–nucleus collisions

The formalism developed in this book is applicable to proton–nucleus collisions, as we saw in Chapter 8. One can therefore test the predictions made in that chapter: in particular we showed that saturation physics predicts a transition of the nuclear modification factor from

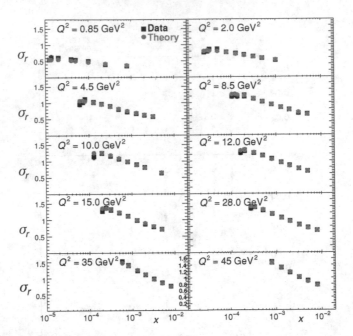

Fig. 9.3. Experimental data for the reduced DIS cross section (squares) in different Q^2 bins with, for comparison, the calculations of Albacete *et al.* (2011) (circles). The data are taken from the H1 and ZEUS collaboration (2010). (With kind permission from Springer Science+Business Media: Albacete *et al.* (2011).) A color version of this figure is available online at www.cambridge.org/9780521112574.

Cronin enhancement to suppression at all produced particle momenta p_T (see Fig. 8.11). This prediction may be compared with the data by means of Fig. 9.4, where we plot the nuclear modification factor $R_{d+\text{Au}}$ for negatively charged hadrons obtained in deuteron–gold ($d+\text{Au}$) collisions at the Relativistic Heavy Ion Collider (RHIC) at Brookhaven National Laboratory by the BRAHMS collaboration. (At RHIC, data was collected for $d+\text{Au}$ collisions instead of pA; we assume that the deuteron is also a dilute parton system, not unlike the proton.) The various different panels in Fig. 9.4 correspond to different pseudo-rapidity values; clearly, suppression sets in as the rapidity η increases, in agreement with the saturation physics prediction.

For dilute–dilute parton-system scattering (say, for pp collisions), we expect that two jets with large transverse momenta $\vec{p}_{1\perp}$ and $\vec{p}_{2\perp}$ are produced back to back, so that $\vec{p}_{2\perp} \approx -\vec{p}_{1\perp}$ as required by momentum conservation if we assume that the other produced particles are few and carry small transverse momenta. In the saturation phase, new processes are possible in which the large value of $\vec{p}_{1\perp}$ is not compensated by a single second jet but instead the momentum is distributed among many particles with average transverse momentum of order Q_s, thus depleting the back-to-back correlation of the jets. A large rapidity interval between the two measured particles only makes this effect stronger, by enhancing extra emissions by powers of rapidity. Figure 9.5 demonstrates that it is likely that this effect has been observed experimentally. Data on neutral pion correlations reported by the PHENIX

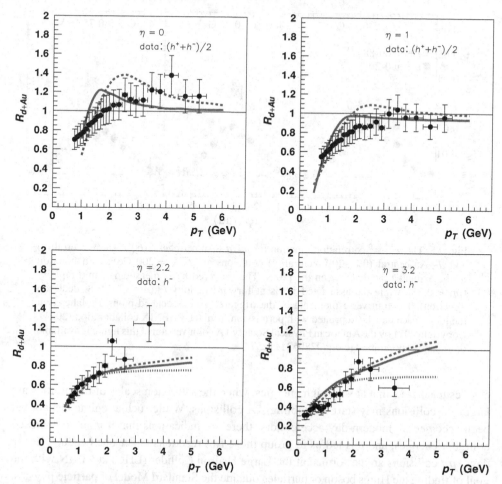

Fig. 9.4. Nuclear modification factor R_{d+Au} of charged particles for different pseudo-rapidities. The data are taken from the BRAHMS collaboration (2004). The theoretical curves and figures are taken from Kharzeev, Kovchegov, and Tuchin (2004). (Reprinted from Kharzeev, Kovchegov and Tuchin (2004), with permission from Elsevier.) A color version of this figure is available online at www.cambridge.org/9780521112574.

collaboration at RHIC are shown. The figure gives the $\pi^0-\pi^0$ correlation function for pp, $d+$Au peripheral, and $d+$Au central collisions as a function of the azimuthal angle $\Delta\phi$ between the pions (the angle in the transverse plane) for three different pairs of values of the pions' transverse momenta. One clearly sees that the back-to-back correlation at $\Delta\phi = \pi$ in the central $d+$Au collisions, where saturation effects should be strongest, is indeed depleted as compared with the pp or peripheral $d+$Au cases.

9.1.3 Proton–proton and heavy ion collisions

Proton–(anti)proton and nucleus–nucleus collisions are indeed very different in the sizes of the colliding particles and in the multiplicity of the produced particles. Still, saturation

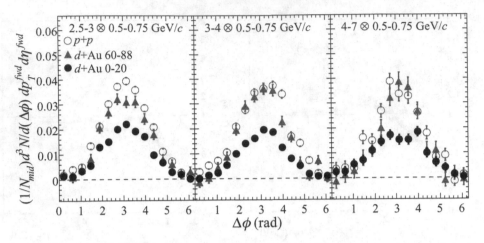

Fig. 9.5. The $\pi^0-\pi^0$ correlation function for *pp*, *d*+Au peripheral (60%−88% centrality) and *d*+Au central (0%−20% centrality) collisions at $\sqrt{s_{NN}} = 200$ GeV, as a function of the azimuthal angle between the pions. The triggered π^0 is measured at mid-rapidity ($|\eta| < 0.35$), while the associated π^0 is at forward rapidity ($3.0 < |\eta| < 3.8$, deuteron direction). The transverse momenta of the triggered and associated pions are labeled at the top of each panel. (Reprinted with permission from the PHENIX collaboration (2011). Copyright 2011 by the American Physical Society.) A color version of this figure is available online at www.cambridge.org/9780521112574.

physics teaches us that at very high energies, when the saturation scales of the protons are large, *pp* collisions may start resembling *AA* collisions. While such a regime has not yet been accessed by modern-day accelerators, there are indications that it may be achieved at higher energies, and this led us to group those two reactions under one heading. These days, *pp* collisions are performed at the Large Hadron Collider (LHC) at CERN with the goal of finding the Higgs boson or particles outside the Standard Model of particle physics. Ultrarelativistic heavy ion (*AA*) collisions are being carried out at RHIC and LHC with the aim of creating a thermal medium of quarks and gluons, the quark–gluon plasma (QGP), and studying its properties (see e.g. the review by Kolb and Heinz (2003)).

The first piece of evidence in favor of saturation physics comes again from geometric scaling: one might expect that geometric scaling in the distribution functions would translate into such a scaling for the produced particle spectra; this conclusion is supported by more detailed calculations. Geometric scaling is observed in both *pp* and *AA* collisions and is depicted here in Fig. 9.6, where we show plots of charged-hadron transverse momentum spectra in *pp* and *AA* collisions as functions of the scaling variable $\tau = (p_T/Q_s)^{2+\lambda}$, where $Q_s = Q_0^{2/(2+\lambda)}(\sqrt{s} \times 10^{-3})^{\lambda/(2+\lambda)}$ with $Q_0 = 1$ GeV, $\sqrt{s}$ measured in GeV, and λ as specified in the figures. In the *AA* case the scaling variable τ also includes a factor $A^{-1/3}$. The quality of scaling is much lower in the *AA* case owing to the quark–gluon plasma (QGP) final-state effects.

Since the saturation scale is the only relevant momentum scale in the problem, the hadron multiplicity produced in a *pp* or *AA* collision per unit transverse area should be

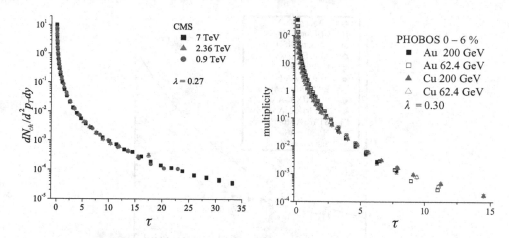

Fig. 9.6. Geometric scaling behavior of the charged hadron spectrum $dN_{ch}/d^2 p_T \, dy$ for proton–proton (left-hand panel) and heavy ion collisions (right-hand panel), plotted as a function of the scaling variable $\tau = (p_T/Q_s)^{2+\lambda}$ for several values of the center-of-mass energy. (Reprinted from McLerran and Praszalowicz (2011) (left) and Praszalowicz (2011) (right), with permission by *Acta Physica Polonica*. The data are as reported by the CMS collaboration (2010a, b) and the PHOBOS collaboration (2002, 2003) respectively. A color version of this figure is available online at www.cambridge.org/9780521112574.

proportional to the saturation scale squared (Gribov, Levin, and Ryskin 1983, McLerran and Venugopalan 1994a),

$$\frac{1}{S_\perp} \frac{dN}{dy} \sim Q_s^2, \tag{9.2}$$

leading to the prediction that the particle multiplicity should grow with Q_s^2 as a power of energy. A compilation of the experimental data on the hadron multiplicity in pp and AA collisions at various energies is shown in Fig. 9.7, demonstrating a power-of-energy growth for both reactions, along with saturation-model fits by Levin and Rezaeian (2011). (The highest-energy pp data point appeared after the saturation prediction.)

Saturation physics also predicts that the average transverse momentum of the produced particles should be proportional to the saturation scale, $\langle p_T \rangle \sim Q_s$ (see e.g. Fig. 8.7) since again it is the only scale in the problem. One expects $\langle p_T \rangle$ to grow with energy and, because of Eq. (9.2), with particle multiplicity. This behavior is demonstrated in Fig. 9.8. The left-hand panel of Fig. 9.8 shows a compilation of the proton–proton and proton–anti-proton collision data collected in several experiments over the years, demonstrating that $\langle p_T \rangle$ does grow with energy. The right-hand panel of Fig. 9.8 shows the growth of $\langle p_T \rangle$ with charged-hadron multiplicity in pp collisions at LHC. It should be stressed that, in the traditional approaches based on high energy pomeron phenomenology, $\langle p_T \rangle$ does not depend on either energy or multiplicity; therefore we consider the observation of these dependences in Fig. 9.8 as a strong argument in favor of the advantage of saturation physics over such models.

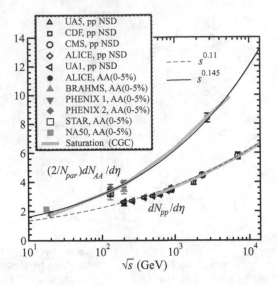

Fig. 9.7. Dependence of hadron multiplicity on energy for *pp* (lower curve and points) and *AA* (upper curve and points) collisions. (Reprinted with permission from Levin and Rezaeian (2011). Copyright 2011 by the American Physical Society.) A color version of this figure is available online at www.cambridge.org/9780521112574.

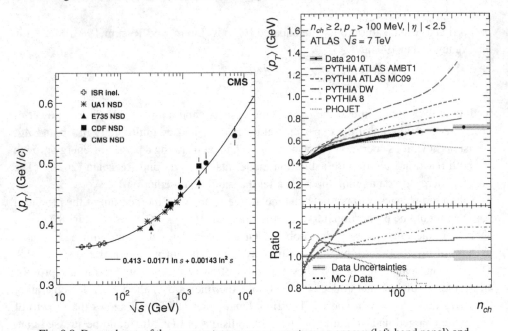

Fig. 9.8. Dependence of the average transverse momentum on energy (left-hand panel) and on multiplicity and hence on centrality (right-hand panel). (The left-hand panel is reprinted with permission from the CMS collaboration (2010b). Copyright 2010 by the American Physical Society. The right-hand panel is reprinted from the ATLAS collaboration (2011), with permission by IOP Publishing.) A color version of this figure is available online at www.cambridge.org/9780521112574.

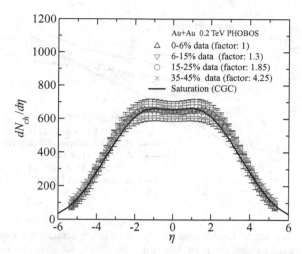

Fig. 9.9. Dependence of hadron multiplicity on rapidity for AA collisions. The experimental data are from the PHOBOS collaboration at RHIC (see PHOBOS collaboration, 2002, 2003). (Reprinted with permission from Levin and Rezaeian (2011). Copyright 2011 by the American Physical Society.) A color version of this figure is available online at www.cambridge.org/9780521112574.

It is difficult to make precise quantitative saturation/CGC physics predictions for the hadron multiplicity and spectra in pp and AA collisions since, mentioned above, these theoretical problems have not been solved. Instead, as a reasonable approximation to the full answer, one may use the k_T-factorization formula (8.45) with the q_T-integral cut off by an upper bound proportional to k_T. This is known as the Kharzeev–Levin–Nardi (KLN) approach (Kharzeev and Nardi 2001, Kharzeev and Levin 2001, Kharzeev, Levin, and Nardi 2005a, b). Predictions based on the KLN approach have been quite successful in describing heavy ion data on multiplicities. A fit of the RHIC multiplicity data plotted as a function of pseudo-rapidity for different centrality bins (denoted by percentages in the legend, along with the appropriate scaling factors) based on the KLN model is shown in Fig. 9.9; clearly the data is well described by the saturation model.

This agreement between the multiplicity data in heavy ion collisions and the saturation predictions is further illustrated in Fig. 9.10, where we show the particle multiplicity per participating nucleon as a function of the collision centrality; N_{part} is the number of nucleons participating in the collision and varies between a few for peripheral collisions to $2A$ for central collisions (for identical nuclei). The three lower lines in the legend of Fig. 9.10 correspond to *predictions* coming from saturation-based models.[1] Clearly all the curves do well; the prediction by Albacete and Dumitru (2011) (the short-dashed curve in Fig. 9.10) based on rcBK evolution for the dipole amplitude combined with the KLN model for particle production matches the data almost perfectly.

---

[1] The three upper curves in the legend of Fig. 9.10 are Monte-Carlo simulations not based on saturation physics; note, however, that the HIJING event generator prediction uses an IR cutoff that grows with energy, which is reminiscent of the saturation scale.

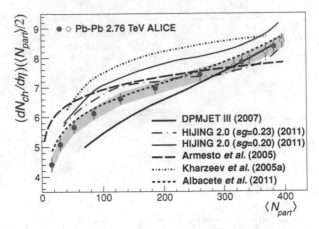

Fig. 9.10. Dependence of hadron multiplicity on collision centrality for AA collisions: the saturation-based models are from Armesto *et al.* (2005), Kharzeev *et al.* (2005a), and Albacete *et al.* (2011), while the no-saturation Monte-Carlo simulations are DPMJET III, Bopp *et al.* (2007); HIJING, Deng *et al.* (2011). (Reprinted with permission from the ALICE collaboration (2011) at the LHC. Copyright 2011 by the American Physical Society.) A color version of this figure is available online at www.cambridge.org/9780521112574.

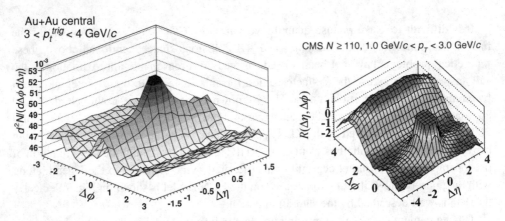

Fig. 9.11. Two-particle correlation function plotted versus pseudo-rapidity interval $\Delta\eta$ and azimuthal angle $\Delta\phi$ between the particles for AA (left-hand panel) and high-multiplicity pp (right-hand panel) collisions. (Left-hand panel: reprinted figure with permission from the STAR collaboration (2009). Copyright 2009 by the American Physical Society. Right-hand panel: reprinted with kind permission from Springer Science+Business Media: CMS collaboration (2010a).) A color version of this figure is available online at www.cambridge.org/9780521112574.

Another possible piece of evidence in favor of saturation/CGC physics is provided by the long-range rapidity correlations between the produced hadrons at small azimuthal angles $\Delta\phi \approx 0$, which have been seen in AA and, more recently, in pp collisions and are shown in Fig. 9.11. Owing to the shape of these correlations they are often referred to as the "ridge". The search for a detailed explanation of them in the saturation framework is still in progress (see Gavin, McLerran, and Moschelli (2009), and Dumitru *et al.* 2008, 2011a, b), but a discussion of this work is beyond of the scope of the present text. However, two facts, inherent properties of the saturation approach, are shown experimentally. First, the correlations are long-range in rapidity. Second, in *pp* collisions the long-range rapidity correlations at $\Delta\phi \approx 0$ appear only in events with high multiplicity ($N \geq 110$ in the right-hand panel of Fig. 9.11): hence high-multiplicity *pp* collisions appear to be similar to the *AA* collisions. This suggests that the source of the correlations shown in Fig. 9.11 is in the properties of dense parton systems.

9.2 Unsolved theoretical problems

In this book we have tried to introduce our readers to the new world of the ideas and methods of high energy QCD. We hope that after reading our book the reader will be able to work in this field, which is in the early stages of development. We conclude the book by outlining unsolved theoretical problems in the field.

1. Impact parameter dependence In Chapters 4 and 5 we showed that the dipole–nucleus forward scattering amplitude $N(\vec{x}_\perp, \vec{b}_\perp, Y)$ resulting either from the GGM model or from BK/JIMWLK evolution does not violate the black-disk limit, so that one always has $N \leq 1$. While this is an improvement over BFKL evolution, some unitarity problems still remain. As we saw in Sec. 3.3.6, the Froissart–Martin bound consists of two ingredients, that the scattering obeys the black-disk limit and that the radius of the black disk grows logarithmically with energy. As follows from Eq. (3.115), the latter property results from QCD having a mass gap (since the lightest particle in the spectrum, the pion, has a nonzero mass). In perturbative QCD one works with gluons, which have zero mass: clearly Eq. (3.115) should no longer apply and the Froissart–Martin bound is violated.

To see this explicitly imagine that we are trying to prove the Froissart–Martin bound for onium–onium scattering. Consider the scattering at large impact parameter $b_\perp$, where the scattering amplitude is small and nonlinear saturation corrections can be neglected. The onium–onium scattering cross section, defined in Eq. (4.85), is then governed by the BFKL evolution (4.87) (in the LLA). At large $b_\perp$ one can show that the general BFKL equation solution in Eq. (4.126) leads to

$$n\left(\vec{x}_{10}, \vec{x}_{1'0'}, \vec{b}_\perp, Y\right) \xrightarrow{b_\perp \gg x_{10}, x_{1'0'}} \int_{-\infty}^{\infty} dv \left(\frac{x_{10}^2 \, x_{1'0'}^2}{b_\perp^4}\right)^{1/2+iv} C(v) \, e^{\bar{\alpha}_s \chi(0,v)Y}, \qquad (9.3)$$

where $C(v)$ is a function of v, the exact form of which is not important to us. Evaluating the v-integral in Eq. (9.3) near the saddle point at $v = 0$, we get

$$n\left(\vec{x}_{10}, \vec{x}_{1'0'}, \vec{b}_\perp, Y\right) \xrightarrow{b_\perp \gg x_{10}, x_{1'0'}} \frac{x_{10} \, x_{1'0'}}{b_\perp^2} \, e^{(\alpha_P - 1)Y}. \qquad (9.4)$$

Using Eq. (9.4) instead of Eq. (3.113), we obtain the black-disk radius $R = b^*$ by requiring that (cf. Eq. (3.114))

$$\frac{x_{10} \, x_{1'0'}}{b^{*2}} \, e^{(\alpha_P - 1) Y} \propto 1 \qquad (9.5)$$

so that

$$b^{*2} \propto s^{\alpha_P - 1}. \qquad (9.6)$$

Thus the radius of interaction increases as a power of s, $R^2 \sim s^{\alpha_P - 1}$. The total cross section increases as a power of energy too,

$$\sigma_{tot} = 2\pi R^2 \propto s^{\alpha_P - 1}. \qquad (9.7)$$

We conclude that nonlinear evolution gives a scattering amplitude that satisfies the black-disk limit but still leads to a violation of the Froissart–Martin bound, owing to the fast growth of the black disk.

This power-like increase in the total cross section was first discussed by Kovner and Wiedemann (2002a, b, 2003). Their conclusions were confirmed by a numerical solution of the BK equation with impact parameter dependence performed by Golec-Biernat and Stasto (2003) (see also Gotsman *et al.* (2004)). Equation (9.7) shows that saturation/CGC physics (or any other perturbative QCD calculation) cannot be trusted in large-impact-parameter scattering and some nonperturbative effects need to come in to make the total cross section satisfy the Froissart–Martin bound. One hopes that, with some minimalistic assumptions about confinement physics included in the evolution, and for a large nucleus, such peripheral nonperturbative effects would give a relatively small contribution to the total cross section, since the perimeter scales as a smaller power of A than the area. The existing data appears to indicate that such a hope is not unfounded.

2. Higher-order corrections to the BFKL, BK, and JIMWLK evolution equations We have briefly discussed the problem of higher-order corrections to the BFKL evolution in Sec. 6.3, outlining possible ways of getting the corrections under control. To improve the precision of the BK/JIMWLK predictions for the phenomenology, one needs to carry out this (or any other) program for calculating higher-order corrections to these nonlinear evolution equations. Would the agreement with the data shown in Sec. 9.1 survive the inclusion of higher-order corrections? Can we devise a systematic way of improving the precision of the nonlinear evolution equations? These are important questions, which need to be studied seriously.

3. Scattering of two dilute systems: BFKL pomeron loops At first sight the scattering of one dilute system of partons on another dilute system of partons (say the scattering of two onia with small sizes) can be described by the exchange of a BFKL pomeron. However, as we discussed in Sec. 3.3.6, such a contribution would violate the black-disk limit at high energy. Owing to the low initial parton density in both onia (so that there is no parameter A), the pomeron fan diagrams of Fig. 3.23 in such a dilute–dilute scattering are not enhanced compared to the pomeron loop diagrams of Fig. 3.24: both kinds of interaction have to be included. The BK/JIMWLK evolution equations assume that one

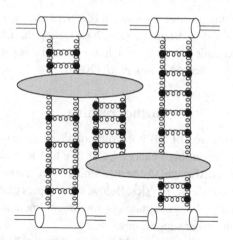

Fig. 9.12. A diagram contributing to the total cross section of nucleus–nucleus scattering in the BFKL pomeron calculus.

scattering particle (the nucleus) is large and thus do not contain the pomeron loop contributions from Fig. 3.24. Nonetheless it is likely that such contributions are essential for the unitarization of onium–onium scattering. This problem is important because the high energy onium–onium scattering process will happen in the next-generation linear colliders (where each colliding lepton could split into a virtual photon, with each γ^* in turn splitting into a $q\bar{q}$ pair). On top of that, unitarization in a dilute–dilute scattering would be "pure QCD" in its nature, since it would not be relying on a large nucleus as one of the scatterers. In spite of many efforts (see Salam 1995, 1996, Navelet and Peschanski 1999, Mueller, Shoshi, and Wong 2005, Levin and Lublinsky 2005b, Iancu and Triantafyllopoulos 2005, Kovner and Lublinsky 2005d, Hatta *et al.* 2006, Altinoluk *et al.* 2009, Levin, Miller, and Prygarin 2008) we are still far from understanding this problem of dilute–dilute scattering.

4. Heavy ion collisions We have already mentioned, in Sec. 8.3, that the calculation of gluon and quark production in AA collisions in the saturation/CGC framework is an important problem for understanding the early-time dynamics of heavy ion collisions. While some progress on this issue has been achieved lately, mainly along the lines of finding a numerical solution, the problem is still an open one. A related problem, which is also open, is the problem of thermalization in heavy ion collisions: one needs to understand how the produced quarks and gluons form a thermal medium (QGP), which is likely to be observed in heavy ion collisions.

Multi-particle correlations in AA collisions are also important, particularly correlations in rapidity since they are sensitive to subtle details of the early-time dynamics and can be measured experimentally. Initial progress on this issue has recently been achieved by Gavin, McLerran, and Moschelli (2009), and by Dumitru *et al.* (2008, 2011a, b).

Another, related, unsolved theoretical problem concerns the calculation of the total nucleus–nucleus scattering cross section in the saturation/CGC framework. One has to sum diagrams fanning out in the directions of both nuclei (Braun 2000b, 2004). A sample diagram contributing to this process is shown in Fig. 9.12. An equation summing all such diagrams has been suggested by Braun (2000a, 2004) using the BFKL pomeron calculus. The solution of such an equation is not known, however, and it is yet to be reproduced using the BK/JIMWLK approach (see Altinoluk *et al.* (2009)).

Further reading

The inevitable space limitations of this book did not allow us to include (or do justice to) everything that is known about high energy QCD. We had to make a selection based on what, in our opinion, is needed to quickly bring a student or a researcher from a neighboring field up to speed in saturation/CGC physics. Below, some topics that did not make it into the book are introduced briefly; they constitute our final recommendation for further reading, complementary to all our earlier suggestions.

Ciafaloni (1988), Catani, Fiorani, and Marchesini (1990a, b), and Marchesini (1995) suggested an evolution equation that, in the framework of a single equation, reproduces both the Q^2 (DGLAP) and x (BFKL) evolutions. This result, known as the CCFM equation, is at the foundation of most Monte Carlo simulations of high energy collisions. While the presentation of CCFM goes beyond the scope of this book, we certainly recommend our readers to learn more about this equation by reading the original papers mentioned above.

We have also omitted any presentation of particle production in the collinear factorization framework; however, the reader may learn the basics of collinear factorization from the textbook by Sterman (1993), with more advanced results along with a presentation of the related jet physics given by Dokshitzer, Diakonov, and Troian (1980), Collins, Soper, and Sterman (1985a, b, 1988a, b), Dokshitzer *et al.* (1991), and Collins (2011).

Finally, in our book we have presented only the basic aspects of the connections between saturation/CGC physics and heavy ion collisions. This is in part due to the fact that understanding AA collisions in saturation physics is a difficult, unsolved, problem. We recommend the reviews by McLerran (2005, 2008, 2009a) as a starting point for the further exploration of the richness of the saturation/CGC dynamics in AA collisions.

Appendix A

Reference formulas

A.1 Dirac matrix element tables

Tables A.1 and A.2 first appeared in Lepage and Brodsky (1980). Following the original authors of these tables we use the notation

$$\vec{p}_\perp \times \vec{p}'_\perp \equiv p^1 p'^2 - p^2 p'^1. \tag{A.1}$$

Note that $\epsilon^{12} = -\epsilon^{21} = 1$, $\epsilon^{11} = \epsilon^{22} = 0$. The spinors in the tables are defined in Eqs. (1.50) and (1.51).

The following formulas will be useful in relating other matrix elements to those tabulated:

$$\bar{v}_{\sigma'}(p')\gamma^\mu v_\sigma(p) = \bar{u}_\sigma(p)\gamma^\mu u_{\sigma'}(p'), \tag{A.2}$$

$$\bar{v}_{\sigma'}(p')\gamma^\mu \gamma^\nu \gamma^\rho v_\sigma(p) = \bar{u}_\sigma(p)\gamma^\rho \gamma^\nu \gamma^\mu u_{\sigma'}(p'), \tag{A.3}$$

$$\bar{v}_{\sigma'}(p')v_\sigma(p) = -\bar{u}_\sigma(p)u_{\sigma'}(p'). \tag{A.4}$$

These formulas allow one to obtain matrix elements constructed only from v-spinors using Table A.1 for matrix elements constructed solely from u-spinors.

For any Dirac spinors ψ and η and a 4×4 matrix Γ, the following is true:

$$\left(\bar{\psi}\Gamma\eta\right)^* = \bar{\eta}\gamma^0\Gamma^\dagger\gamma^0\psi. \tag{A.5}$$

This allows one to construct matrix elements of the type

$$\bar{u}_\sigma(p)\Gamma v_{\sigma'}(p') = [\bar{v}_{\sigma'}(p')\gamma^0\Gamma^\dagger\gamma^0 u_\sigma(p)]^* \tag{A.6}$$

from those tabulated in Table A.2. As

$$\gamma^0\gamma^\mu\gamma^0 = (\gamma^\mu)^\dagger \tag{A.7}$$

in the case $\Gamma = \gamma^\mu$, Eq. (A.6) gives

$$\bar{u}_\sigma(p)\gamma^\mu v_{\sigma'}(p') = \left(\bar{v}_{\sigma'}(p')\gamma^\mu u_\sigma(p)\right)^*. \tag{A.8}$$

A.2 Some useful integrals

Below we list several integrals used throughout the book. We leave their derivation as an exercise for the reader. The Green function of the Laplace equation in two dimensions is

$$\int \frac{d^2 q_\perp}{q_\perp^2} e^{i\vec{q}_\perp \cdot \vec{x}_\perp} = 2\pi \ln \frac{1}{x_\perp \Lambda}, \tag{A.9}$$

Table A.1. *Dirac matrix elements constructed from u-spinors only. Table reprinted with permission from Lepage and Brodsky (1980). Copyright 1980 by the American Physical Society.*

Matrix element	Value
$\dfrac{\bar{u}_{\sigma'}(p')}{\sqrt{p'^+}}\gamma^+\dfrac{u_\sigma(p)}{\sqrt{p^+}}$	$2\delta_{\sigma\sigma'}$
$\dfrac{\bar{u}_{\sigma'}(p')}{\sqrt{p'^+}}\gamma^-\dfrac{u_\sigma(p)}{\sqrt{p^+}}$	$\delta_{\sigma\sigma'}\dfrac{2}{p^+p'^+}(\vec{p}_\perp\cdot\vec{p}'_\perp-i\sigma\vec{p}_\perp\times\vec{p}'_\perp+m^2)$ $-\delta_{\sigma,-\sigma}\sigma\dfrac{2m}{p^+p'^+}\left[(p'^1+i\sigma p'^2)-(p^1+i\sigma p^2)\right]$
$\dfrac{\bar{u}_{\sigma'}(p')}{\sqrt{p'^+}}\gamma_\perp^i\dfrac{u_\sigma(p)}{\sqrt{p^+}}$	$\delta_{\sigma\sigma'}\left(\dfrac{p'^i_\perp-i\sigma\epsilon^{ij}p'^j_\perp}{p'^+}+\dfrac{p^i_\perp+i\sigma\epsilon^{ij}p^j_\perp}{p^+}\right)$ $-\delta_{\sigma,-\sigma}\sigma m\left(\dfrac{p'^+-p^+}{p'^+p^+}\right)(\delta^{i1}+i\sigma\delta^{i2})$
$\dfrac{\bar{u}_{\sigma'}(p')}{\sqrt{p'^+}}\dfrac{u_\sigma(p)}{\sqrt{p^+}}$	$\delta_{\sigma\sigma'}m\dfrac{p^++p'^+}{p^+p'^+}-\delta_{\sigma,-\sigma}\sigma\left(\dfrac{p'^1+i\sigma p'^2}{p'^+}-\dfrac{p^1+i\sigma p^2}{p^+}\right)$
$\dfrac{\bar{u}_{\sigma'}(p')}{\sqrt{p'^+}}\gamma^-\gamma^+\gamma^-\dfrac{u_\sigma(p)}{\sqrt{p^+}}$	$4\dfrac{\bar{u}_{\sigma'}(p')}{\sqrt{p'^+}}\gamma^-\dfrac{u_\sigma(p)}{\sqrt{p^+}}$
$\dfrac{\bar{u}_{\sigma'}(p')}{\sqrt{p'^+}}\gamma^-\gamma^+\gamma_\perp^i\dfrac{u_\sigma(p)}{\sqrt{p^+}}$	$\delta_{\sigma\sigma'}4\dfrac{p'^i_\perp-i\epsilon^{ij}\sigma p'^j_\perp}{p'^+}+\delta_{\sigma,-\sigma'}\sigma\dfrac{4m}{p'^+}(\delta^{i1}+i\sigma\delta^{i2})$
$\dfrac{\bar{u}_{\sigma'}(p')}{\sqrt{p'^+}}\gamma_\perp^i\gamma^+\gamma^-\dfrac{u_\sigma(p)}{\sqrt{p^+}}$	$\delta_{\sigma\sigma'}4\dfrac{p^i_\perp+i\epsilon^{ij}\sigma p^j_\perp}{p^+}-\delta_{\sigma,-\sigma'}\sigma\dfrac{4m}{p^+}(\delta^{i1}+i\sigma\delta^{i2})$
$\dfrac{\bar{u}_{\sigma'}(p')}{\sqrt{p'^+}}\gamma_\perp^i\gamma^+\gamma_\perp^j\dfrac{u_\sigma(p)}{\sqrt{p^+}}$	$\delta_{\sigma\sigma'}2(\delta^{ij}+i\sigma\epsilon^{ij})$

where Λ is the IR cutoff on the integration. Taking the transverse gradient of Eq. (A.9) yields

$$\int d^2q_\perp e^{i\vec{q}_\perp\cdot\vec{x}_\perp}\frac{\vec{q}_\perp}{q_\perp^2}=2\pi i\frac{\vec{x}_\perp}{x_\perp^2}. \tag{A.10}$$

Here is a variation of Eq. (A.9), for a massive Green function:

$$\int \frac{d^2q_\perp}{q_\perp^2+m^2}e^{i\vec{q}_\perp\cdot\vec{x}_\perp}=2\pi K_0(mx_\perp). \tag{A.11}$$

Table A.2. *Dirac matrix elements constructed from u- and v-spinors. Table reprinted with permission from Lepage and Brodsky (1980). Copyright 1980 by the American Physical Society.*

Matrix element	Value
$\dfrac{\bar{v}_{\sigma'}(p')}{\sqrt{p'^+}}\gamma^+\dfrac{u_\sigma(p)}{\sqrt{p^+}}$	$2\delta_{\sigma,-\sigma'}$
$\dfrac{\bar{v}_{\sigma'}(p')}{\sqrt{p'^+}}\gamma^-\dfrac{u_\sigma(p)}{\sqrt{p^+}}$	$\delta_{\sigma,-\sigma'}\dfrac{2}{p^+p'^+}(\vec{p}_\perp\cdot\vec{p}'_\perp-i\sigma\vec{p}_\perp\times\vec{p}'_\perp-m^2)$ $-\delta_{\sigma\sigma'}\sigma\dfrac{2m}{p^+p'^+}\left[(p'^1+i\sigma p'^2)+(p^1+i\sigma p^2)\right]$
$\dfrac{\bar{v}_{\sigma'}(p')}{\sqrt{p'^+}}\gamma^i_\perp\dfrac{u_\sigma(p)}{\sqrt{p^+}}$	$\delta_{\sigma,-\sigma'}\left(\dfrac{p'^i_\perp-i\sigma\epsilon^{ij}p'^j_\perp}{p'^+}+\dfrac{p^i_\perp+i\sigma\epsilon^{ij}p^j_\perp}{p^+}\right)$ $-\delta_{\sigma\sigma'}\sigma m\left(\dfrac{p'^++p^+}{p'^+p^+}\right)(\delta^{i1}+i\sigma\delta^{i2})$
$\dfrac{\bar{v}_{\sigma'}(p')u_\sigma(p)}{\sqrt{p'^+}\sqrt{p^+}}$	$\delta_{\sigma,-\sigma}m\dfrac{p'^+-p'^+}{p^+p'^+}-\delta_{\sigma\sigma'}\sigma\left(\dfrac{p'^1+i\sigma p'^2}{p'^+}-\dfrac{p^1+i\sigma p^2}{p^+}\right)$
$\dfrac{\bar{v}_{\sigma'}(p')}{\sqrt{p'^+}}\gamma^-\gamma^+\gamma^-\dfrac{u_\sigma(p)}{\sqrt{p^+}}$	$4\dfrac{\bar{v}_{\sigma'}(p')}{\sqrt{p'^+}}\gamma^-\dfrac{u_\sigma(p)}{\sqrt{p^+}}$
$\dfrac{\bar{v}_{\sigma'}(p')}{\sqrt{p'^+}}\gamma^-\gamma^+\gamma^i_\perp\dfrac{u_\sigma(p)}{\sqrt{p^+}}$	$\delta_{\sigma,-\sigma'}4\dfrac{p'^i_\perp-i\epsilon^{ij}\sigma p'^j_\perp}{p'^+}+\delta_{\sigma\sigma'}\sigma\dfrac{4m}{p'^+}(\delta^{i1}+i\sigma\delta^{i2})$
$\dfrac{\bar{v}_{\sigma'}(p')}{\sqrt{p'^+}}\gamma^i_\perp\gamma^+\gamma^-\dfrac{u_\sigma(p)}{\sqrt{p^+}}$	$\delta_{\sigma,-\sigma'}4\dfrac{p^i_\perp+i\epsilon^{ij}\sigma p^j_\perp}{p^+}-\delta_{\sigma\sigma'}\sigma\dfrac{4m}{p^+}(\delta^{i1}+i\sigma\delta^{i2})$
$\dfrac{\bar{v}_{\sigma'}(p')}{\sqrt{p'^+}}\gamma^i_\perp\gamma^+\gamma^j_\perp\dfrac{u_\sigma(p)}{\sqrt{p^+}}$	$\delta_{\sigma,-\sigma'}2(\delta^{ij}+i\sigma\epsilon^{ij})$

Equations (A.10) and (A.9) can be used to show that

$$\int d^2y_\perp\frac{\vec{y}_\perp\cdot(\vec{y}_\perp+\vec{x}_\perp)}{y_\perp^2(\vec{y}_\perp+\vec{x}_\perp)^2}=2\pi\ln\frac{1}{x_\perp\Lambda}.\tag{A.12}$$

Several angular integrals are useful too:

$$\int_0^{2\pi}\frac{d\varphi_q}{(\vec{q}_\perp-\vec{l}_\perp)^2}=\frac{2\pi}{|l_\perp^2-q_\perp^2|},\tag{A.13}$$

where φ_q is the angle between $\vec{q}_\perp$ and $\vec{l}_\perp$;

$$\int_0^{2\pi} d\varphi_q \frac{\vec{q}_\perp - \vec{l}_\perp}{(\vec{q}_\perp - \vec{l}_\perp)^2} = -2\pi \theta(l_\perp - q_\perp)\frac{\vec{l}_\perp}{l_\perp^2}; \tag{A.14}$$

$$\int_0^{2\pi} \frac{d\varphi_q}{q_\perp^2 + (\vec{q}_\perp - \vec{l}_\perp)^2} = \frac{2\pi}{\sqrt{4q_\perp^4 + l_\perp^4}}; \tag{A.15}$$

$$\int_0^{2\pi} d\varphi \, e^{iz\sin\phi - in\varphi} = 2\pi J_n(z), \tag{A.16}$$

for integer $n \geq 0$;

$$\int_0^\infty dk \, k^{\lambda-1} J_\nu(kx) = 2^{\lambda-1} x^{-\lambda} \frac{\Gamma\left(\frac{1}{2}(\nu+\lambda)\right)}{\Gamma\left(\frac{1}{2}(2+\nu-\lambda)\right)}. \tag{A.17}$$

The integral (A.17) converges for real $x > 0$ and for $-\text{Re}\,\nu < \text{Re}\,\lambda < 3/2$, but it can be analytically continued outside this region of λ. A useful special case is $\nu = 0$:

$$\int_0^\infty dk \, k^{\lambda-1} J_0(kx) = 2^{\lambda-1} x^{-\lambda} \frac{\Gamma\left(\frac{1}{2}\lambda\right)}{\Gamma\left(1 - \frac{1}{2}\lambda\right)}. \tag{A.18}$$

One can also show that

$$\int_0^\infty dk \, k^{\lambda-1}[1 - J_0(kx)] = -2^{\lambda-1} x^{-\lambda} \frac{\Gamma\left(\frac{1}{2}\lambda\right)}{\Gamma\left(1 - \frac{1}{2}\lambda\right)}. \tag{A.19}$$

This integral converges for real x and for $-2 < \text{Re}\,\lambda < 0$: it can also be analytically continued outside this region of λ.

A.3 Another useful integral

Let us find here the integral

$$I_{dip} = \int d^2x_2 \frac{x_{10}^2}{x_{20}^2 x_{21}^2}, \tag{A.20}$$

which is very useful in the Mueller's dipole model. We first note that one can write

$$d^2x_2 = 2\pi x_{02} x_{12} \int_0^\infty dk k \, J_0(kx_{10}) \, J_0(kx_{20}) \, J_0(kx_{21}), \tag{A.21}$$

where the right-hand side is non-zero only if there exists a triangle with sides x_{10}, x_{20}, and x_{21}. Equation (A.20) becomes

$$I_{dip} = 2\pi x_{10}^2 \int\limits_0^\infty dk\, k J_0(kx_{10}) \int\limits_\rho^\infty \frac{dx_{20}}{x_{20}} J_0(kx_{02}) \int\limits_\rho^\infty \frac{dx_{21}}{x_{21}} J_0(kx_{12}), \qquad (A.22)$$

where we have inserted a UV regulator ρ into the x_{20}- and x_{21}- integrals. Writing

$$\int\limits_\rho^\infty \frac{dx}{x} J_0(kx) = \lim_{\epsilon \to 0} \left\{ \int\limits_0^\infty dx\, x^{\epsilon-1} J_0(kx) - \int\limits_0^\rho dx\, x^{\epsilon-1} J_0(kx) \right\}, \qquad (A.23)$$

we use Eq. (A.18) to perform the first integral on the right while putting $J_0(kx) = 1$ in the second integral before integrating over x, thus neglecting higher powers of the UV regulator ρ. Expanding the result in ϵ and taking the limit $\epsilon \to 0$ yields

$$\int\limits_\rho^\infty \frac{dx}{x} J_0(kx) = \ln \frac{2}{k\rho} - \gamma_E + O(\rho). \qquad (A.24)$$

Substituting Eq. (A.24) into Eq. (A.22) we obtain

$$I_{dip} = 2\pi x_{10}^2 \int\limits_0^\infty dk\, k J_0(kx_{10}) \left(\ln \frac{2}{k\rho} - \gamma_E \right)^2. \qquad (A.25)$$

Using (for $x_{10} > 0$)

$$\int\limits_0^\infty dk\, k J_0(kx_{10}) = 0, \qquad (A.26)$$

$$\int\limits_0^\infty dk\, k J_0(kx_{10}) \ln k = \lim_{\epsilon \to 0} \frac{\partial}{\partial \epsilon} \int\limits_0^\infty dk\, k^{1+\epsilon} J_0(kx_{10}) = -\frac{1}{x_{10}^2}, \qquad (A.27)$$

and

$$\int\limits_0^\infty dk\, k J_0(kx_{10}) \ln^2 k = \lim_{\epsilon \to 0} \frac{\partial^2}{\partial \epsilon^2} \int\limits_0^\infty dk\, k^{1+\epsilon} J_0(kx_{10}) = \frac{2}{x_{10}^2} \left(\ln \frac{x_{10}}{2} + \gamma_E \right), \qquad (A.28)$$

all of which follow from Eq. (A.18), we can rewrite Eq. (A.25) as

$$I_{dip} = 4\pi \ln \frac{x_{01}}{\rho}, \qquad (A.29)$$

which is used in arriving at the last line of Eq. (4.64) in the main text.

Appendix B

Dispersion relations, analyticity, and unitarity of the scattering amplitude

Our analysis of high energy scattering amplitudes cannot be complete, since the problem of quark (and gluon) confinement in QCD has not been solved. It is clear that the solution of this problem lies beyond the realm of perturbative QCD. At the same time, nonperturbative physics may affect the scattering amplitudes (though maybe to a lesser extent than one would naively expect, since the saturation dynamics described in this book tends to suppress nonperturbative effects). Therefore, it would be very instructive to summarize the properties of the scattering amplitudes in the perturbative and nonperturbative approaches to any field theory.

First, any scattering amplitude should be a relativistic invariant (a scalar with respect to Lorenz transformations) and, because of this, it can depend only on variables that are relativistic invariants, namely on quantities such as $(p_i - p_j)^2$, where p_i^μ and p_j^μ are the four-momenta of external lines labeled i and j. In the case of a $2 \to 2$ scattering amplitude we have three such invariants, given by the Mandelstam variables (Mandelstam 1958)

$$s = (p_A + p_B)^2 = m_A^2 + m_B^2 + 2p_B \cdot p_A,$$

$$u = (p_C - p_B)^2 = m_C^2 + m_B^2 - 2p_C \cdot p_B, \tag{B.1}$$

$$t = (p_A - p_C)^2 = m_C^2 + m_A^2 - 2p_C \cdot p_A,$$

with

$$s + u + t = m_A^2 + m_B^2 + m_C^2 + m_D^2. \tag{B.2}$$

The process is illustrated in the left-hand panel of Fig. B.1, where the notation is explained as well.

The second basic principle is the unitarity of the S-matrix: $S^\dagger S = I$ where I is the identity operator. This translates into the following equation for the T-matrix, defined by $S = I + iT$:

$$i(T^\dagger - T) = T^\dagger T. \tag{B.3}$$

Below we will rewrite Eq. (B.3) as a condition on scattering amplitudes.

The presentation of the material in this appendix is based mainly on the books Chew (1961, 1966), Roman (1969), Schweber (1961), and Weinberg (1996), vol. 1.

B.1 Crossing symmetry and dispersion relations

It turns out that when calculating any amplitude using Feynman diagrams in a field theory one always obtains a function that is analytic in its Lorentz-invariant arguments. In the case of a

312

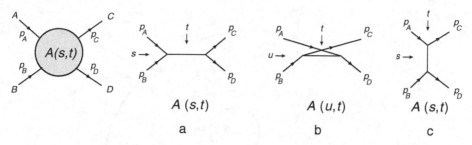

Fig. B.1. Scattering amplitude for a $2 \to 2$ process and tree-level diagrams for the process in the ϕ^3 theory in three different channels, s, u, and t, corresponding respectively to parts a, b, and c of the figure.

$2 \to 2$ scattering amplitude these arguments are the Mandelstam variables s, t, u. The singularities of the scattering amplitude are located only at the real values of these Lorentz-invariant variables. These singularities are closely related to the physical processes: they correspond to the production thresholds for physical particles (this is known as the Landau principle (Landau 1959, 1960)).

Relations between the scattering amplitudes for different processes may be obtained using *crossing symmetry*. This symmetry allows one to use only one function (say $A(s, t)$) to describe three different processes: $A + B \to C + D$, in the kinematic region where $s > 0$ and $t < 0$; $\bar{C} + B \to \bar{A} + D$, for $u > 0$ and $t < 0$; and $A + \bar{C} \to B + \bar{D}$, for $t > 0$ and $s < 0$. (Here $\bar{C}$, $\bar{D}$, and $\bar{A}$ denote antiparticles in complex-field theories and particles in real-field theories: in both cases the four-momenta are inverted under crossing symmetry transformations, e.g., $p_{\bar{A}}^{\mu} = -p_A^{\mu}$.)

The crossing symmetry can be illustrated using the example of simple tree-level graphs in ϕ^3 theory with the Lagrangian of Eq. (1.71). In Figs. B.1 a, b, c we plot the diagrams for the s-, t- and u-channel contributions respectively, all corresponding to the same process $A + B \to C + D$. Indeed, the diagrams of Fig. B.1 lead to the following expressions for the scattering amplitudes:

$$A(s, t; \text{Fig. B1a}) = \frac{\lambda^2}{m^2 - (p_A + p_B)^2 - i\epsilon} = \frac{\lambda^2}{m^2 - s - i\epsilon}$$

$$A(u, t; \text{Fig. B1b}) = \frac{\lambda^2}{m^2 - (-p_C + p_B)^2 - i\epsilon} = \frac{\lambda^2}{m^2 - u - i\epsilon} \tag{B.4}$$

$$A(t, s; \text{Fig. B1c}) = \frac{\lambda^2}{m^2 - (p_A + -p_C)^2 - i\epsilon} = \frac{\lambda^2}{m^2 - t - i\epsilon}.$$

It is clear that Fig. B.1a describes the process $A + B \to C + D$, while the diagram of Fig. B.1b can be viewed either as describing the same process as the diagram of Fig. B.1a but with s replaced by $u = 4m^2 - s - t$ or it can be viewed as the tree-level diagram for the process $\bar{C} + B \to \bar{A} + D$, with the invariant s defined now as $s = (p_{\bar{C}} + p_B)^2$ if we assume that $p_{\bar{C}}^{\mu} = -p_C^{\mu}$. Defining

$$A(s, t) = A_{A+B \to C+D}(p_A, p_B, p_C, p_D) \tag{B.5}$$

we see that the relation between the amplitudes in diagrams Figs. B.1a, b that describes their crossing symmetry is

$$A_{\bar{C}+B \to \bar{A}+D}(p_{\bar{C}}, p_B, p_{\bar{A}}, p_D) = A_{A+B \to C+D}(-p_C, p_B, -p_A, p_D) = A(u, t). \tag{B.6}$$

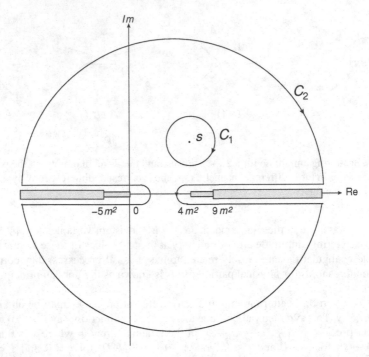

Fig. B.2. The singularities of the scattering amplitude $A(s, t)$, shown in the complex plane of the variable s. They are mainly given by branch cuts that start at the production thresholds for two, three, and more particles. For the sake of simplicity we do not show the pole contributions of Fig. B.1.

Similarly, for the amplitude resulting from the diagram Fig. B.1c we have

$$A_{A+\bar{C}\to B+\bar{D}}\,(p_A, p_{\bar{C}}, p_B, p_{\bar{D}}) = A_{A+B\to C+D}\,(p_A, -p_C, -p_B, p_D) = A(t, s). \qquad \text{(B.7)}$$

Therefore, the scattering amplitude $A(s, t)$ as a function of the variables s and t is able to describe all three processes.

The analyticity of the scattering amplitude gives more detailed information about the amplitude. Indeed, owing to Cauchy's theorem the amplitude, being an analytical function, can be written in the form (see Fig. B.2)

$$A(s, t) = \frac{1}{2\pi i} \oint_{C_1} ds' \frac{A(s', t)}{s' - s} = \frac{1}{2\pi i} \oint_{C_2} ds' \frac{A(s', t)}{s' - s}. \qquad \text{(B.8)}$$

The contours C_1 and C_2 are shown in Fig. B.2, and s is taken somewhere in the complex plane away from the real axis. The singularities of $A(s, t)$ are also shown in Fig. B.2: as mentioned before, they are confined to the real s-axis, and are typically branch cuts starting at the particle production thresholds. Since the amplitude does not have singularities at complex values of s, we can stretch the contour of integration C_1 to C_2 without modifying the value of the integral, as is reflected in Eq. (B.8). One can see that the integration over contour C_2 in Eq. (B.8) can be rewritten in the form

$$A(s,t) = \frac{1}{2\pi i} \left\{ \int\limits_{s_{min}}^{\infty} ds' \frac{\mathrm{Disc}_s \, A(s',t)}{s'-s} + \int\limits_{u_{min}}^{\infty} du' \frac{\mathrm{Disc}_u \, A(u',t)}{u'-u} \right\}$$

$$+ \frac{1}{2\pi i} \oint\limits_{\text{large circle}} ds' \frac{A(s',t)}{s'-s}, \tag{B.9}$$

where

$$\mathrm{Disc}_s \, A\,(s',t) = \lim_{\epsilon \to 0} \left[A(s'+i\epsilon,t) - A(s'-i\epsilon,t) \right] \tag{B.10}$$

and

$$\mathrm{Disc}_u \, A\,(u',t) = \lim_{\epsilon \to 0} \left[A(u'+i\epsilon,t) - A(u'-i\epsilon,t) \right]. \tag{B.11}$$

In the second term on the right-hand side of Eq. (B.9) we have changed the integration variable to $u' = 4m^2 - t - s'$; note that $A(u,t)$ is no longer equal to $A(s,t)$ with s replaced by u: rather $A(u,t) = A(s = 4m^2 - t - u, t)$. Typically the limits of integration in Eq. (B.9) would be $s_{min} = 4m^2$ and $u_{min} = 4m^2 - t$, and we are keeping t fixed and real. (If the amplitude has poles on the real axis for $0 < s < 4m^2$, as is the case for the ϕ^3-theory amplitudes given by Eq. (B.4), the contributions of such poles has to be included in the right-hand side of Eq. (B.9) by appropriately lowering s_{min}.)

The amplitude $A(s,t)$ has no imaginary part (no branch cuts corresponding to particle production) along the real axis between $s = 0$ (corresponding to $u = 4m^2$, $t = 0$) and $s = 4m^2$. Therefore it is a real function of s and t in this interval and, as can be shown, is in fact a real function of s and t in the whole region of its analyticity. We thus conclude that $A(s'-i\epsilon,t) = A^*(s'+i\epsilon,t)$, such that

$$\mathrm{Disc}_s \, A(s',t) = 2i \, \mathrm{Im} \, A(s,t). \tag{B.12}$$

Similarly,

$$\mathrm{Disc}_u \, A(u',t) = 2i \, \mathrm{Im} \, A(u,t). \tag{B.13}$$

(It should be stressed that, owing to the optical theorem, which follows from Eq. (B.3), Im $A(u,t)$ and Im $A(s,t)$ are directly related to physical processes.)

One can show that the contribution to the right-hand side of Eq. (B.9) coming from the integral over the large circle vanishes as we stretch the radius of the circle to infinity (see e.g. Weinberg (1996), vol. 1):

$$\frac{1}{2\pi i} \oint\limits_{\text{large circle}} ds' \frac{A(s',t)}{s'-s} \longrightarrow 0. \tag{B.14}$$

Neglecting this last term on the right of Eq. (B.9) and employing Eqs. (B.12) and (B.13) we finally obtain the following *dispersion relation*:

$$A(s,t) = \frac{1}{\pi} \left\{ \int\limits_{s_{min}}^{\infty} ds' \frac{\mathrm{Im}_s \, A(s',t)}{s'-s} + \int\limits_{u_{min}}^{\infty} du' \frac{\mathrm{Im}_u \, A(u',t)}{u'-u} \right\}. \tag{B.15}$$

Equation (B.15) allows one to reconstruct the full amplitude if its imaginary part is known.

For example, the tree-level diagrams in Fig. B.1 yield

$$\text{Im}_s \, A\big(s', t; \text{Fig. B1a}\big) = \pi \lambda^2 \delta \big(m^2 - s'\big), \tag{B.16a}$$

$$\text{Im}_u \, A\big(u', t; \text{Fig. B1b}\big) = \pi \lambda^2 \delta \big(m^2 - u'\big). \tag{B.16b}$$

Substituting each of these imaginary parts into the right-hand side of Eq. (B.15) yields the appropriate amplitude after straightforward integration over the delta functions.

Note that a dispersion relation in the form Eq. (B.15) cannot be used in QCD since we know that QCD amplitudes grow as the energy s at large s (see e.g. Eq. (3.17)), making the integrals in Eq. (B.15) divergent. Therefore, we have to alter Eq. (B.15) by subtracting, for example, the amplitude $A(s = 0, t)$ obtained by putting $s = 0$ in Eq. (B.15). Doing this, we obtain the *subtracted dispersion relation*

$$A(s, t) = A(s = 0, t) + \frac{1}{\pi} \left\{ s \int_{s_{min}}^{+\infty} ds' \frac{\text{Im}_s \, A\big(s', t\big)}{s'(s' - s)} \right.$$

$$\left. + [u - u(s = 0)] \int_{u_{min}}^{+\infty} du' \frac{\text{Im}_u \, A\big(u', t\big)}{[u' - u(s = 0)](u' - u)} \right\}. \tag{B.17}$$

Finally, subtracting $s \partial_s A(s = 0, t)$ from Eq. (B.17) (with $A(s, t)$ again given by Eq. (B.15)) we obtain the *double-subtracted dispersion relation*

$$A(s, t) = A(s = 0, t) + s \partial_s A(s = 0, t) + \frac{1}{\pi} \left\{ s^2 \int_{s_{min}}^{+\infty} ds' \frac{\text{Im}_s \, A\big(s', t\big)}{s'^2(s' - s)} \right.$$

$$\left. + [u - u(s = 0)]^2 \int_{u_{min}}^{+\infty} du' \frac{\text{Im}_u \, A\big(u', t\big)}{[u' - u(s = 0)]^2(u' - u)} \right\}. \tag{B.18}$$

This is exactly the dispersion relation used in Eq. (3.43). Note that in perturbative QCD $A(s = 0, t) = 0$.

B.2 Unitarity and the Froissart–Martin bound

The unitarity constraint (B.3) can be written in terms of scattering amplitudes as (see e.g. Peskin and Schroeder (1995))

$$M(k_1, k_2 \to k_1, k_2) - M^*(k_1, k_2 \to k_1, k_2)$$

$$= i \sum_{n=2}^{\infty} \int \prod_{i=1}^{n} \frac{d^3 q_i}{(2\pi)^3 2 E_{q_i}} |M(k_1, k_2 \to q_1, \dots, q_n)|^2 (2\pi)^4 \delta^4 \left(k_1 + k_2 - \sum_{j=1}^{n} q_j \right), \tag{B.19}$$

where $M(k_1, k_2 \to q_1, \dots, q_n)$ is the $2 \to n$ scattering amplitude for the scattering of two particles with momenta k_1, k_2 into n particles with momenta $q_1, \dots, q_n$, and $M(k_1, k_2 \to k_1, k_2)$ is the forward scattering amplitude; E_{q_i} is the energy of a particle with momentum q_i.

Let us consider the case of high energy scattering, where k_1^+ and k_2^- are very large and so are $q_1^+ \approx k_1^+$ and $q_2^- \approx k_2^-$. Separating the *elastic* $2 \to 2$ contribution from the *inelastic* contributions ($2 \to 3, 2 \to 4$, etc.) on the right-hand side of Eq. (B.19), and integrating over the delta-function in that contribution, yields

$$2 \, \text{Im} \, A(k_1, k_2 \to k_1, k_2) = \int \frac{d^2 q_\perp}{(2\pi)^2} |A(k_1, k_2 \to q_1, q_2)|^2 + \text{inelastic terms}, \tag{B.20}$$

where q is the momentum transfer four-vector, defined by

$$q = q_1 - k_1 = k_2 - q_2, \tag{B.21}$$

and we also define a new rescaled scattering amplitude

$$A(k_1, k_2 \to q_1, q_2) \equiv \frac{M(k_1, k_2 \to q_1, q_2)}{2\sqrt{2E_{k_1} 2E_{k_2} 2E_{q_1} 2E_{q_2}}} \approx \frac{M(k_1, k_2 \to q_1, q_2)}{2k_1^+ k_2^-}. \tag{B.22}$$

Since both the incoming and outgoing particles are on mass shell the momentum transfer q has only two free components, which we choose to be transverse and over which we integrated in Eq. (B.20).

The optical theorem then states that the total scattering cross section is given by (again, see e.g. Peskin and Schroeder (1995))

$$\sigma_{tot} = 2 \operatorname{Im} A(k_1, k_2 \to k_1, k_2) \tag{B.23}$$

so that Eq. (B.20) simply implies that

$$\sigma_{tot} = \sigma_{el} + \sigma_{inel}, \tag{B.24}$$

where σ_{el} is the elastic $2 \to 2$ cross section and σ_{inel} is the total inelastic cross section.

As we have seen above, in general the elastic amplitude $A(k_1, k_2 \to q_1, q_2)$ can be written as a function of the Mandelstam variables s and t. However, for our purposes it is convenient to go to impact parameter ($\vec{b}_\perp$) space, using

$$A(k_1, k_2 \to q_1, q_2) = \int d^2b \, e^{-i\vec{q}_\perp \cdot \vec{b}_\perp} A(s, \vec{b}_\perp), \tag{B.25}$$

which, when applied in Eq. (B.20) yields

$$2 \operatorname{Im} A(s, \vec{b}_\perp) = |A(s, \vec{b}_\perp)|^2 + \text{inelastic terms}. \tag{B.26}$$

In arriving at Eq. (B.26) we have used the fact that the forward amplitude corresponds to the case of zero momentum transfer, $t = 0$, or, equivalently, $q_\perp = 0$, such that

$$A(k_1, k_2 \to k_1, k_2) = \int d^2b \, A(s, \vec{b}_\perp). \tag{B.27}$$

Note that the total cross section in impact parameter space is

$$\sigma_{tot} = 2 \int d^2b \operatorname{Im} A(s, \vec{b}_\perp). \tag{B.28}$$

We also see immediately from Eq. (B.26) that the elastic cross section is given by

$$\sigma_{el} = \int d^2b \, |A(s, \vec{b}_\perp)|^2. \tag{B.29}$$

Relating the inelastic terms in Eq. (B.26) to the corresponding cross section yields

$$2 \operatorname{Im} A(s, \vec{b}_\perp) = |A(s, \vec{b}_\perp)|^2 + \frac{d\sigma_{inel}}{d^2b}. \tag{B.30}$$

The simple nonnegativity condition

$$\frac{d\sigma_{inel}}{d^2b} \geq 0 \tag{B.31}$$

used in Eq. (B.30) yields

$$\operatorname{Im} A(s, \vec{b}_\perp) \leq 2. \tag{B.32}$$

This is an important condition, which follows from unitarity. When used in Eq. (B.28) it yields an upper bound for the total cross section:

$$\sigma_{tot} = 2 \int d^2b \, \text{Im} A(s, \vec{b}_\perp) \le 4 \int d^2b = 4\pi R^2, \tag{B.33}$$

where R is the radius of the region in $b_\perp$-space where the interactions are sufficiently strong (the radius of the "black disk").

Parametrizing the forward scattering amplitude by (as follows from $S = I + iT$)

$$A(s, \vec{b}_\perp) = i\left[1 - S(s, \vec{b}_\perp)\right], \tag{B.34}$$

with $S(s, \vec{b}_\perp)$ the forward matrix element of the S-matrix, we see that the constraint (B.32) and the nonnegativity of the total cross section σ_{tot} together lead to $|\text{Re} S(s, \vec{b}_\perp)| \le 1$.

Using Eq. (B.34) in Eqs. (B.28), (B.29), and (B.30) yields

$$\sigma_{tot} = 2 \int d^2b \left[1 - \text{Re} \, S(s, \vec{b}_\perp)\right], \tag{B.35a}$$

$$\sigma_{el} = \int d^2b \left|1 - S(s, \vec{b}_\perp)\right|^2, \tag{B.35b}$$

$$\sigma_{inel} = \int d^2b \left[1 - |S(s, \vec{b}_\perp)|^2\right]. \tag{B.35c}$$

In high energy scattering the bound on the total cross section is even stronger than Eq. (B.33). At very high energies inelastic processes dominate, so that $\sigma_{inel} \ge \sigma_{el}$, which leads to

$$\text{Re} \, S(s, \vec{b}_\perp) \ge 0. \tag{B.36}$$

With the help of Eq. (B.34) we obtain

$$\text{Im} A(s, \vec{b}_\perp) \le 1, \tag{B.37}$$

which is a stronger constraint than (B.32). Equation (B.37) leads to

$$\sigma_{tot} = 2 \int d^2b \, \text{Im} \, A(s, \vec{b}_\perp) \le 2\pi R^2. \tag{B.38}$$

This is the bound used in the text in Eq. (3.112). (For a derivation of this result in nonrelativistic quantum mechanics see Landau and Lifshitz (1958), vol. 3, Chapter 131.) Using the estimate (3.115) for the typical interaction range, i.e.,

$$R = b^* \sim \frac{\Delta}{2m_\pi} \ln s, \tag{B.39}$$

in Eq. (B.38) yields the Froissart–Martin bound (3.116)

$$\sigma_{tot} \le \frac{\pi \Delta^2}{2m_\pi^2} \ln^2 s \tag{B.40}$$

(Froissart 1961, Martin 1969, Lukaszuk and Martin 1967).

References

The references for a given first-mentioned author are in date order.

Abramovsky, V. A., Gribov, V. N., and Kancheli, O. V. (1973), Character of inclusive spectra and fluctuations produced in inelastic procesesses by multi-Pomeron exchange, *Sov. J. Nucl. Phys.* **18**, 308 [*Yad. Fiz.* **18** (1973) 595].

Abramowicz, H., and Dainton, J. B. (1996), On the energy dependence of the deep inelastic diffractive cross-section, *J. Phys.* **G22**, 911.

Abramowicz, H., and Caldwell, A. (1999), HERA collider physics, *Rev. Mod. Phys.* **71**, 1275.

Albacete, J. L., Armesto, N., Kovner, A., Salgado, C. A., and Wiedemann, U. A. (2004), Energy dependence of the Cronin effect from nonlinear QCD evolution, *Phys. Rev. Lett.* **92**, 082 001.

Albacete, J. L., Armesto, N., Milhano, J. G., Salgado, C. A., and Wiedemann, U. A. (2005), Numerical analysis of the Balitsky–Kovchegov equation with running coupling: dependence of the saturation scale on nuclear size and rapidity, *Phys. Rev.* **D71**, 014 003.

Albacete, J. L., and Kovchegov, Y. V. (2007a), Baryon stopping in proton–nucleus collisions, *Nucl. Phys.* **A781**, 122.

Albacete, J. L., and Kovchegov, Y. V. (2007b), Solving high energy evolution equation including running coupling corrections, *Phys. Rev.* **D75**, 125 021.

Albacete, J. L., Armesto, N., Milhano, J. G., and Salgado, C. A. (2009), Non-linear QCD meets data: a global analysis of lepton–proton scattering with running coupling BK evolution, *Phys. Rev.* **D80**, 034 031.

Albacete, J. L., Armesto, N., Milhano, J. G., Quiroga Arias, C., and Salgado, C. A. (2011), AAMQS: a non-linear QCD analysis of new HERA data at small-x including heavy quarks, *Eur. Phys. J.* **C71**, 1705.

Albacete, J. L., and Dumitru, A. (2011). A model for gluon production in heavy-ion collisions at the LHC with rcBK unintegrated gluon densities, arXiv:1011.5161 [hep-ph].

ALICE collaboration Aamodt, K. *et al.* (2011), *Phys. Rev. Lett.* **106**, 032 301.

Altarelli, G., and Parisi, G. (1977), Asymptotic freedom in parton language, *Nucl. Phys.* **B126**, 298.

Altarelli, G., Ball, R. D., and Forte, S. (2006), Perturbatively stable resummed small x evolution kernels, *Nucl. Phys.* **B742**, 1.

Altinoluk, T., Kovner, A., Lublinsky, M., and Peressutti, J. (2009), QCD reggeon field theory for every day: Pomeron loops included, *JHEP* **0903**, 109.

Amsler, C., *et al.* (Particle Data Group) (2008), *Phys. Lett.* **B667**, 1, and 2009 partial update for the 2010 edition.

Armesto, N., Salgado, C. A., and Wiedemann, U. A. (2005), Relating high-energy hepton–hadron, proton–nucleus and nucleus–nucleus collisions through geometric scaling, *Phys. Rev. Lett.* **94**, 022 002 [arXiv/0407018[hep-ph]].

ATLAS collaboration, Aad, G., *et al.* (2011). *New J. Phys.* **13**, 053 033.

Avsar, E., Stasto, A. M., Triantafyllopoulos, D. N., and Zaslavsky, D. (2011), Next-to-leading and resummed BFKL evolution with saturation boundary, arXiv:1107.1252 [hep-ph].

Ayala, A. L., Gay Ducati, M. B., and Levin, E. M. (1996), Unitarity boundary for deep inelastic structure functions, *Phys. Lett.* **B388**, 188.

Ayala, A. L., Gay Ducati, M. B., and Levin, E. M. (1997), QCD evolution of the gluon density in a nucleus, *Nucl. Phys.* **B493**, 305.

Ayala, A. L., Gay Ducati, M. B., and Levin, E. M. (1998), Parton densities in a nucleon, *Nucl. Phys.* **B511**, 355.

Baier, R., Dokshitzer, Y. L., Mueller, A. H., Peigne, S., and Schiff, D. (1997), Radiative energy loss and $p(T)$ broadening of high-energy partons in nuclei, *Nucl. Phys.* **B484**, 265.

Baier, R., Kovner, A., and Wiedemann, U. A. (2003), Saturation and parton level Cronin effect: enhancement versus suppression of gluon production in pA and AA collisions, *Phys. Rev.* **B68**, 054 009.

Baier, R., Mueller, A. H., and Schiff, D. (2004), Saturation and shadowing in high-energy proton nucleus dilepton production, *Nucl. Phys.* **A741**, 358.

Balitsky, I. I., and Lipatov, L. N. (1978), The Pomeranchuk singularity in quantum chromodynamics, *Sov. J. Nucl. Phys.* **28**, 822 [*Yad. Fiz.* **28** 1597].

Balitsky, I. I. (1996), Operator expansion for high-energy scattering, *Nucl. Phys.* **B463**, 99.

Balitsky, I. I. (1998), Scattering of shock waves in QCD, *Phys. Rev.* **D70**, 114 030 [arXiv:hep-ph/0409314].

Balitsky, I. I. (1999a), Factorization for high-energy scattering, *Phys. Rev. Lett.* **81**, 2024.

Balitsky, I. I. (1999b), Factorization and high-energy effective action, *Phys. Rev.* **D60**, 014 020.

Balitsky, I. I., and Belitsky, A. V. (2002), Nonlinear evolution in high density QCD, *Nucl. Phys.* **B629**, 290.

Balitsky, I. I. (2004), Scattering of shock waves in QCD, *Phys. Rev.* **D70**, 114 030 [hep-ph/0409314].

Balitsky, I. I. (2007), Quark contribution to the small-x evolution of color dipole, *Phys. Rev.* **D75**, 014 001.

Balitsky, I. I., and Chirilli, G. A. (2008), Next-to-leading order evolution of color dipoles, *Phys. Rev.* **D77**, 014 019 [arXiv:0710.4330 [hep-ph]].

Barone, V. and Predazzi, E. (2002), *High Energy Particle Diffraction*, Springer.

Bartels, J. (1980), High-energy behavior in a nonabelian gauge theory. 2. First corrections to $T(n \to m)$ beyond the leading LNS approximation, *Nucl. Phys.* **B175**, 365.

Bartels, J., Schuler, G. A., and J. Blumlein, J. (1991), A numerical study of the small x behavior of deep inelastic structure functions in QCD, *Z. Phys.* **C50**, 91; *Nucl. Phys. Proc. Suppl.* **18C**, 147.

Bartels, J., and Levin, E. (1992), Solutions to the Gribov–Levin–Ryskin equation in the nonperturbative region, *Nucl. Phys.* **B387**, 617.

Bartels, J. (1993a), Unitarity corrections to the Lipatov pomeron and the small x region in deep inelastic scattering in QCD, *Phys. Lett.* **B298**, 204.

Bartels, J. (1993b), A note on the infrared cutoff dependence of the BFKL pomeron, *J. Phys.* **G19**, 1611.

Bartels, J., and Wusthoff, M. (1995), The triple Regge limit of diffractive dissociation in deep inelastic scattering, *Z. Phys.* **C66**, 157.

Bartels, J., Lotter, H., and Vogt, M. (1996), A numerical estimate of the small-k_T region in the BFKL pomeron, *Phys. Lett.* **B373**, 215.

Bartels, J., and Ryskin, M. G. (1997), The space–time picture of the Wee partons and the AGK cutting rules in perturbative QCD, *Z. Phys.* **C76**, 241.

Bartels, J., Lipatov, L. N., and Vacca, G. P. (2000). A new odderon solution in perturbative QCD, *Phys. Lett.* **B477**, 178.

Bartels, J., and Kowalski, H. (2001), Diffraction at HERA and the confinement problem, *Eur. Phys. J.* **C19**, 693.

Bartels, J., Braun, M., and Vacca, G. P. (2005), Pomeron vertices in perturbative QCD in diffractive scattering, *Eur. Phys. J.* **C40**, 419.

Bartels, J., Lipatov, L. N., and Vacca, G. P. (2005), Interactions of Reggeized gluons in the Moebius representation, *Nucl. Phys.* **B706**, 391.

Bartels, J., and Kutak, K. (2008), A momentum space analysis of the triple pomeron vertex in pQCD, *Eur. Phys. J.* **C53**, 533.

Bartels, J., Salvadore, M., and Vacca, G. P. (2008), Inclusive 1-jet production cross section at small x in QCD: multiple interactions, *JHEP* **0806**, 032.

Bertsch, G., Brodsky, S. J., Goldhaber, A. S., and Gunion, J. F. (1981), Diffraction excitation in QCD, *Phys. Rev. Lett.* **47**, 297.

Beuf, G., and Peschanski, R. (2007), Universality of QCD traveling-wave with running coupling, *Phys. Rev.* **D75**, 114 001.

Beuf, G. (2010), Universal behavior of the gluon saturation scale at high energy including full NLL BFKL effects, arXiv:1008.0498 [hep-ph].

Bialas, A., Navelet, H., and Peschanski, R. B. (1998), High-mass diffraction in the QCD dipole picture, *Phys. Lett.* **B427**, 147.

Bialas, A., Navelet, H., and Peschanski, R. B. (1999), Diffraction at small M^2/Q^2 in the QCD dipole picture, *Eur. Phys. J.* **C8**, 643.

Bjorken, J. D. (1969), Asymptotic sum rules at infinite momentum, *Phys. Rev.* **179**, 1547.

Bjorken, J. D., and Paschos, E. A. (1969), Inelastic electron–proton and gamma–proton scattering, and the structure of the nucleon, *Phys. Rev.* **185**, 1975.

Bjorken, J. D., Kogut, J. B., and Soper, D. E. (1971), Quantum electrodynamics at infinite momentum: scattering from an external field, *Phys. Rev.* **D3**, 1382.

Bjorken, J. D., and Kogut, J. B. (1973), Correspondence arguments for high-energy collisions, *Phys. Rev.* **D8**, 1341.

Blaettel, B., Baym, G., Frankfurt, L. L., and Strikman, M. (1993), How transparent are hadrons to pions?, *Phys. Rev. Lett.* **70**, 896.

Blaizot, J. P., Gelis, F., and Venugopalan, R. (2004). High-energy pA collisions in the color glass condensate approach. 2. Quark production, *Nucl. Phys.* **A743**, 57.

Blaizot, J. P., and Mehtar-Tani, Y. (2009), The classical field created in early stages of high energy nucleus–nucleus collisions, *Nucl. Phys.* **A818**, 97.

Blaizot, J. P., Lappi, T., and Mehtar-Tani, Y. (2010), On the gluon spectrum in the glasma, *Nucl. Phys.* **A846**, 62.

Bondarenko, S., Kozlov, M., and Levin, E. (2003), QCD saturation in the semiclassical approach, *Nucl. Phys.* **A727**, 139.

Bopp, F. W., Engel, R., Ranft, J., and Roesler, S. (2007), Inclusive distributions at the LHC as predicted from the DPMJET-III model with chain fusion, arXiv:0706.3875 [hep-ph].

Boreskov, K. G., Kaidalov, A. B., and Kancheli, O. V. (2006), Strong interactions at high energies in the Reggeon approach, *Phys. Atom. Nucl.* **69**, 1765.

BRAHMS collaboration, Bearden, I. G., *et al.* (2002), Pseudorapidity distributions of charged particles from Au+Au collisions at the maximum RHIC energy, *Phys. Rev. Lett.* **88**, 202 301.

BRAHMS collaboration, Arsene, I., *et al.* (2004), *Phys. Rev. Lett.* **93**, 242 303.

Bramson, M. (1983), Convergence of solutions of the Kolmogorov equation to travelling waves, *Mem. Am. Math. Soc.* **44**, 285.

Braun, M. A. (1995), Reggeized gluons with a running coupling constant, *Phys. Lett.* **B348**, 190.

Braun, M. A., and Vacca, G. P. V. (1997), Triple pomeron vertex in the limit $N_c \to \infty$, *Eur. Phys. J.* **C6**, 147.

Braun, M. (2000a), Structure function of the nucleus in the perturbative QCD with $N_c \to \infty$ (BFKL pomeron fan diagrams), *Eur. Phys. J.* **C16**, 337.

Braun, M. (2000b), Nucleus nucleus scattering in perturbative QCD with $N_c \to \infty$, *Phys. Lett.* **B483**, 115.

Braun, M. A. (2000c), Inclusive jet production on the nucleus in the parturbative QCD with $N_c \to \infty$, *Phys. Lett.* **B483**, 105 [arXiv:hep-ph/00030037].

Braun, M. (2004), Nucleus nucleus interaction in the perturbative QCD, *Eur. Phys. J.* **C33**, 113.

Braun, M. (2006), Conformal invariant pomeron interaction in the perturbative QCD with large N_c, *Phys. Lett.* **B632**, 297.

Braun, M. A. (2008), Single and double inclusive cross-sections for nucleus–nucleus collisions in the perturbative QCD, *Eur. Phys. J.* **C55**, 377.

Brodsky, S. J., and Farrar, G. R. (1973), Scaling laws at large transverse momentum, *Phys. Rev. Lett.* **31**, 1153.

Brodsky, S. J., Lepage, G. P., and Mackenzie, P. B. (1983), On the elimination of scale ambiguities in perturbative quantum chromodynamics, *Phys. Rev.* **D28**, 228.

Brodsky, S. J., and Lepage, G. P. (1989), Exclusive processes in quantum chromodynamics, *Adv. Ser. Direct. High Energy Phys.* **5**, 93.

Brodsky, S. J., Frankfurt, L., Gunion, J. F., Mueller, A. H., and Strikman, M. (1994), Diffractive leptoproduction of vector mesons in QCD, *Phys. Rev.* **D50**, 3134.

Brodsky, S. J., Pauli, H. C., and Pinsky, S. S. (1998), Quantum chromodynamics and other field theories on the light cone, *Phys. Rept.* **301**, 299.

Brodsky, S. J., Fadin, V. S., Kim, V. T., Lipatov, L. N., and Pivovarov, G. B. (1999), The QCD pomeron with optimal renormalization, *JETP Lett.* **70**, 155.

Buchmuller, W., Gehrmann, T., and Hebecker, A. (1997), Inclusive and diffractive structure functions at small x, *Nucl. Phys.* **B537**, 477.

Buchmuller, W., McDermott, M. F., and Hebecker, A. (1999), High-p_T jets in diffractive electroproduction, *Phys. Lett.* **B410**, 304.

Callan, C. G., and Gross, D. J. (1969), High energy electroproduction and the constitution of the electric current, *Phys. Rev. Lett.* **22**, 156.

Capua, M., Grothe, M., Ivanov, D. Y., and Kapishin, M. N. (2008), Summary from the working group on "Diffraction and vector mesons" at DIS 2008, arXiv:0901.2409 [hep-ph].

Catani, S., Ciafaloni, M., and Hautmann, F. (1990), Gluon contributions to small x heavy flavor production, *Phys. Lett.* **B242**, 97.

Catani, S., Fiorani, F., and Marchesini, G. (1990a), QCD coherence in initial state radiation, *Phys. Lett.* **B234**, 339.

Catani, S., Fiorani, F., and Marchesini, G. (1990b), Small x behavior of initial state radiation in perturbative QCD, *Nucl. Phys.* **B336**, 18.

Catani, S., Ciafaloni, M., and Hautmann, F. (1991), High-energy factorization and small x heavy flavor production, *Nucl. Phys.* **B366**, 135.

Chen, Z., and Mueller, A. H. (1995), The dipole picture of high-energy scattering, the BFKL equation and many gluon compound states, *Nucl. Phys.* **B451**, 579.

Chew, G. (1961), *S-Matrix Theory of Strong Interactions*, W. A. Benjamin.

Chew, G. (1966), *The Analytic S-Matrix*, W. A. Benjamin.

Christ, N. H., Hasslacher, B., and Mueller, A. H. (1972), Light cone behavior of perturbation theory, *Phys. Rev.* **D6**, 3543.

Ciafaloni, M. (1988), Coherence effects in initial jets at small q^2/s, *Nucl. Phys.* **B96**, 49.

Ciafaloni, M. and Camici, G. (1998), Energy scale(s) and next-to-leading BFKL equation, *Phys. Lett.* **B430**, 349.

Ciafaloni, M., Colferai, D., and Salam, G. P. (1999a), Renormalization group improved small-x equation, *Phys. Rev.* **D60**, 114 036.

Ciafaloni, M., Colferai, D., and Salam, G. P. (1999b) A collinear model for small-x physics, *JHEP* **9910**, 017.

Ciafaloni, M., Colferai, D., Salam, G. P., and Stasto, A. M. (2003a), Renormalisation group improved small-x Green's function, *Phys. Rev.* **D68**, 114 003.

Ciafaloni, M., Colferai, D., Salam, G. P., and Stasto, A. M. (2003b), Tunneling transition to the pomeron regime, *Phys. Lett.* **B541**, 314.

Ciafaloni, M. (2005), Small-x QCD: the (improved) perturbative picture, *Nucl. Phys. Proc. Suppl.* **146**, 129.

CMS collaboration, Khachatryan, V., *et al.* (2010a), *JHEP* **1009**, 091 [arXiv:1009.4122 [hep-ex]].

CMS collaboration, Khachatryan, V., *et al.* (2010b), *Phys. Rev. Lett.* **105**, 022 002.

Collins, J. C., Soper, D. E., and Sterman, G. (1985a), Factorization for short distance hadron–hadron scattering, *Nucl. Phys.* **B261**, 104.

Collins, J. C., Soper, D. E., and Sterman, G. (1985b), Transverse momentum distribution in Drell–Yan pair and W and Z boson production, *Nucl. Phys.* **B250**, 199.

Collins, J. C., Soper, D. E., and Sterman, G. (1988a), Factorization of hard processes in QCD, *Adv. Ser. Direct. High Energy Phys.* **5**, 1.

Collins, J. C., Soper, D. E., and Sterman, G. (1988b), Soft gluons and factorization, *Nucl. Phys.* **B308**, 833.

Collins, J. C., and Kwiecinski, J. (1990), Shadowing in gluon distributions in the small x region, *Nucl. Phys.* **B335**, 89.

Collins, J. C., and Ellis, R. K. (1991), Heavy quark production in very high-energy hadron collisions, *Nucl. Phys.* **B360**, 3.

Collins, J. C. (2011), *Foundations of Perturbative QCD*, Cambridge University Press.

Collins, P. D. B. (1977), *An Introduction to Regge Theory and High-Energy Physics*, Cambridge University Press.

Courant, R. and Hilbert, D. (1953), *Methods of Mathematical Physics*, Interscience.

Cronin, J. W., Frisch, H. J., Shochet, M. J., *et al.* (1975), Production of hadrons with large transverse momentum at 200 GeV, 300 GeV, and 400 GeV, *Phys. Rev.* **D11**, 3105.

CTEQ collaboration, Brock, R., *et al.* (1995), Handbook of perturbative QCD, *Rev. Mod. Phys.* **67**, 157.

Del Duca, V. (1995), An introduction to the perturbative QCD pomeron and to jet physics at large rapidities, *Scientifica Acta* **10** (2), 91–139 [arXiv:hep-ph/9503226].

Deng, W. T., Wang, X. N., and Xu, R. (2011), Gluon shadowing and hadron production in heavy-ion collisions at LHC, *Phys. Lett.* **B701**, 133 [arXiv:1011.5907 [nucl-th]].

Devenish, R., and Cooper-Sarkar, A. (2004), *Deep Inelastic Scattering*, Oxford University Press.

DeWitt, B. S. (1967), Quantum theory of gravity. II. The manifestly covariant theory, *Phys. Rev.* **162**, 1195.

Diaz Saez, B., and Levin, E. (2011), Violation of the geometric scaling behaviour of the amplitude for running QCD coupling in the saturation region, arXiv:1106.6257 [hep-ph].

Dokshitzer, Y. L. (1977), Calculation of the structure functions for deep inelastic scattering and $e^+ e^-$ annihilation by perturbation theory in quantum chromodynamics, *Sov. Phys. JETP* **46**, 641 [*Zh. Eksp. Teor. Fiz.* **73**, 1216 (1977)].

Dokshitzer, Y. L., Diakonov, D., and Troian, S. I. (1980), Hard processes in quantum chromodynamics, *Phys. Rept.* **58**, 269.

Dokshitzer, Y. L., Khoze, V. A., Mueller, A. H., and Troian, S. I. (1991), *Basics of Perturbative QCD*, Éditions Frontières.

Dokshitzer, Y. L., and Shirkov, D. V. (1995). On exact account of heavy quark thresholds in hard processes, *Z. Phys.* **C67**, 449.

Dokshitzer, Y. L., and Kharzeev, D. E. (2004). The Gribov conception of quantum chromodynamics, *Ann. Rev. Nucl. Part. Sci.* **54**, 487 [arXiv:hep-ph/0404216].

Dominguez, F., Marquet, C., Xiao, B. W., and Yuan, F. (2011). Universality of unintegrated gluon distributions at small x, *Phys. Rev.* **D83**, 105 005.

Donnachie, A., and Landshoff, P. V. (1992). Total cross-sections, *Phys. Lett.* **B296**, 227.

Donnachie, S., Dosch, G., and Landshoff, P. (2005), *Pomeron Physics and QCD*, Cambridge University Press.

Dumitru, A., and Jalilian-Marian, J. (2002). Forward quark jets from protons shattering the colored glass, *Phys. Rev. Lett.* **89**, 022 301.

Dumitru, A., and McLerran, L. D. (2002). How protons shatter colored glass, *Nucl. Phys.* **A700**, 492.

Dumitru, A., Gelis, F., McLerran, L., and Venugopalan, R. (2008), Glasma flux tubes and the near side ridge phenomenon at RHIC, *Nucl. Phys.* **A810**, 91.

Dumitru, A., Dusling, K., Gelis, F., Jalilian-Marian, J., Lappi, T., and Venugopalan, R. (2011a), The ridge in proton–proton collisions at the LHC, *Phys. Lett.* **B697**, 21.

Dumitru, A., Jalilian-Marian, J., Lappi, T., Schenke, B., and Venugopalan, R. (2011b), Renormalization group evolution of multi-gluon correlators in high energy QCD [arXiv:1108.4764 [hep-ph]].

Ellis, J., Kowalski, H., and Ross, D. A. (2008), Evidence for the discrete asymptotically-free BFKL pomeron from HERA data, *Phys. Lett.* **B668**, 51.

Ellis, R. K., Kunszt, Z., and Levin, E. (1994), The evolution of parton distributions at small x, *Nucl. Phys.* **B420**, 517.

Ellis, R. K., Stirling, W. J., and Webber, B. R. (1996), *QCD and Collider Physics*, Cambridge University Press.

Ewerz, C. (2003), The odderon in quantum chromodynamics, arXiv:hep-ph/0306137.

Faddeev, L. D., and Popov, V. N. (1967), Feynman diagrams for the Yang–Mills field, *Phys. Lett.* **B25**, 29.

Fadin, V. S., Kuraev, E. A., and Lipatov, L. N. (1975), On the Pomeranchuk singularity in asymptotically free theories, *Phys. Lett.* **B60**, 50.

Fadin, V. S., Kuraev, E. A., and Lipatov, L. N. (1976), Multi-Reggeon processes in the Yang–Mills theory, *Sov. Phys. JETP* **44**, 443 [*Zh. Eksp. Teor. Fiz.* **71**, 840].

Fadin, V. S., Kuraev, E. A., and Lipatov, L. N. (1977). The Pomeranchuk singularity in nonabelian gauge theories, *Sov. Phys. JETP* **45**, 199 [*Zh. Eksp. Teor. Fiz.* **72**, 377].

Fadin, V. S., and Lipatov, L. N. (1989), High-energy production of gluons in a quasi-multi-Regge kinematics, *Sov. J. Nucl. Phys.* **50**, 712 [*Yad. Fiz.* **50**, 1144.]

Fadin, V. S., Kotsky, M. I., and Fiore, R. (1995), Gluon reggeization in QCD in the next-to-leading order, *Phys. Lett.* **B359**, 181.

Fadin, V. S., Fiore, R., and Kotsky, M. I. (1996), Gluon Regge trajectory in the two-loop approximation, *Phys. Lett.* **B387**, 593 [hep-ph/9605357].

Fadin, V. S., and Fiore, R. (1998), The generalized nonforward BFKL equation and the bootstrap condition for the gluon Reggeization in the NLLA, *Phys. Lett.* **B440**, 359.

Fadin, V. S., and Lipatov, L. N. (1998), BFKL pomeron in the next-to-leading approximation, *Phys. Lett.* **B429**, 127.

Feinberg, E. L., and Pomeranchuk, I. (1956), High energy inelastic diffraction phenomena, *Nuovo Cimento Suppl.* **3**, 652.

Ferreiro, E., Iancu, E., Leonidov, A., and McLerran, L. (2002), Nonlinear gluon evolution in the color glass condensate. II, *Nucl. Phys.* **A703**, 489.

Feynman, R. P. (1963), Quantum theory of gravitation, *Acta Phys. Polon.* **24**, 697.

Feynman, R. P. (1969), Very high-energy collisions of hadrons, *Phys. Rev. Lett.* **23**, 1415.

Feynman, R. P. (1972), *Photon–Hadron Interactions*, Reading.

Fisher, R. A. (1937), The wave of advance of advantageous genes, *Ann. Eugen.* **7**, 355.

Forshaw, J. R., and Ross, D. A. (1997), *Quantum Chromodynamics and the Pomeron*, Cambridge University Press.

Franco, V., and Glauber, R. J. (1966), High-energy deuteron cross-sections, *Phys. Rev.* **142**, 1195.

Frankfurt, L. L., and Strikman, M. I. (1988), Hard nuclear processes and microscopic nuclear structure, *Phys. Rept.* **160**, 235.

Frankfurt, L. L., Miller, G. A., and Strikman, M. (1993), Coherent nuclear diffractive production of mini-jets: illuminating color transparency, *Phys. Lett.* **B304**, 1.

Froissart, M. (1961), Asymptotic behavior and subtractions in the Mandelstam representation, *Phys. Rev.* **123**, 1053.

Fritzsch, H., Gell-Mann, M., and Leutwyler, H. (1973), Advantages of the color octet gluon picture, *Phys. Lett.* **B47**, 365.

Gavin, S., McLerran, L., and Moschelli, G. (2009), Long range correlations and the soft ridge in relativistic nuclear collisions, *Phys. Rev.* **C79**, 051 902.

Gardi, E., Kuokkanen, J., Rummukainen, K., and Weigert, H. (2007), Running coupling and power corrections in nonlinear evolution at the high-energy limit, *Nucl. Phys.* **A784**, 282.

Gelis, F., and Jalilian-Marian, J. (2002a), Photon production in high-energy proton nucleus collisions, *Phys. Rev.* **D66**, 014 021.

Gelis, F., and Jalilian-Marian, J. (2002b), Dilepton production from the color glass condensate, *Phys. Rev.* **D66**, 094 014.

Gelis, F., Lappi, T., and Venugopalan, R. (2007), High energy scattering in quantum chromodynamics, *Int. J. Mod. Phys.* **E16**, 2595.

Gelis, F., Lappi, T., and Venugopalan, R. (2008a), High energy factorization in nucleus–nucleus collisions, *Phys. Rev.* **D78**, 054 019.

Gelis, F., Lappi, T., and Venugopalan, R. (2008b), High energy factorization in nucleus–nucleus collisions. II. Multigluon correlations, *Phys. Rev.* **D78**, 054 020.

Gelis, F., Lappi, T., and Venugopalan R. (2009), High energy factorization in nucleus–nucleus collisions. 3. Long range rapidity correlations, *Phys. Rev.* **D79**, 094 017.

Gelis, F., Iancu, E., Jalilian-Marian, J., and Venugopalan, R. (2010), The color glass condensate, arXiv:1002.0333 [hep-ph].

Georgi, H., and Politzer, H. D. (1974), Electroproduction scaling in an asymptotically free theory of strong interactions, *Phys. Rev.* **D9**, 416.

Glauber, R. J. (1955), Cross-sections in deuterium at high-energies, *Phys. Rev.* **100**, 242.

Glauber, R. J., and Matthiae, G. (1970), High-energy scattering of protons by nuclei, *Nucl. Phys.* **B21**, 135.

Golec-Biernat, K. J., and Wusthoff, M. (1999a), Saturation in diffractive deep inelastic scattering, *Phys. Rev.* **D60**, 114 023.

Golec-Biernat, K. J., and Wusthoff, M. (1999b), Saturation effects in deep inelastic scattering at low Q^2 and its implications on diffraction, *Phys. Rev.* **D59**, 014 017.

Golec-Biernat, K. J., and Wusthoff, M. (2001), Diffractive parton distributions from the saturation model, *Eur. Phys. J.* **C20**, 313.

Golec-Biernat, K. J., Motyka, L., and Stasto, A. M. (2002). Diffusion into infrared and unitarization of the BFKL pomeron, *Phys. Rev.* **D65**, 074 037.

Golec-Biernat, K. J., and Stasto, A. M. (2003), On solutions of the Balitsky–Kovchegov equation with impact parameter, *Nucl. Phys.* **B668**, 345 [arXiv:hep-ph/0306279].

Golec-Biernat, K. J., and Marquet, C. (2005), Testing saturation with diffractive jet production in deep inelastic scattering, *Phys. Rev.* **D71**, 114 005.

Golec-Biernat, K. J., and Luszczak, A. (2009). Dipole model analysis of the newest diffractive deep inelastic scattering data, *Phys. Rev.* **D79**, 114 010.

Good, M. L., and Walker, W. D. (1960), Diffraction dissociation of beam particles, *Phys. Rev.* **120**, 1857.

Gorshkov, V. G., Gribov, V. N., Lipatov, L. N., and Frolov, G. V. (1968), Doubly logarithmic asymptotic behavior in quantum electrodynamics, *Sov. J. Nucl. Phys.* **6**, 95 [*Yad. Fiz.* **6**, 129 (1967)].

Gotsman, E., Kozlov, M., Levin, E., Maor, U., and Naftali, E. (2004), Towards a new global QCD analysis: solution to the nonlinear equation at arbitrary impact parameter, *Nucl. Phys.* **A742**, 55 [arXiv:hep-ph/0401021].

Gotsman, E., Levin, E., Maor, U., and Naftali, E. (2005), A modified Balitsky–Kovchegov equation, *Nucl. Phys.* **A750**, 391 [arXiv:hep-ph/0411242].

Gradshteyn, I. S., and Ryzhik, I. M. (1994). *Table of Integrals, Series, and Products*, fifth edition, Academic Press.

Green, M. B., Schwarz, J. H., and Witten, E. (1987), *Superstring theory*, Vol. 1, *Introduction*, Cambridge University Press.

Gribov, V. N., Ioffe, B. L., and Pomeranchuk, I. Y. (1966), What is the range of interactions at high-energies?, *Sov. J. Nucl. Phys.* **2**, 549 [*Yad. Fiz.* **2**, 768 (1965)].

Gribov, V. N. (1968), A Reggeon diagram technique, *Sov. Phys. JETP* **26**, 414 [*Zh. Eksp. Teor. Fiz.* **53**, 654 (1967)].

Gribov, V. N. (1969a), Inelastic processes at super high-energies and the problem of nuclear cross-sections, *Sov. J. Nucl. Phys.* **9**, 369 [*Yad. Fiz.* **9**, 640].

Gribov, V. N. (1969b), Glauber corrections and the interaction between high-energy hadrons and nuclei, *Sov. Phys. JETP* **29**, 483 [*Zh. Eksp. Teor. Fiz.* **56**, 892].

Gribov, V. N. (1970), Interaction of gamma quanta and electrons with nuclei at high-energies, *Sov. Phys. JETP* **30**, 709 [*Zh. Eksp. Teor. Fiz.* **57**, 1306 (1969)].

Gribov, V. N., and Lipatov, L. N. (1972), Deep inelastic Ep scattering in perturbation theory, *Sov. J. Nucl. Phys.* **15**, 438 [*Yad. Fiz.* **15**, 781].

Gribov, V. N. (1973), Space–time description of hadron interactions at high energies, in *Proc. Moscow 1 ITEP School*, Vol. 1 "Elementary particles", 65 [arXiv:hep-ph/0006158].

Gribov, V. N. (1978), Quantization of non-Abelian gauge theories, *Nucl. Phys.* **B139**, 1.

Gribov, L. V., Levin, E. M., and Ryskin, M. G. (1983), Semihard processes in QCD, *Phys. Rept.* **100**, 1.

Gross, D. J., and Wilczek, F. (1973). Ultraviolet behavior of non-abelian theories. *Phys. Rev. Lett.* **30**, 1343.

Gross, D. J. and Wilczek, F. (1974), Asymptotically free gauge theories. 2, *Phys. Rev.* **D9**, 980.

Gubser, S. S. (2011), Conformal symmetry and the Balitsky–Kovchegov equation, arXiv:1102.4040 [hep-th].

Gunion, J. F., and Bertsch, G. (1982), Hadronization by color bremsstrahlung, *Phys. Rev.* **D25**, 746.

H1 collaboration, Adloff, C., *et al.* (1997), Inclusive measurement of diffractive deep inelastic ep scattering, *Z. Phys.* **C76**, 613.

H1 and ZEUS collaboration, Glasman, C., *et al.* (2008), Precision measurements of alpha(s) at HERA, *J. Phys. Conf. Ser.* **110**, 022 013 [arXiv:0709.4426 [hep-ex]].

H1 and ZEUS collaborations, Kapishin, M., *et al.* (2008), Diffraction and vector meson production at HERA, *Fizika* **B17**, 131.

H1 and ZEUS collaboration, Aaron, F. D., *et al.* (2010), Combined measurement and QCD analysis of the inclusive $e^+ -$ p scattering cross sections at HERA, *JHEP* **1001**, 109.

Halzen, F., and Martin, A. D. (1984), *Quarks and Leptons: An Introductory Course in Modern Particle Physics*, Wiley.

Hatta, Y., Iancu, E., Itakura, K., and McLerran, L. (2005a), Odderon in the color glass condensate, *Nucl. Phys.* **A760**, 172.

Hatta, Y., Iancu, E., McLerran, L., and Stasto, A. (2005b), Color dipoles from bremsstrahlung in QCD evolution at high energy, *Nucl. Phys.* **A762**, 272.

Hatta, Y., Iancu, E., Marquet, C., Soyez, G., and Triantafyllopoulos, D. N. (2006), Diffusive scaling and the high-energy limit of deep inelastic scattering in QCD at large N_c, *Nucl. Phys.* **A773**, 95.

Hatta, Y., Iancu, E., McLerran, L., Stasto, A., and Triantafyllopoulos, D. N. (2006), Effective Hamiltonian for QCD evolution at high energy, *Nucl. Phys.* **A764**, 423.

Hebecker, A. (2000), Diffraction in deep inelastic scattering, *Phys. Rept.* **331**, 1.

Heiselberg, H., Baym, G., Blaettel, B., Frankfurt, L. L., and Strikman, M. (1991), Color transparency, color opacity, and fluctuations in nuclear collisions, *Phys. Rev. Lett.* **67**, 2946.

Heisenberg, W. (1939), *Z. Phys.* **113**, 61.

Heisenberg, W. (1952), Production of mesons as a shock wave problem, *Z. Phys.* **133**, 65.

Hentschinski, M., Weigert, H., and Schafer, A. (2006), Extension of the color glass condensate approach to diffractive reactions, *Phys. Rev.* **D73**, 051 501.

Iancu, E., Leonidov, A., and McLerran, L. D. (2001a), The renormalization group equation for the color glass condensate, *Phys. Lett.* **B510**, 133.

Iancu, E., Leonidov, A., and McLerran, L. D. (2001b), Nonlinear gluon evolution in the color glass condensate. I, *Nucl. Phys.* **A692**, 583.

Iancu, E., and McLerran, L. D. (2001), Saturation and universality in QCD at small x, *Phys. Lett.* **B510**, 145.

Iancu, E., Itakura, K., and McLerran, L. (2002), Geometric scaling above the saturation scale, *Nucl. Phys.* **A708**, 327.

Iancu, E., and Venugopalan, R. (2003), The color glass condensate and high energy scattering in QCD, in *Quark Gluon Plasma*, eds. Hwa, R. C., and Wang, X. N. World Scientific.

Iancu, E., and Triantafyllopoulos, D. N. (2005), A Langevin equation for high energy evolution with pomeron loops, *Nucl. Phys.* **A756**, 419.

Ioffe, B. L. (1969), Space–time picture of photon and neutrino scattering and electroproduction cross-section asymptotics, *Phys. Lett.* **B30**, 123.

Ioffe, B. L., Fadin, V. S., and Lipatov, L. N. (2010), *Quantum Chromodynamics: Perturbative and Nonperturbative Aspects*, Cambridge University Press.

Jackson, J. D. (1998), *Classical Electrodynamics*, John Wiley & Sons.

Jalilian-Marian, J., Kovner, A., McLerran, L. D., and Weigert, H. (1997a), The intrinsic glue distribution at very small x, *Phys. Rev.* **D55**, 5414.

Jalilian-Marian, J., Kovner, A., McLerran, L. D., and Weigert, H. (1997b), The BFKL equation from the Wilson renormalization group, *Nucl. Phys.* **B504**, 415.

Jalilian-Marian, J., Kovner, A., Leonidov, A., and Weigert, H. (1999a), The Wilson renormalization group for low x physics: towards the high density regime, *Phys. Rev.* **D59**, 014 014.

Jalilian-Marian, J., Kovner, A., and Weigert, H. (1999b), The Wilson renormalization group for low x physics: gluon evolution at finite parton density, *Phys. Rev.* **D59**, 014 015.

Jalilian-Marian, J., and Kovchegov, Y. V. (2004), Inclusive two-gluon and valence quark–gluon production in DIS and pA, *Phys. Rev.* **D70**, 114 017 [Erratum: *ibid* **D71**, 079 901].

Jalilian-Marian, J., and Kovchegov, Y. V. (2006), Saturation physics and deuteron gold collisions at RHIC, *Prog. Part. Nucl. Phys.* **56**, 104.

Jaroszewicz, T. (1980). Infrared divergences and Regge behavior in QCD, *Acta Phys. Polon.* **B11**, 965.

Katz, U. F. (2000), Deep inelastic positron proton scattering in the high-momentum-transfer regime of HERA, in *Springer Tracts Mod. Phys.* Vol. 168, pp. 1.

Kharzeev, D., and Levin, E. (2001), Manifestations of high density QCD in the first RHIC data, *Phys. Lett.* **B523**, 79.

Kharzeev, D., and Nardi, M. (2001), Hadron production in nuclear collisions at RHIC and high density QCD, *Phys. Lett.* **B507**, 121.

Kharzeev, D., Kovchegov, Y. V., and Levin, E. (2002), Instantons in the saturation environment, *Nucl. Phys.* **A699**, 745.

Kharzeev, D., Kovchegov, Y. V., and Tuchin, K. (2003). Cronin effect and high p_T suppression in pA collisions, *Phys. Rev.* **D68**, 094 013.

Kharzeev, D., Levin, E., and McLerran, L. (2003), Parton saturation and N_{part} scaling of semi-hard processes in QCD, *Phys. Lett.* **B561**, 93.

Kharzeev, D., Kovchegov, Y. V., and Tuchin, K. (2004). *Phys. Lett.* **B599**, 23.

Kharzeev, D., Levin, E., and Nardi, M. (2004), QCD saturation and deuteron nucleus collisions, *Nucl. Phys.* **A730**, 448 [Erratum: *ibid.* **A743**, 329].

Kharzeev, D., Levin, E., and McLerran, L. (2005), Jet azimuthal correlations and parton saturation in the color glass condensate, *Nucl. Phys.* **A748**, 627.

Kharzeev, D., Levin, E., and Nardi, M. (2005a), Color glass condensate at the LHC: hadron multiplicities in pp, pA and AA collisions, *Nucl. Phys.* **A747**, 609.

Kharzeev, D., Levin, E., and Nardi, M. (2005b), The onset of classical QCD dynamics in relativistic heavy ion collisions, *Phys. Rev.* **C71**, 054 903.

Khoze, V. A., Martin, A. D., Ryskin, M. G., and Stirling, W. J. (2004), The spread of the gluon k_t-distribution and the determination of the saturation scale at hadron colliders in resummed NLL BFKL, *Phys. Rev.* **D70**, 074 013.

Khriplovich, I. B. (1969), Green's functions in theories with non-abelian gauge group, *Yad. Fiz.* **10**, 409.

Kolb, P. F., and Heinz, U. W. (2003), Hydrodynamic description of ultrarelativistic heavy ion collisions, invited review, in *Quark Gluon Plasma*, Vol. 3, pp. 634–714, eds. Hwa, R. C., and Wang, X. N. World Scientific [arXiv:nucl-th/0305084].

Kolmogorov, A., Petrovsky, I., and Piskunov, N. (1937), Study of the diffusion equation with growth of the quantity of matter and its application to a biology problem, *Moscow Univ. Bull. Math.* **A1**, 1.

Kopeliovich, B. Z., Lapidus, L., and Zamolodchikov, A. (1981), Dynamics of color in hadron diffraction on nuclei, *JETP Lett.* **33**, 595 [*Pisma Zh. Eksp. Teor. Fiz.* **33**, 612].

Kopeliovich, B. Z., Tarasov, A. V., and Schafer, A. (1999), Bremsstrahlung of a quark propagating through a nucleus, *Phys. Rev.* **C59**, 1609.

Kopeliovich, B. Z., Raufeisen, J., Tarasov, A. V., and Johnson, M. B. (2003), Nuclear effects in the Drell–Yan process at very high-energies, *Phys. Rev.* **C67**, 014 903.

Kopeliovich, B. Z., Potashnikova, I. K., and Schmidt, I. (2007), Diffraction in QCD, *Braz. J. Phys.* **37**, 473.

Korchemsky, G. P. (1999), Conformal bootstrap for the BFKL pomeron, *Nucl. Phys.* **B550**, 397.

Korchemsky, G. P., Kotanski, J., and Manashov, A. N. (2002), Solution of the multi-Reggeon compound state problem in multicolor QCD, *Phys. Rev. Lett.* **88**, 122 002.

Korchemsky, G. P., Kotanski, J., and Manashov, A. N. (2004), Multi-reggeon compound states and resummed anomalous dimensions in QCD, *Phys. Lett.* **B583**, 121.

Kotikov, A. V., and Lipatov, L. N. (2000), NLO corrections to the BFKL equation in QCD and in supersymmetric gauge theories, *Nucl. Phys.* **B582**, 19.

Kovchegov, Y. V. (1996), Non-Abelian Weizsaecker–Williams field and a two-dimensional effective color charge density for a very large nucleus, *Phys. Rev.* **D54**, 5463.

Kovchegov, Y. V. (1997), Quantum structure of the non-Abelian Weizsacker–Williams field for a very large nucleus, *Phys. Rev.* **D55**, 5445.

Kovchegov, Y. V., and Rischke, D. H. (1997), Classical gluon radiation in ultrarelativistic nucleus nucleus collisions, *Phys. Rev.* **C56**, 1084.

Kovchegov, Y. V., and Mueller, A. H. (1998a), Gluon production in current nucleus and nucleon nucleus collisions in a quasi-classical approximation, *Nucl. Phys.* **B529**, 451.

Kovchegov, Y. V., and Mueller, A. H. (1998b), Running coupling effects in BFKL evolution, *Phys. Lett.* **B439**, 428.

Kovchegov, Y. V. (1999), Small-x F_2 structure function of a nucleus including multiple pomeron exchanges, *Phys. Rev.* **60**, 034 008.

Kovchegov, Y. V., and McLerran, L. D. (1999), Diffractive structure function in a quasi-classical approximation, *Phys. Rev.* **D60**, 054 025.

Kovchegov, Y. V. (2000), Unitarization of the BFKL pomeron on a nucleus, *Phys. Rev.* **61**, 074 018.

Kovchegov, Y. V., and Levin, E. (2000), Diffractive dissociation including multiple pomeron exchanges in high parton density QCD, *Nucl. Phys.* **B577**, 221.

Kovchegov, Y. V. (2001), Diffractive gluon production in proton nucleus collisions and in DIS, *Phys. Rev.* **D64**, 114 016. [Erratum: *ibid.* **D68**, 039 901 (2003)].

Kovchegov, Y. V., and Tuchin, K. (2002), Inclusive gluon production in DIS at high parton density, *Phys. Rev.* **D65**, 074 026.

Kovchegov, Y. V., Szymanowski, L., and Wallon, S. (2004), Perturbative odderon in the dipole model, *Phys. Lett.* **B586**, 267.

Kovchegov, Y. V., and Tuchin, K. (2006), Production of $\bar{q}q$ pairs in proton–nucleus collisions at high energies, *Phys. Rev.* **D74**, 054 014 [arXiv:hep-ph/0603055].

Kovchegov, Y. V., and Weigert, H. (2007a), Triumvirate of running couplings in small-x evolution, *Nucl. Phys.* **A784**, 188.

Kovchegov, Y. V., and Weigert, H. (2007b), Quark loop contribution to BFKL evolution: running coupling and leading-N_f NLO intercept, *Nucl. Phys.* **A789**, 260.

Kovchegov, Y. V., Kuokkanen, J., Rummukainen, K., and Weigert, H. (2009), Subleading-N_c corrections in non-linear small-x evolution, *Nucl. Phys.* **A823**, 47.

Kovner, A., McLerran, L. D., and Weigert, H. (1995a), Gluon production from non-Abelian Weizsacker–Williams fields in nucleus–nucleus collisions, *Phys. Rev.* **D52**, 6231.

Kovner, A., McLerran, L. D., and Weigert, H. (1995b), Gluon production at high transverse momentum in the McLerran–Venugopalan model of nuclear structure functions, *Phys. Rev.* **D52**, 3809.

Kovner, A., and Milhano, J. G. (2000), Vector potential versus color charge density in low x evolution, *Phys. Rev.* **D61**, 014 012.

Kovner, A., Milhano, J. G., and Weigert, H. (2000), Relating different approaches to nonlinear QCD evolution at finite gluon density, *Phys. Rev.* **D62**, 114 005.

Kovner, A., and Wiedemann, U. A. (2001), Eikonal evolution and gluon radiation, *Phys. Rev.* **D64**, 114 002.

Kovner, A., and Wiedemann, U. A.(2002a), Nonlinear QCD evolution: saturation without unitarization, *Phys. Rev.* **D66**, 051 502.

Kovner, A., and Wiedemann, U. A. (2002b), Perturbative saturation and the soft pomeron, *Phys. Rev.* **D66**, 034 031.

Kovner, A., and Wiedemann, U. A. (2003), No Froissart bound from gluon saturation, *Phys. Lett.* **B551**, 311.

Kovner, A., and Lublinsky, M. (2005a), In pursuit of Pomeron loops: the JIMWLK equation and the Wess–Zumino term, *Phys. Rev.* **D71**, 085 004.

Kovner, A., and Lublinsky, M. (2005b), From target to projectile and back again: selfduality of high energy evolution, *Phys. Rev. Lett.* **94**, 181 603.

Kovner, A., and Lublinsky, M. (2005c), Dense–dilute duality at work: dipoles of the target, *Phys. Rev.* **72**, 074 023.

Kovner, A., and Lublinsky, M. (2005d), Remarks on high energy evolution, *JHEP* **0503**, 001.

Kovner, A., and Lublinsky, M. (2005e), In pursuit of pomeron loops: the JIMWLK equation and the Wess–Zumino term, *Phys. Rev.* **D71**, 085 004.

Kovner, A., and Lublinsky, M. (2006), One gluon, two gluon: multigluon production via high energy evolution, *JHEP* **0611**, 083.

Kovner, A., Lublinsky, M., and Weigert, H. (2006), Treading on the cut: semi inclusive observables at high energy, *Phys. Rev.* **D74**, 114 023.

Kovner, A., and Lublinsky, M. (2007), Odderon and seven Pomerons: QCD Reggeon field theory from JIMWLK evolution, *JHEP* **0702**, 058 [arXiv:hep-ph/0512316].

Krasnitz, A., and Venugopalan, R. (2000), The initial energy density of gluons produced in very high-energy nuclear collisions, *Phys. Rev. Lett.* **84**, 4309.

Krasnitz, A., Nara, Y., and Venugopalan, R. (2001), Coherent gluon production in very high-energy heavy ion collisions, *Phys. Rev. Lett.* **87**, 192 302.

Krasnitz, A., and Venugopalan, R. (2001), The initial gluon multiplicity in heavy ion collisions, *Phys. Rev. Lett.* **86**, 1717.

Krasnitz, A., Nara, Y., and Venugopalan, R. (2003a), Gluon production in the color glass condensate model of collisions of ultrarelativistic finite nuclei, *Nucl. Phys.* **A717**, 268.

Krasnitz, A., Nara, Y., and Venugopalan, R. (2003b). Classical gluodynamics of high-energy nuclear collisions: an erratumn and an update, *Nucl. Phys.* **A727**, 427.

Krasnitz, A., Nara, Y., and Venugopalan, R. (2003c). Gluon production in the color glass condensate model of collisions of ultrarelativistic finite nuclei, *Nucl. Phys.* **A717**, 268.

Kwiecinski, J., and Praszalowicz, M. (1980), Three gluon integral equation and odd c-singlet Regge singularities in QCD, *Phys. Lett.* **B94**, 413.

Kwiecinski, J., and Stasto, A. M. (2002), Geometric scaling and QCD evolution, *Phys. Rev.* **D66**, 014 013.

Laenen, E., Levin, E., and Shuvaev, A. G. (1994), Anomalous dimensions of high twist operators in QCD at $N \to 1$ and large Q^2, *Nucl. Phys.* **B419**, 39.

Laenen, E., and Levin, E. (1995), A new evolution equation, *Nucl. Phys.* **B451**, 207.

Landau, L. D., and Pomeranchuk, I. Y. (1955), On point interactions in quantum electrodynamics, *Dokl. Akad. Nauk Ser. Fiz.* **102**, 489.

Landau, L. D., Abrikosov, A., and Halatnikov, L. (1956), On the quantum theory of fields, *Nuovo Cim. Suppl.* **3**, 80.

Landau, L. D., and Lifshitz, E. M. (1958). *Quantum Mechanics, Non-Relativistic Theory*, Vols. 2 and 3, Pergamon Press.

Landau, L. D. (1959), On analytic properties of vertex parts in quantum field theory, *Nucl. Phys.* **13**, 181.

Landau, L. D. (1960), in *Theoretical Physics in the Twentieth Century: A Memorial Volume to Wolfgang Pauli*, pp. 245–247, eds. Fierz, M., and Weisskopf, V. F., Interscience.

Lappi, T. (2003), Production of gluons in the classical field model for heavy ion collisions, *Phys. Rev.* **C67**, 054 903.

Laycock, P. (2009), Diffraction at H1 and Zeus, arXiv:0906.1525 [hep-ex].

Lepage, G. P., and Brodsky, S. J. (1980), Exclusive processes in perturbative quantum chromodynamics, *Phys. Rev.* **D22**, 2157.

Levin, E., and Ryskin, M. G. (1981), Production of hadrons with large transverse momenta on nuclei in framework of QCD, *Sov. J. Nucl. Phys.* **33**, 901 [*Yad. Fiz.* **33**, 1673].

Levin, E., and Ryskin, M. G. (1985), Deep inelastic scattering on nuclei at small x, *Sov. J. Nucl. Phys.* **41**, 300 [*Yad. Fiz.* **41**, 472].

Levin, E., and Ryskin, M. G. (1987), Diffraction dissociation on nuclei in the QCD leading logarithmic approximation, *Sov. J. Nucl. Phys* **45**, 150 [*Yad. Fiz.* **45** 234].

Levin, E., and Ryskin, M. G. (1990), High-energy hadron collisions in QCD, *Phys. Rept.* **189**, 267.

Levin, E., Ryskin, M. G., Shabelski, Yu. M., and Shuvaev, A. G. (1991), Heavy quark production in semihard nucleon interactions, *Sov. J. Nucl. Phys.* **53**, 657 [*Yad. Fiz.* **53**, 1059].

Levin, E., Ryskin, M. G., and Shuvaev, A. G. (1992), Anomalous dimension of the twist four gluon operator and pomeron cuts in deep inelastic scattering, *Nucl. Phys.* **B387**, 589.

Levin, E., and Wusthoff, M. (1994), Photon diffractive dissociation in deep inelastic scattering, *Phys. Rev.* **D50**, 4306.

Levin, E. (1995), Renormalons at low x, *Nucl. Phys.* **B453**, 303.

Levin, E. (1999), The BFKL high energy asymptotic in the next-to-leading approximation, *Nucl. Phys.* **B545**, 481.

Levin, E., and Tuchin, K. (2000), Solution to the evolution equation for high parton density QCD, *Nucl. Phys.* **B573**, 833.

Levin, E., and Lublinsky, M. (2001), Non-linear evolution and high energy diffractive production, *Phys. Lett.* **B521**, 233.

Levin, E., and Tuchin, K. (2001), Nonlinear evolution and saturation for heavy nuclei in DIS, *Nucl. Phys.* **A693**, 787.

Levin, E., and Lublinsky, M. (2002a), Diffractive dissociation from non-linear evolution in DIS on nuclei, *Nucl. Phys.* **A712**, 95.

Levin, E., and Lublinsky, M. (2002b), Diffractive dissociation and saturation scale from non-linear evolution in high energy DIS, *Eur. Phys. J.* **C22**, 647.

Levin, E., and Lublinsky, M. (2004), A linear evolution for nonlinear dynamics and correlations in realistic nuclei, *Nucl. Phys.* **A730**, 191.

Levin, E., and Lublinsky, M. (2005a), Balitsky's hierarchy from Mueller's dipole model and more about target correlations, *Phys. Lett.* **B607**, 131.

Levin, E., and Lublinsky, M. (2005b), Towards a symmetric approach to high energy evolution: generating functional with Pomeron loops, *Nucl. Phys.* **A763**, 172.

Levin, E., Miller, J., and Prygarin, A. (2008), Summing Pomeron loops in the dipole approach, *Nucl. Phys.* **A806**, 245.

Levin, E., and Prygarin, A. (2008), Inclusive gluon production in the dipole approach: AGK cutting rules, *Phys. Rev.* **C78**, 065 202.

Levin, E., and Rezaeian, A. H. (2011), Gluon saturation and energy dependence of hadron multiplicity in pp and AA collisions at the LHC, *Phys. Rev.* **D83**, 114 001.

Lipatov, L. N. (1974), The parton model and perturbation theory, *Sov. J. Nucl. Phys.* **20**, 94 [*Yad. Fiz.* **20**, 181].

Lipatov, L. N. (1976), Reggeization of the vector meson and the vacuum singularity in nonabelian gauge theories, *Sov. J. Nucl. Phys.* **23**, 338 [*Yad. Fiz.* **23**, 642].

Lipatov, L. N. (1986), The bare Pomeron in quantum chromodynamics, *Sov. Phys. JETP* **63**, 904 [*Zh. Eksp. Teor. Fiz.* **90**, 1536].

Lipatov, L. N. (1989), in *Perturbative Quantum Chromodynamics*, ed. Mueller, A. H., World Scientific.

Lipatov, L. N. (1997), Small-x physics in perturbative QCD, *Phys. Rept.* **286**, 131.

Lipatov, L. N. (1999), Hamiltonian for Reggeon interactions in QCD, *Phys. Rept.* **320**, 249.

Lipatov, L. N. (2009), Integrability of scattering amplitudes in $N = 4$ SUSY, arXiv:0902.1444 [hep-th].

Low, F. F. (1975), A model of the bare Pomeron, *Phys. Rev.* **D12**, 163.

Luo, M., Qiu, J. W., and Sterman, G. (1994a), Anomalous nuclear enhancement in deeply inelastic scattering and photoproduction, *Phys. Rev.* **50**, 1951.

Luo, M., Qiu, J. W., and Sterman, G. (1994b), Twist four nuclear parton distributions from photoproduction, *Phys. Rev.* **D49**, 4493.

Lukaszuk, L., and Martin, A. (1967), Absolute upper bounds for $\pi\pi$ scattering, *Nuovo Cim.* **A52**, 122.

Lukaszuk, L., and Nicolescu, B. (1973), A possible interpretation of pp rising total cross-sections, *Lett. Nuovo Cim.* **8**, 405.

Mandelstam, S. (1958), Determination of the pion – nucleon scattering amplitude from dispersion relations and unitarity: general theory, *Phys. Rev.* **112**, 1344.

Marage, P. (2009), Vector meson production at HERA, *Int. J. Mod. Phys.* **A24**, 237.

Marchesini, G. (1995), QCD coherence in the structure function and associated distributions at small x, *Nucl. Phys.* **B445**, 49.

Marquet, C. (2005), A QCD dipole formalism for forward-gluon production, *Nucl. Phys.* **B705**, 319.

Marquet, C., and Schoeffel, L. (2006), Geometric scaling in diffractive deep inelastic scattering, *Phys. Lett.* **B639**, 471.

Marquet, C. (2007), Forward inclusive dijet production and azimuthal correlations in pA collisions, *Nucl. Phys.* **A796**, 41.

Marquet, C., Peschanski, R. B., and Soyez, G. (2007), Exclusive vector meson production at HERA from QCD with saturation, *Phys. Rev.* **D76**, 034011.

Martin, A. (1969), *Scattering Theory: Unitarity, Analyticity and Crossing*, Springer-Verlag.

McLerran, L., and Venugopalan, R. (1994a), Computing quark and gluon distribution functions for very large nuclei, *Phys. Rev.* **D49**, 2233.

McLerran, L., and Venugopalan, R. (1994b), Gluon distribution functions for very large nuclei at small transverse momentum, *Phys. Rev.* **D49**, 3352.

McLerran, L., and Venugopalan, R. (1994c), Green's functions in the color field of a large nucleus, *Phys. Rev.* **D50**, 2225.

McLerran, L. (2005), The color glass condensate and RHIC, *Nucl. Phys.* **A752**, 355.

McLerran, L. (2008), From AGS-SPS and onwards to the LHC, *J. Phys.* **G35**, 104001.

McLerran, L. (2009a), Theoretical concepts for ultra-relativistic heavy ion collisions, arXiv:0911.2987 [hep-ph].

McLerran, L. (2009b), The phase diagram of QCD and some issues of large N_c, *Nucl. Phys. Proc. Suppl.* **95**, 275.

McLerran, L., and Praszalowicz, M. (2010), Saturation and scaling of multiplicity, mean p_T and p_T distributions from 200 GeV $< \sqrt{s} < 7$ TeV, *Acta Phys. Polon.* **B41**, 1917.

McLerran, L., and Praszalowicz, M. (2011), Saturation and scaling of multiplicity, mean p_T and p_T distributions from 200 GeV $< \sqrt{s} < 7$ TeV, Addendum, *Acta Phys. Polon.* **B42**, 99.

Moschelli, G., Gavin, G., and McLerran, L. (2009), Long range untriggered two particle correlations, *Eur. Phys. J.* **C62**, 277.

Mueller, A. H. (1970), O(2,1) analysis of single particle spectra at high energy, *Phys. Rev.* **D2**, 2963.

Mueller, A. H. (1981), Perturbative QCD at high energies, *Phys. Rept.* **73**, 237.

Mueller, A. H. (1985), On the structure of infrared renormalons in physical processes at high-energies, *Nucl. Phys.* **B250**, 327.

Mueller, A. H., and Qiu, J. W. (1986), Gluon recombination and shadowing at small values of x, *Nucl. Phys.* **B268**, 427.

Mueller, A. H. (1990), Small x behavior and parton saturation: a QCD model, *Nucl. Phys.* **B335**, 115.

Mueller, A. H. (1992), The QCD perturbation series, talk given at the workshop "QCD: 20 years later", Aachen, 9–13 June 1992.

Mueller, A. H. (1994), Soft gluons in the infinite momentum wave function and the BFKL pomeron, *Nucl. Phys.* **B415**, 373.

Mueller, A. H., and Patel, B. (1994), Single and double BFKL pomeron exchange and a dipole picture of high-energy hard processes, *Nucl. Phys.* **B425**, 471.

Mueller, A. H. (1995), Unitarity and the BFKL pomeron, *Nucl. Phys.* **B437**, 107.

Mueller, A. H. (2001), A simple derivation of the JIMWLK equation, *Phys. Lett.* **B523**, 243.

Mueller, A. H., and Triantafyllopoulos, D. N. (2002), The energy dependence of the saturation momentum, *Nucl. Phys.* **B640**, 331.

Mueller, A. H., (2002), Gluon distributions and color charge correlations in a saturated light cone wave function, *Nucl. Phys.* **B643**, 501.

Mueller, A. H. (2003), Nuclear A-dependence near the saturation boundary, *Nucl. Phys.* **A724**, 223.

Mueller, A. H., and Shoshi, A. I. (2004), Small-x physics beyond the Kovchegov equation, *Nucl. Phys.* **B692**, 175.

Mueller, A. H., Shoshi, A. I., and Wong, S. M. H. (2005), Extension of the JIMWLK equation in the low gluon density region, *Nucl. Phys.* **B715**, 440.

Munier, S., Stasto, A. M., and Mueller, A. H. (2001), Impact parameter dependent S-matrix for dipole proton scattering from diffractive meson electroproduction, *Nucl. Phys.* **B603**, 427.

Munier, S., and Peschanski, R. B. (2003). Geometric scaling as traveling waves, *Phys. Rev. Lett.* **91**, 232001.

Munier, S., and Peschanski, R. B. (2004a), Traveling wave fronts and the transition to saturation, *Phys. Rev.* **D69**, 034008.

Munier, S., and Peschanski, R. B. (2004b), Universality and tree structure of high-energy QCD, *Phys. Rev.* **D70**, 077503.

Munier, S., and Shoshi, A. (2004), Diffractive photon dissociation in the saturation regime from the Good and Walker picture, *Phys. Rev.* **D69**, 074022.

Nardi, M. (2005), Hadronic multiplicity at RHIC (and LHC), *J. Phys. Conf. Ser.* **5**, 148.

Narison, S. (2002), *QCD as a Theory of Hadrons (from Partons to Confinement)*, Cambridge University Press.

Navelet, H., and Peschanski, R. B. (1997), Conformal invariance and the exact solution of BFKL equations, *Nucl. Phys.* **B507**, 353.

Navelet, H., and Peschanski, R. B. (1999), The elastic QCD dipole amplitude at one loop, *Phys. Rev. Lett.* **82**, 1370.

Navelet, H., and Peschanski, R. B. (2001), Unifying approach to hard diffraction, *Phys. Rev. Lett.* **87**, 252002.

Nikolaev, N. N., and Zakharov, V. I. (1975), Parton model and deep inelastic scattering on nuclei, *Phys. Lett.* **B55**, 397.

Nikolaev, N. N., and Zakharov, B. G. (1991), Colour transparency and scaling properties of nuclear shadowing in deep inelastic scattering, *Z. Phys.* **C49**, 607.

Nikolaev, N. N., and Zakharov, B. G. (1994), The triple pomeron regime and the structure function of the pomeron in the diffractive deep inelastic scattering at very small x, *Z. Phys.* **C64**, 631.

Nikolaev, N. N., Zakharov, B. G., and Zoller, V. R. (1994), The s channel approach to Lipatov's pomeron and hadronic cross-sections, *JETP Lett.* **59**, 6.

Nussinov, S. (1976), A perturbative recipe for quark gluon theories and some of its applications, *Phys. Rev.* **D14**, 246.

Nussinov, S. (2008), Is the Froissart bound relevant for the total pp cross section at $s = (14 \text{ TeV})^2$?, arXiv:0805.1540 [hep-ph].

Peschanski, R. B., Royon, C., and Schoeffel, L. (2005), Confronting next-leading BFKL kernels with proton structure function data, *Nucl. Phys.* **B716**, 401.

Peskin, M. E., and Schroeder, D. V. (1995) *An Introduction to Quantum Field Theory*, Perseus Books.

PHENIX collaboration, Adare, A., *et al.* (2011), *Phys. Rev. Lett.* **107**, 172301 [arXiv:1105.5112 [nucl-ex]].

PHOBOS collaboration, Back, B. B., *et al.* (2002), *Phys. Rev.* **C65**, 061901(R) [nucl-ex/0201005].

PHOBOS collaboration, Back, B. B., *et al.* (2003), *Phys. Rev. Lett.* **91**, 052303 [nucl-ex/0210015].

Politzer, H. D. (1973), Reliable perturbative results for strong interactions? *Phys. Rev. Lett.* **30**, 1346.

Polyanin, A. D. (2002). *Handbook of Linear Partial Differential Equations for Engineers and Scientists*, Chapman & Hall/CRC.

Polyanin, A. D., and Zaitsev, V. F. (2004), *Handbook of Nonlinear Partial Differential Equations*, Chapman & Hall/CRC.

Pomeranchuk, I. Y. (1958). *Sov. Phys.* **7**, 499.

Praszalowicz, M. (2011), Geometrical scaling in hadronic collisions, *Acta Phys. Polon.* **B42**, 1557.

Pumplin, J. (1973), Eikonal models for diffraction dissociation on nuclei, *Phys. Rev.* **D8**, 2899.

Regge, T. (1959), Introduction to complex orbital momenta, *Nuovo Cim.* **14**, 951.

Regge, T. (1960), Bound states, shadow states and Mandelstam representation, *Nuovo Cim.* **18**, 947.

Risken, H. (1989), *The Fokker–Planck Equation (Methods of Solution and Applications)*, Springer-Verlag.

Roberts, R. G. (1993), *The Structure of the Proton: Deep Inelastic Scattering*, Cambridge University Press.

Roman, P. (1969), *Introduction to Quantum Field Theory*, John Wiley & Sons.

Ross, D. A. (1998), The effect of higher order corrections to the BFKL equation on the perturbative pomeron, *Phys. Lett.* **B431**, 161.

Rummukainen, K., and Weigert, H. (2004), Universal features of JIMWLK and BK evolution at small-x, *Nucl. Phys.* **A739**, 183.

Ryskin, M. G. (1993), Diffractive J/Ψ electroproduction in LLA QCD, *Z. Phys.* **C57**, 89.

Salam, G. P. (1995), Multiplicity distribution of color dipoles at small x, *Nucl. Phys.* **B449**, 589.

Salam, G. P. (1996), Studies of unitarity at small x using the dipole formulation, *Nucl. Phys.* **B461**, 512.

Salam, G. P. (1998), A resummation of large sub-leading corrections at small x, *JHEP* **9807**, 019.

Salam, G. P. (1999), An introduction to leading and next-to-leading BFKL, *Acta Phys. Polon.* **B30**, 3679.

Schmidt, C. R. (1999), Rapidity-separation dependence and the large next-to-leading corrections to the BFKL equation, *Phys. Rev.* **D60**, 074003.

Schweber, S. S. (1961). *An Introduction to Relativistic Quantum Field Theory*, Row, Peterson and Co.

Schwimmer, A. (1975), Inelastic rescattering and high-energy reactions on nuclei, *Nucl. Phys.* **94**, 445.

Stasto, A. M., Golec-Biernat, K., and Kwiecinski, J. (2001), *Phys. Rev. Lett.* **86**, 596.

STAR collaboration, Adams, J., *et al.* (2003), *Phys. Rev. Lett.* **91**, 072304.

STAR collaboration, Abelev, B. I., *et al.* (2009), Long range rapidity correlations and jet production in high energy nuclear collisions, *Phys. Rev.* **C80**, 064912.

Sterman, G. (1993). *An Introduction to Quantum Field Theory*, Cambridge University Press.

't Hooft, G. (1972), Unpublished.

't Hooft, G. (1974), A two-dimensional model for mesons, *Nucl. Phys.* **B75**, 461.

't Hooft, G. (1979), Can we make sense out of quantum chromodynamics?, *Subnucl. Ser.* **15**, 943.

van Saarloos, W. (2003), Front propagation into unstable states, *Phys. Rept.* **386**, 29.

Wiedemann, U. A. (2008), Heavy-ion collisions: selected topics, in *Proc. European School of High-Energy Physics 2007, Trest*, pp. 277–306.

Weigert, H. (2002), Unitarity at small Bjorken x, *Nucl. Phys.* **A703**, 823.

Weigert, H. (2005), Evolution at small x_{Bj}: the color glass condensate, *Prog. Part. Nucl. Phys.* **55**, 461.

Weigert, H. (2007), A compact introduction to evolution at small x and the color glass condensate, *Nucl. Phys.* **A783**, 165.

Weinberg, S. (1973), Nonabelian gauge theories of the strong interactions, *Phys. Rev. Lett.* **31**, 494.

Weinberg, S. (1996), *The Quantum Theory of Fields*, Vols. 1–3, Cambridge University Press.

White, C. D., and Thorne, R. S. (2007), A global fit to scattering data with NLL BFKL resummations, *Phys. Rev.* **D75**, 034 005.

Wilson, K. G. (1974), Confinement of quarks, *Phys. Rev.* **D10**, 2445.

Witten, E. (1979), Baryons in the $1/N$ expansion, *Nucl. Phys.* **B160**, 57.

Woods, R. D., and Saxon, D. S. (1954), Diffuse surface optical model for nucleon-nuclei scattering, *Phys. Rev.* **95**, 577.

Wusthoff, M., and Martin, A. D. (1999), The QCD description of diffractive processes, *J. Phys.* **G25**, R309.

Yang, C. N., and Mills, R. L. (1954), Conservation of isotopic spin and isotopic gauge invariance, *Phys. Rev.* **96**, 191.

Yndurain, F. J. (2006), *The Theory of Quark and Gluon Interactions*, fourth edition Springer.

Zakharov, V. I. (1992), QCD perturbative expansions in large orders, *Nucl. Phys.* **B385**, 452.

Zamolodchikov, A. B., Kopeliovich, B. Z., and Lapidus, L. I. (1981), Dynamics of colour in hadron diffraction by nuclei, *JETP Lett.* **33**, 595 [*Pisma Zh. Eksp. Teor. Fiz.* **33**, 612].

ZEUS collaboration, Breitweg, J., *et al.* (1999), Measurement of the diffractive cross-section in deep inelastic scattering using ZEUS 1994 data, *Eur. Phys. J.*, **C6**, 43.

ZEUS collaboration, Chekanov, S., *et al.* (2003), A ZEUS next-to-leading-order QCD analysis of data on deep inelastic scattering, *Phys. Rev.* **D67**, 012 007 [arXiv:hep-ex/0208023].

Zhang, W. M., and Harindranath, A. (1993), Light front QCD. 2: Two component theory, *Phys. Rev.* **D48**, 4881.

Index

Printed in the United States
by Baker & Taylor Publisher Services